Essentials of
Soil Mechanics
and Foundations

Essentials of
Soil Mechanics
and Foundations

David F. McCarthy, *P.E.*

Reston Publishing Company, Inc.
A Prentice-Hall Company
Reston, Virginia

Library of Congress Cataloging in Publication Data

McCarthy, David F
   Essentials of soil mechanics and foundations.

   Bibliography: p. 477
   Includes index.
   1. Soil mechanics. 2. Foundations. I. Title.
TA710.M29   624'.15   76-53771
ISBN 0-87909-221-1

© 1977 by Reston Publishing Company, Inc.
A Prentice-Hall Company
Reston, Virginia 22090

10 9 8 7 6 5 4 3

Printed in the United States of America

Contents

Appendix 1

Selected Soil Testing Procedures   423

Appendix 2

Sample Specification for Test Borings   455

Appendix 3

Typical Earthwork Specifications   463

Preface

This text has been developed as an educational instrument to introduce the theory and application of soil mechanics to students intending to enter the planning, design, or field phases of the construction profession.

In its function as an introductory level text, the material is limited to basic principles. Advanced topics are beyond its intended scope. Advanced mathematics have been avoided. Detail has been restricted to ensure that major concepts do not become masked or lost. For the student, the text establishes the importance of soil mechanics to the design and construction profession. The practitioner and teacher will find that the methods presented for analysis and evaluation apply to many practical problems.

Illustrations and photographs provided by contractors and manufacturers to complement the written text have been acknowledged where presented. The opportunity to use them is greatly appreciated. Soiltest, Inc. and Acker Drill Co., Inc. deserve special mention for the considerable material made available during the text's development. Many diagrams and photographs relating to soil-boring methods appear in Will Acker's book, *Basic Procedures for Soil Sampling and Core Drilling*, a recommended reference where detailed information on drilling and boring techniques is required. Important to the manuscript, and deserving of acknowledgement, are the reviews, suggestions, and materials provided by former colleagues at Dames and Moore and at Empire Soil Investigations—Thomsen Associates.

The preparation and editing of a technical manuscript requires time and patience. To this end, the skill of Mrs. Patricia Klossner was essential and not unappreciated. The efforts of Mrs. Sue Fox Hill and Miss Laurene Allen, involved during the manuscript's formative stage, require acknowledgement. A special note of gratitude is due Mrs. Grace Messere of Mohawk Valley Community College for her interest and assistance throughout the various stages of the book's development. Also deserving of recognition are the patience and understanding of the author's family, whose lifestyle was uncomplainingly altered throughout the text's gestation period.

David F. McCarthy

Units

Soil mechanics is a field where U.S. Customary units (also referred to as British units) and customary metric units have been commonly used in the past. The international system, SI units (the international metric system) is proposed as the system for eventual worldwide usage. However, to date, application of the SI system within the technology and construction professions in English-speaking countries has been limited. Sole use of the SI system in a related textbook is not yet warranted. The material in this text has been presented primarily in the units historically familiar to North America, but SI equivalents are also included in discussions and tabulations of data and soil properties.

For reference, a listing of conversions between common U.S. Customary and SI units is shown below.

U.S. Customary	SI
one inch (in.)	25.4 millimeters (mm)
one inch	2.54 centimeters (cm)
one inch	0.0254 meter (m)
one foot (ft)	0.305 meter (m)
3.28 feet	1.0 meter (m)
one yard (yd)	0.91 meter (m)
one mile (mi)	1610 meters (1.61 kilometers)
one sq inch (in.^{2})	6.45 sq cm (cm^2)
one sq foot (ft^2)	0.093 sq meter (m^2)
one acre	4047 sq meters (m^2)
one sq mile (mi^2)	2.59 sq kilometers (km^2)

one cubic inch (in.^{3})	16.4 cubic centimeters (cm^3)
one cubic foot (ft^3)	0.028 cubic meter (m^3)
one cubic yard (yd^3)	0.765 cubic meter (m^3)

one pound (lb)	454 grams (g)
one pound	0.454 kilogram (kg)
one ton (2000 pounds, or 2 kips)	907.2 kilograms (kg)

one pound force (lbf)	4.45 newtons (N)
one slug	14.59 kilograms (kg)

one pound per sq foot (psf)	47.9 newtons per sq meter (N/m^2) or 0.048 kN/m^2
one pound per sq inch (psi)	6.9 kilonewtons per sq meter (kN/m^2)
one ton per sq foot	95.8 kN/m^2
one pound per cu foot (pcf)	157 newtons per cu meter (N/m^3) or 0.157 kN/m^3 or 15.95 kgf/m^3

one gallon per minute (gal/min)	0.0038 cu meter per minute

Other Terms and Relationships

one pascal (Pa) = 1 newton per sq meter
one kilogram force (kgf) = 9.81 newtons
one bar = 1.02 kg/cm^2 = 10^5 Pa

one pound per sq foot = 4.8 × 10^{-4} bar = 48 Pa
one ton per sq foot (tsf) = 1.09 kg/cm^2 = 1.07 bar = 95.8 kN/m^2

one dyne = 1 × 10^{-5} newtons

one poise = $\dfrac{\text{dyne-sec}}{\text{cm}^2}$

Nomenclature

The tabulation that follows indicates the nomenclature used throughout this text. Generally, the notation and symbols are standard for the soil mechanics profession. Some dual use of nomenclature does exist (the use of the same symbol for different factors). This practice is continued, for certain nomenclature has become well established in soil mechanics use, and it is felt not advisable to arbitrarily assign new notation. Fortunately, repeated nomenclature only occurs with different topics, and the potential for confusion is remote.

A	area (in.^{2}, ft^2, cm^2, m^2)
A_{tip}	end area of pile tip (ft^2, m^2)
$\overset{\circ}{A}$	Angstrom unit (1 $\overset{\circ}{A}$ = 10^{-8} cm)
AASHO	American Association of State Highway Officials
ASCE	American Society of Civil Engineers
ASTM	American Society for Testing and Materials
a_v	coefficient of compressibility (ft^2/lb, or cm^2/gm)
a_1, a_2	shape factors for rectangular footings (dimensionless)
B	footing width, pile or pile group width (ft, m)
C_c	compression index (dimensionless)
C_r	slope of recompression curve from consolidation test (dimensionless)
C_u	uniformity coefficient for sands (dimensionless)
c	unit cohesion or shear strength of cohesive soils (lb/ft^2, kip/ft^2, kg/cm^2, N/m^2)
c_h	coefficient of consolidation for horizontal drainage (ft^2/day, month, or year; cm^2/min, hour, day, or month)

c_v	coefficient of consolidation for vertical drainage (ft^2/day, month, or year; cm^2/min, hour, day, month)
D, d	diameter of soil particles (in., cm, mm)
D_c	critical depth for pile penetration (ft, m)
D_f	depth of footing (ft, m)
D_r	relative density of cohesionless soil (percent)
D_{10}	effective particle size, diameter of the 10 percent finer size (cm, mm)
D_{60}	particle size corresponding to 60 percent finer (cm, mm)
E	modulus of elasticity ($lb/in.^2$, lb/ft^2, N/m^2)
e	void ratio (dimensionless)
Eff.	efficiency of pile driving hammer (decimal, fraction, percent)
F	force (lb, kip, ton, kg, N)
FS and SF	factor of safety or safety factor (dimensionless)
f	skin friction or adhesion between pile and soil (lb/ft^2, kip/ft^2, N/m^2)
G_w	specific gravity of water (dimensionless)
G_s	specific gravity of soil solids (dimensionless)
g	acceleration of gravity (ft/sec^2, cm/sec^2)
H	height or thickness (ft, m)
H_{dr}	longest drainage path
H_w, h	height of water, head of water (ft, m)
ΔH	settlement (in., cm)
h_c	height of capillary rise
I	electrical current (amperes)
I_B, I_W	influence factor for determining subsurface stress (dimensionless)
i	hydraulic gradient (dimensionless)
K	lateral earth pressure coefficient (dimensionless)
K_o	lateral earth pressure coefficient for at-rest condition (dimensionless)
K	permeability (ft^2, m^2)
k	coefficient of permeability (ft/min, day; cm/sec, min)
K_v	modulus of subgrade reaction (kip/ft^3, N/m^3)
ksf	kips per square foot (one kip equals 1,000 lb)
L, l	length (ft, m)
LI	liquidity index (dimensionless)
LL	liquid limit (percent)
M_s	slope stability factor (dimensionless)
m_v	coefficient of volume compressibility (ft^2/lb, cm^2/kg, m^2/N)
N	normal force (lb, kip, ton, kg, N)

N	blow count from standard penetration test
N_c, N_γ, N_q	bearing capacity factors for foundation design (dimensionless)
$N_{cs}, N_{\gamma s}$	bearing capacity factors for footings on slopes (dimensionless)
N_d	number of equipotential drops in flow net analysis (dimensionless)
N_f	number of flow channels in flow net analysis (dimensionless)
n	porosity (dimensionless)
PI	plasticity index (percent)
PL	plastic limit (percent)
p_w	water pressure (lb/ft^2, gm/cm^2, N/m^2)
pcf	pounds per cubic foot (unit weight or density) (lb/ft^3)
psf	pounds per square foot (pressure) (lb/ft^2)
Q	total water flow (ft^3, cm^3, m^3, gal)
Q, Q_t	total load (lb, kip, ton, kg, N)
$q_{ultimate}, q_{ult}$	ultimate unit soil bearing pressure (lb/ft^2, kip/ft^2, ton/ft^2, kg/cm^2, N/m^2)
q_{design}	design value of soil bearing pressure (lb/ft^2, kip/ft^2, kg/cm^2, N/m^2)
q_c	static cone resistance (kg/cm^2)
q	volume of water flowing per unit time (gal/min, ft^3/min, cm^3/min)
R	correction factor to obtain bearing capacity of inclined foundation (dimensionless)
R_H	hydraulic radius (ft)
r	radius (ft, m)
S	degree of saturation (percent)
S	set, penetration of driven pile per hammer blow (in., cm)
S_t	sensitivity (dimensionless)
S_o	elastic compression of pile (in., cm)
SPT	standard penetration test blow count (blows)
T_h	time factor for horizontal drainage (dimensionless)
T_s	value of surface tension (lb/ft, dyne/cm)
T_v	time factor for vertical drainage (dimensionless)
t	time (sec, min, hour, day, month, year)
U	percentage of consolidation (percent)
u	neutral stress due to water pressure (lb/ft^2, kg/cm^2, N/m^2)
V	volume (ft^3, cm^3, m^3)
V, v	velocity (ft/s, m/s)

V	electrical potential difference in soil resistance studies (V, volts)
W	weight (lb, gm, kg)
W	width (ft, m)
W_h	weight of pile hammer ram (lb, ton, kg)
w	water content (percent)
z	depth or vertical distance (ft, m)
α (alpha)	factor expressing ratio of adhesion to cohesion for clay (dimensionless)
β (beta)	soil resistivity (ohm-ft)
β	skin friction factor for piles embedded in clay (dimensionless)
η (eta)	fluid viscosity, poises $\dfrac{\text{dyne-sec}}{\text{cm}^2}, \dfrac{\text{N-sec}}{\text{m}^2}, \dfrac{\text{lb-sec}}{\text{ft.}^2}$
η	coefficient for silt and clay soils applied when calculating foundation settlement from static cone penetration data (dimensionless)
γ (gamma)	unit weight (lb/ft^3, gm/cm^3, N/m^3)
γ_{dry}	dry unit weight of soil
γ_{total}, γ_t	total unit weight of soil including water content
γ_{sat}	unit weight of saturated soil
γ_{sub}	submerged unit weight of soil
γ_w	unit weight of water
λ (lambda)	friction capacity coefficient in pile foundation analysis (dimensionless)
μ (mu)	micron (one micron = 0.0001 cm)
ϕ (phi)	angle of internal friction (deg)
ρ (rho)	density (gm/cm^3, slugs/ft^3)
σ (sigma)	normal stress (lb/in.^{2}, lb/ft^2, kip/ft^2, ton/ft^2, kg/cm^2, N/m^2)
$\sigma_1, \sigma_2, \sigma_3$	principal (normal) stresses (lb/in.^{2}, lb/ft^2, kip/ft^2, kg/cm^2, N/m^2)
σ_h	horizontal stress (lb/ft^2, kg/cm^2, N/m^2)
σ_v	vertical stress or overburden stress (lb/ft^2, kg/cm^2, N/m^2)
$\bar{\sigma}_v$	effective vertical stress, effective overburden stress (lb/ft^2, kg/cm^2, N/m^2)
σ_{v_0}	original overburden stress (lb/ft^2, kg/cm^2, N/m^2)
τ (tau)	shear stress (lb/ft^2, kg/cm^2, N/m^2)
θ (theta)	inclination angle of inclined foundation or inclined load on footing (deg)
θ	angle of slope in earth dams
$\tan \delta$ (delta)	coefficient of friction (dimensionless)

Essentials of
Soil Mechanics
and Foundations

CHAPTER 1
Origin and Characteristics of Soil Deposits

The earth's crust is composed of soil and rock. Rock can be defined as a natural aggregate of minerals that are connected by strong and permanent attractive forces; for this reason, rock is often considered a consolidated material. Soil may be defined as the unconsolidated sediments and deposits of solid particles that have resulted from the disintegration of rock. In the construction and engineering professions, however, soil is also assumed to include the residue of vegetable and animal life, including industrial wastes.

Soil is a *particulate* material, which means that a soil mass consists of an accumulation of individual particles that are bonded together by mechanical or attractive means, though not as strongly as for rock. In soil (and in rock), voids exist between particles, and the voids may be filled with a liquid, usually water, or gas, usually air. As a result, soils are often referred to as a three-phase material or system (solids plus liquid plus gas).

Rock: The Source of Soils

Most of the nonorganic materials that are identified as soil originated from rock as the parent material. Rock types are classified into three major categories—igneous, sedimentary, or metamorphic—depending on the origin or method of formation. The type of soil that develops is related to the method of rock formation and to the minerals included in the rock.

Igneous rock resulted from the cooling and hardening of molten rock called *magma*, which originated deep within the earth. Molten magma escaping through volcanoes and fissures in the earth's crust and forming at or near the surface of the earth cooled quickly. As a result of rapid

1

cooling, the mineral components solidified in small crystals and possessed a fine texture. In some situations, the cooling was so rapid that a crystal-free, glassy texture resulted. The molten materials that cooled rapidly at or near the earth's surface are called extrusive or volcanic rock types and include the *basalts, rhyolites,* and *andesites.*

Molten rock trapped deep below the surface of the earth cooled slowly. The mineral components formed in large crystals, and coarse-textured rocks resulted. These rocks are classified as intrusive or plutonic types and include the *granites,* the most common, as well as the *syenites, diorites,* and *gabbros.*

Many of the chemical combinations in igneous rocks are unstable in the environment existing at the earth's surface. Upon exposure to air, water, chemicals in solution in water, freezing temperatures, varying temperatures, and wind, the rock minerals break down to the soil types existing today. Rock whose chief mineral is quartz or orthoclase (high silica content) decomposes to predominantly sandy or gravelly soil with a little clay.[1] Granites and rhyolites are in this category. Because of the high silica content, these rocks are classified as acidic.

Rocks whose minerals contain iron, magnesium, calcium, or sodium, but little silica, such as the gabbros, diabases, and basalts, are classified as basic rocks. These rocks decompose to the fine-textured silt and clay soils.

Generally, the acidic rocks are light-colored, while the basic rocks are very dark. Intermediate colors reflect an intermediate chemical composition. Rock types intermediate between acidic and basic include the *syenetes, trachytes, diorites,* and *andesites.* Because of their mineral components, diorite and andesite easily break down into the fine-textured soils.

The clay portion of fine-textured soil is the result of primary rock minerals decomposing to form secondary minerals. The clays are *not* small fragments of the original minerals that existed in the parent rock. Because of this change, the properties and behavior of clay soils are different from the gravel, sand, and silt soils, which *are* still composed of the primary rock minerals.

Rocks that are *acidic* (not *basic*) such as the granites, are considered to be good construction materials.

Sedimentary rocks are formed from accumulated deposits of soil particles or remains of certain organisms that have become hardened by pressure or cemented by minerals. Pressure sufficient to harden or

---

[1]According to the engineering definition, coarser soils include sands and gravels and are particles larger than 0.074 mm. The fine textured silt and clay soils are smaller than 0.074 mm. The 0.074-mm size is close to the smallest particle size observable with the unaided eye under normal conditions. Most clay particles are smaller than .002 mm.

solidify a soil deposit can result from the weight of great thicknesses of overlying soil or from glaciers. Under this pressure, the deposit is compacted and consolidated, and strong attractive bonds are established. Cementing minerals such as silica, calcium carbonate, and the iron oxides are abundant in soil as a result of rock weathering, and when dissolved in the water circulating through a soil deposit, these minerals precipitate out onto the soil particles. Other cementing may be obtained from within the mass by solution or chemical change of materials. Sedimentary-type rocks include the limestones and dolomites, shales, sandstone, conglomerate, and breccia.

Geologic conditions in past historic times have had a very significant effect on the location and type of sedimentary rocks that exist across North America today. In earliest prehistoric times, most of what is now the United States was under water. Gradually, land near the existing west coast and in the northeast rose, but what are the central and southern areas of the United States, and the Atlantic coast region remained beneath shallow seas (Fig. 1-1). Accumulated sediments in these shallow seas eventually became the limestones, shales, and sandstones that underlie these regions today. As a consequence of the layered manner of soil deposition, many sedimentary rocks are easily recognized today because of their stratified formation (Fig. 1-2).

Shales are formed predominantly from deposited clay and silt soils. The degree of hardness varies, depending on the type of soil minerals, the bonding that developed, and the presence of foreign materials. The hardness is generally due to external pressures and particle bonds that resulted, and is not due to cementing minerals. Many shales are relatively stable when exposed to the environment, but some shales expand or delaminate (layers separate) after contact with water or air. Weathering breaks down shale to fragments of varying size. In turn, these fragments may be quickly reduced back to clay-particle sizes. The properties of shale are quite important to the construction industry, for it is estimated that shale represents approximately 50 percent of the rock that is exposed at the earth's surface or closest to the surface under the soil cover. Sound shale can provide a good foundation material, but its use as a construction material is questionable because of its tendency to break down under handling and abuse.

Sandstone is predominantly quartz cemented together with silica, calcium carbonate, or iron compounds. These materials are generally considered good construction materials. *Conglomerate* (cemented sand and gravel) and *breccia* (cemented rock fragments) are similar to sandstone.

Limestone is predominantly crystalline calcium carbonate (calcite) formed under water. This material can result from chemical transformation of other soil minerals in solution and from the remains of marine

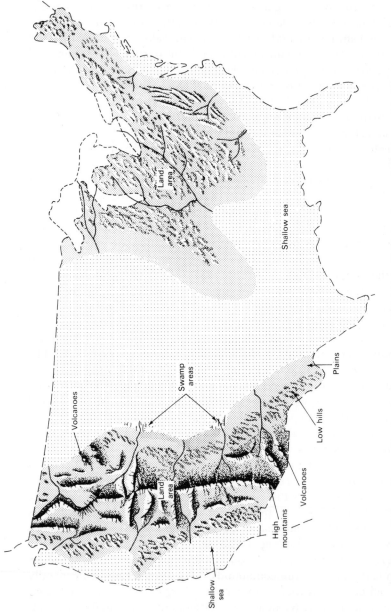

Figure 1-1. Generalized geographic map of the United States for period approximately 100 million years ago. Sediments deposited in shallow seas became sedimentary rock which exists in such areas today. (Source: U.S. Department of the Interior, Geological Survey)

Figure 1-2. *Sedimentary rock in central New York state showing stratified formation.*

organisms and action of plant life. Because of their sedimentary nature, limestones frequently include impurities such as clays and organic material. The degree of hardness and durability of limestones varies. Some limestone formations are very sound, but some are very soluble with the result that the formation contains many cavities.

Dolomite is a variety of limestone, but harder and more durable. *Marl* and *chalk* are softer forms of limestone.

Weathering of limestones can produce a soil that includes a large range of particle sizes, but the fine-grained soils predominate. Limestone is a good foundation material, provided that the formation is sound and free of cavities. Sound limestone is considered a good construction material.

Frequently, because of the deposition of the original rock-forming sediments, sandstone and shale, or limestone and shale, and sometimes sandstone, limestone, and shale are interbedded. Upon weathering, the resulting soil will have the characteristics of the predominant rock type.

Metamorphic rock results when igneous or sedimentary rocks are subject to combinations of heat, pressure, and plastic flow so that the original rock structure and mineral composition is changed. Plastic flow for rock refers to slow viscous movement and rearrangement within the rock mass as it changes and adjusts to the pressures created by external forces. Under these conditions, limestone is changed to *marble,* and sandstone to *quartzite.* When subject to pressure and plastic flow, shale is transformed to *slate* or *phyllite.* Higher levels of these factors change the shale, or slate, to *schist.* Because of the processes affecting their formation, slates and schists become foliated rocks (that is, layered or banded as in a folio). *Gneiss,* similarly, is banded or foliated rock resulting from metamorphosis of sedimentary rock or basalt or granite. In spite of their possibly different origin, a distinction between gneisses and schists is not great. Frequently, these two rock types grade or blend together.

Metamorphic rocks formed from sound igneous or sedimentary rocks are good materials for construction. But schist, gneiss, and slate are questionable construction materials because of their foliated or banded structure and resulting planes of weakness.

Upon weathering, some metamorphic rocks break down to soil types comparable to that which would be derived from the original igneous or sedimentary-type rock. Others reflect the changes brought about by metamorphism. Gneiss and schist decompose to silt-sand mixtures with mica. Soils from slates and phyllites are more clayey. Soils derived from marble are similar to those resulting from limestone. Decomposition of quartzite generally produces sands and gravels.

The process of rock changing to soil, soil changing to rock, and alteration of rock is a continuous and simultaneously occurring process. The process of change or alteration takes place over long periods of time, and there is *not* a set sequence to the order in which changes occur.

Rock types are many. To establish a proper perspective, the construction industry's concern is generally not with the rock names but with the properties. In-place properties such as hardness and possible presence of fractures or fissures affect drilling, blasting, and excavation operations. The suitability for use as a foundation for structures is related to strength, durability, presence of cavities, or fractures and fissures. The commercial value of excavated and crushed rock for fill and as an ingredient of concrete is influenced by soundness and durability. In a general way, desirable and undesirable properties have been associated with the different rock types.

Soil Categories

Soils can be grouped into two broad categories—residual or transported—depending on the method of deposition.

Residual soils have formed from the weathering of rock and remain at the location of their origin. The weathering process may be attributed to mechanical weathering or chemical and solution weathering. Mechanical weathering refers to the effects of wind, rain, running water, and tectonic forces (such as earthquakes). Chemical and solution weathering is rock decomposition due to chemical changes in the rock minerals brought about by exposure to the atmosphere, temperature changes, water, or other materials. The rate of weathering is generally greater in warm humid regions than in cool dry regions.

Residual soils generally include particles having a wide range of sizes, shapes, and composition, depending on the amount and type of weathering and the minerals in the parent rock. The thickness of residual soil existing at any particular location is affected by the rate of rock weathering and the presence or lack of erosive forces to carry the soil away after it is formed.

Transported soils are those materials that have been moved from their place of origin. Transportation may have resulted from the effects of gravity, wind, water, glaciers, or man—either singularly or in combination. Soil particles are often segregated according to size by, or during, the transportation process. The method of transportation and deposition has significant effect on the properties of the resulting soil mass, as discussed in subsequent sections of this chapter.

GRAVITY AND WIND-TRANSPORTED SOILS / Gravity is generally capable of transporting materials only limited distances, such as down a hill or mountain slope, with the result that there is little change in the soil material brought about by the transportation procedure.

Wind can move small particles by rolling or carrying them. Soils carried by wind and subsequently deposited are designated *aeolian* deposits. Particles of small sand sizes can be rolled and carried short distances. Accumulations of such wind-deposited sands are dunes. Dunes are characterized by low hill and ridge formation. They generally occur in sandy desert areas, and on the downwind side of bodies of water having sandy beaches. Dune material is a good source of sand for some construction purposes, but if the particles are of uniform size and very weathered and rounded, the sand may not be highly suitable for all construction purposes.

Fine-textured soils, the silts and clays, can be carried great distances by wind. Silt soils in arid regions have no moisture to bond the particles together and are very susceptible to the effects of wind. Clay, however, has sufficient bonding or cohesion to withstand the eroding effects of wind. Deposits of wind-blown silts laid down in a loose condition that has been retained because of particle bonding or cementing minerals is classified as *loess*. Significant loess deposits are found in North America, Europe, and Asia.

In the United States, great thicknesses of loess exist to the east of the Mississippi River and northeast of the Missouri River. With these materials, accumulations have built up slowly, and grasses growing at the surface could keep pace with the rate of deposit. The resulting rootholes and grass channels that remain have created a soil that has a high porosity and cleavage in the vertical direction. Natural and man-made cuts in this material will stand with nearly vertical slopes. However, if the soil is exposed to excessive water (becomes saturated or inundated), or is subject to severe ground vibrations, the soil's stable structure can be broken down. Subsidence or settlement results. Consequently, loess formations should be considered as poor foundation soils unless they can be protected from the effects of water and vibrations.

Volcanic eruptions have also produced "wind-transported" soils. The volcanic ash carried into the air with the escaping gases are small fragments of igneous rock. The soil type expected to result will be related to the mineral characteristics of the igneous rock, as discussed earlier. Generally, remains of volcanic ash deposits are limited. Because of surface deposition, they are quickly affected by other weathering agents.

GLACIAL DEPOSITS / Much of Canada and the northern United States in North America, and northern Europe have been subjected to the past effects of massive moving sheets of ice, the continental glaciers. The period of glaciation is referred to as the Great Ice Age, and geologists estimate that it covered the span of time that extended from about one million years ago to about 10,000 years ago. Some areas were covered once, whereas some other areas may have been covered as often as five different times. Glaciers grew and advanced over the land when climatic conditions permitted or contributed to the formation of ice (Fig. 1-3). Glacial advance ceased when melting at its limits equalled the rate of growth. When the rate of melting exceeded the rate of growth, because of climatic or other changes, the glaciers receded or shrunk. Generally glacier growth or shrinkage and movement were very slow.

Considerable quantities of soil have been moved and deposited by or because of glacial action. Such soils are glacial deposits. But although glaciers moved vast quantities of soil and created a surface topography, the major topographical features such as mountain ranges or plains areas are not the direct result of glaciers. Indirectly, however, major topographical features have very likely been affected by the continental glaciers. The glaciers were several thousand feet thick in many areas and this had two very significant effects: the tremendous amount of water taken to form the glaciers lowered the level of the sea by some 400 to 500 feet (125 to 150 m), and the tremendous weight of the glaciers caused the land beneath them to depress. As a result of the lowering of the sea level, the continental limits of North America extended beyond its current limits. It is probable that much glacially eroded soil was dropped or washed to areas that are now under the sea. As a result of the land's being depressed, areas of low elevation were flooded over. After the glaciers retreated and the areas were relieved of their weight, the land rose in elevation. Sedimentary rock and soil deposits that had been carried to and formed in the low flooded areas were elevated to create new land surfaces. Many lake areas became diminished in size

Figure 1-3. *Map of North America indicating areas that have been covered with glacial ice.*

or disappeared entirely. Great Salt Lake in Utah, for instance, is the shrunken remain of a 20,000-square-mile (50,000-km^2) glacial lake which once had a depth in excess of 1000 feet (300 m).

As a glacier grew or advanced, it gathered and pushed soil ahead of it or enveloped and gathered the soil into itself. All sizes of soil particles were picked up and mixed together, there being no sorting according to size. Some of the material picked up was subsequently dropped during the advance, either under the glacier or in front, and then overrun by the continued movement. When a glacier stopped advancing, soil being pushed by the glacier and soil being freed by the melting process accumulated in front of the glacier. When the glacier receded, all soil trapped in the melting ice was dropped. Such direct glacial deposits are a heterogeneous mixture of all soil sizes and are termed *glacial till*.

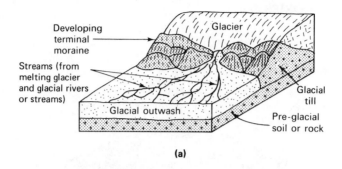

(a)

(b)

Figure 1-4. (a) The development of a terminal moraine and outwash plain in front of a glacier; (b) aerial photo of glacier showing developing moraine and outwash area.

The land form or topographic surface resulting after a glacier receded is called a *ground moraine* or *till plain*. The hills and ridges of soil that formed at the front of the glacier and marked its farthest advance are *terminal moraines* (Fig. 1-4). *Recessional moraines* are hills or ridges that represent deposits along the front of a glacier where it made temporary stops during the recession process. Soil dropped along the side of a glacier as it moved through a valley is termed a *lateral moraine*. Long low hills of till that extend in the direction of the glacial movement are called *drumlins*.

Where the till material was dropped under the glacier and overrun, it became very dense and compact, and can provide excellent foundation support. The suitability of a till material for construction purposes, such as for a compacted earth fill, depends on the quantity and range of sizes of the soil particles. Till deposits that have a preponderance of coarse soils are good construction materials, while deposits containing an excess of silt and clay materials are, generally, relatively difficult to work with.

Where an area was subjected to repeated glacial action, original deposits could be overlain by the more recent glacial soils, or the original deposits may have been moved and redeposited by the more recent glacier as a new land form. As a result, the original source of a material may be difficult to determine. For some situations, however, the color of the soil may provide information as to its source. With reference to glacial deposits in the northern United States, material gathered by glaciers from the Hudson Bay area in Canada is gray, whereas soils picked up from glaciers originating in the area northwest of the Great Lakes are red in color as a result of the high iron content in the original soils of that area.

Even while the glaciers covered a land area, there were streams and rivers of water flowing on the surface of the glaciers and in subterranean tunnels eroded within the glaciers. These flowing bodies of water carried soil picked up from the land surface or eroded from the glaciers (Fig. 1-5). Much such soil was ultimately carried to the front of the glacier, but much was also dropped along the routes of flow, and where the water became trapped within the glacier or between a glacier and valley wall.

Soils deposited by the surface and subsurface glacial rivers, and those which remained in the form of long winding ridges, are called *eskers*. These deposits are, usually, mostly coarse-grained soils. Although coarse, the range of particle sizes can vary considerably in cross section or over short distances. As a result, eskers can provide a good source of coarse-grained soils for construction purposes, but the uniformity of gradation should not be expected to be constant at different locations within the esker. Eskers frequently followed along locations

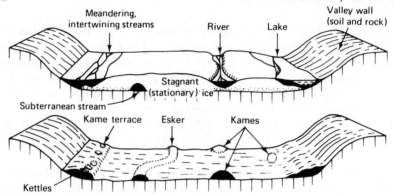

Figure 1-5.   *Flowing water on a glacier and resulting effect on land form.*

of low ground elevation, such as river valleys. As a result, they frequently represent paths of good foundation soils across areas where poor soil deposits, such as fine-grained soil sediments or marshes, have subsequently accumulated. Such paths of coarse-grained soils ideally suit the requirements for highway subgrades because of their good drainage characteristics, low susceptibility to frost heave, and ease of handling during the construction process.

Kames provide soil deposits somewhat similar to eskers, being the remains of material dropped along the boundaries of a glacier and a valley wall (kame terrace), or in holes in a glacier. The resulting land form has the shape of knobs or small hills. As for eskers, the gradation of soils from kames should be expected to be variable.

In some areas where kames exist, the surface topography is further affected by depressions called *kettle holes*. Kettle holes apparently formed when great chunks of ice remained behind a receding glacier and became buried by the glacial soil. When the ice melted, a hole or depression in the soil surface resulted.

RIVER DEPOSITS / Flowing bodies of water are capable of moving considerable volumes of soil by carrying the particles in suspension or skipping them along the river bottom. The largest-diameter particles that can be carried in suspension are related to the square of the velocity of the flowing water. Coarser particles being carried in suspension are dropped when a decrease in the water velocity occurs, as when the river deepens, widens, or changes direction. Finer particles remain in suspension to be deposited in quieter waters. Thus, river deposits are segregated according to size. For most rivers the volume of water that flowed and the volume of soil transported were variables that changed as land drainage forms changed or as seasonal variations in precipitation occurred.

All soils carried and deposited by rivers are classified as *alluvial deposits*. However, glacial soils carried by rivers created from melting glacial waters and subsequently sorted and deposited according to size are referred to as *glacial fluvial deposits* or *stratified drift*.

Glacially created rivers were developed from water escaping at the edge of a glacier or from trapped water breaking through a recessional moraine left by a receding glacier. Upon leaving the glacier or moraine area, the flowing water rapidly fanned out over a broad area of land, temporarily flooding it. With the resulting decrease in the flow velocity, the larger soil particles dropped out, forming fan- or delta-shaped flat beds of predominantly sand and gravel soils. Overlapping deltas of coarse soils spread over broad areas create land forms classified as *outwash plains*. The finer soil particles remained in suspension in the escaping water. Subsequently these too settled out where the velocity of flow slowed or the water became ponded.

At locations where a heavily loaded natural or glacial river broadened or encountered flatter terrain so that its velocity decreased, coarse soil particles dropped out to form submerged spreading triangular-shaped deposits termed *alluvial fans*. The alluvial fans are good sources of sand and gravel for construction purposes.

Rivers flowing through broad flat valleys have often overflowed their banks during periods of flooding. When this occurred, the overflow velocity quickly diminished, and the heavier gravel and sand particles dropped out in the vicinity of the bank, forming low ridges termed *natural levees*. The broad lowland areas on either side of the river were also flooded over, but the materials dropped in the areas were the finer-grained soils. These are *flood plain deposits* (Fig. 1-6).

Where rivers bend or curve to change direction, the velocity of the flowing water can vary considerably between the inside and outside edges. Erosion may take place along the outside, while deposition takes place on the inside. The deposits are the coarser soils, generally sorted according to size. Constant erosion along the outside of a bend while the inside is being built up with sediments causes the river to migrate laterally. The old river locations provide good sources of coarse soils for construction and are good foundation sites.

When a river shortcuts a large bend when eroding a new route, the

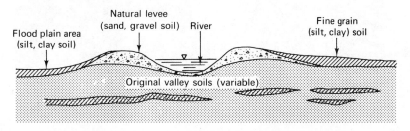

Figure 1-6. *Cross section of flood plain deposit.*

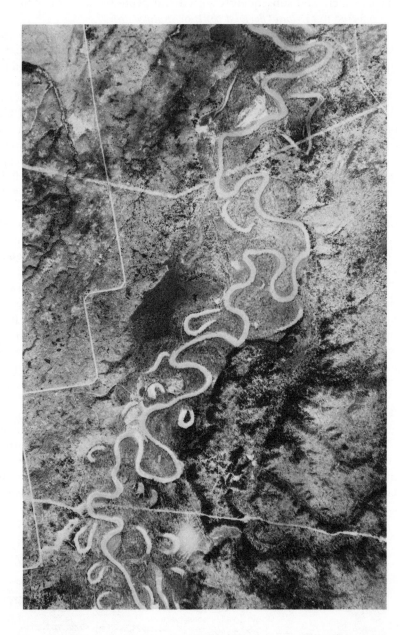

Figure 1-7.   *Aerial photo of meandering river and forming oxbow lakes.*

old channel left behind is cut off from new flow, and the trapped water forms an *oxbow lake* (Fig. 1-7). Such lakes eventually fill with predominantly fine-grained soils carried by low-velocity flood waters or surface runoff. These areas become poor foundation sites.

Lake areas were created in natural basins in the topography, or, if in glacial areas, in depressed reservoirs created between a terminal (or recessional) moraine and a retreating glacier. Natural lakes and glacial lakes often covered vast areas of land. The coarsest soil particles (sands) carried by rivers feeding into the lake would fall out of suspension quickly after entering the lake area because of the sharp decrease of velocity. Such coarse soil deposits are termed *lake deltas*, because of the resulting shape of the deposit, and are good sources of sand and gravel for construction purposes. Fine-grained particles remained in suspension and were carried to the body of the lake where they eventually settled out. The larger fine sand and silt particles settled out first while the smaller clay particles continued to remain in suspension. After the waters in the lake quieted, as in periods of little or no flow into the lake, the clay settled out. Alternating layers of these fine-grained deposits built up as the variation of water flowing into the lake area continued (Fig. 1-8). If the lake area were extremely large, the coarser silt particles might settle out in areas close to the shore while the clays settled out in the quiet central areas of the lake. Frequently, the basin eventually filled with soil, or drained and left the lake deposit if the trapped water created an outlet. Soil formations remaining at the locations of former lake areas are termed *lacustrine deposits*. When the alternating layers are less than $\frac{1}{8}$ inch (3 mm) in thickness, the resulting deposit of fine sand-silt and clay is known as *laminated clay*. Thicknesses greater than $\frac{1}{8}$ inch are termed *varved clays*. Unless such lake sediments were subject to the weight of a new glacier during a new glacial advance or to other overburden pressures, they are weak and compressible and make poor

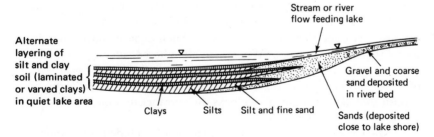

Figure 1-8. *Cross section of soil deposits in a lake area, indicating typical alternate layering of silt and clay soil.*

foundations. However, the deltas of coarse soils dropped when the flow-ing waters entered the lake area provide good foundation support and soils for construction use.

At some former glacial lake locations, the silt-clay soil near the surface has become firm because of drying (dessication). This can give the illusion of a strong deposit with good foundation capabilities. Such an area may be suitable for carrying roadways and light structures, but foundations for larger and heavier structures generally can not be satis-factorily supported.

Where flowing waters carried fine-grained soils to ocean or sea water areas, the smaller clay particles flocculated in the presence of the salt water.[2] The silt and clay then settled out of suspension at about the same rate, creating deposits of *marine clays*. In spite of their classifica-tions, marine clays may actually consist of more silt than clay. The marine clays are typically gray or blue-gray in color. Frequently, these deposits contain the shells or remains of shells from marine life. Much of the seaboard area of the northeastern United States and southeast Canada has such deposits. Because of the method of deposition, marine clay deposits are generally weak and compressible, and are poor founda-tion materials. Where the land covered by the marine clays has sub-sequently elevated to be above sea level, there is an additional danger that the sodium (from the sea water) that reacted with the clays to cause them to flocculate may have been leached from the soil by percolating rain water. The result is that the clays suffer a loss in strength and be-come sensitive to disturbance, and land areas that had been stable be-come unstable.

BEACH DEPOSITS / Ocean beach deposits are predominantly sand materials, and are constantly being changed by the erosive and redistributing effects of currents and wave action. These same currents and wave action keep silt and clay particles in suspension and carry them to the deeper, quieter, off-shore areas, where they eventually settle out. Long ridges of sand that form slightly offshore are termed *bars* (Fig. 1-9). When the formation that develops includes two or more submerged ridges, they are *longshore bars*. Such deposits have been built up by the breaking waves. When a small exposed ridge forms offshore from a gently sloping beach, the formation is termed an *off-shore bar* or *island bar* (also called barrier beach or barrier island). A *bar barrier* is a deposit that almost completely blocks the entrance to a bay. A bar that continues out into open water is a *split*.

Marine sands are somewhat weathered (rounded and smooth), particularly the larger particles, and the deposit at a particular loca-

---

[2]Flocculation may be defined as the development of an attraction between and bonding of individual particles to form larger particles.

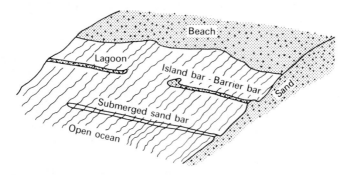

Figure 1-9. *Typical sand bar deposits along a shoreline.*

tion may consist of particles of uniform size. Such factors, plus the corrosion potential due to salinity, may affect the usefulness for certain construction purposes, such as making quality concrete. Generally, however, such sands are potentially good sources of material for construction purposes, particularly for waterfront and marine structures, because excavation is generally uncomplicated and transportation will be economical if barges or hydraulic pumping techniques (through pipelines) can be used.

SWAMP AND MARSH DEPOSITS / Swamps and marshes develop in stagnated areas where limited depths of water accumulate, or where periodic inundating and drying occurs because of fluctuations in the ground water level, and where vegetation has the chance to grow. The soils that subsequently form on the surface of swamp areas are generally of high organic content (from the decaying vegetation) and soft and odoriferous. Accumulations of decomposed or partially decomposed aquatic plants in swamp or marsh areas are termed *muck* or *peat*. Muck, geologically older than peat, is almost fully decomposed vegetation and is relatively dense. Peat includes partially decomposed vegetation and is normally spongy and relatively light. These materials are generally weak and highly compressible. Muck or peat deposits may be buried beneath the ground surface if the marsh area is subsequently overlain by glacial or alluvial materials.

In an early geologic period of time, when much of the present United States was covered over by shallow seas, an eastern section of the continent was generally a low swampy area where heavy vegetation flourished (Fig. 1-10). Due to subsequent geologic changes, these lush forest areas slowly became flooded and covered over. But while being destroyed, vast accumulations of forest vegetation were also being deposited. Subsequently, heat and pressure changed these deposits to the extensive coal beds found throughout Pennsylvania, West Virginia, Tennessee, Illinois, and Kentucky.

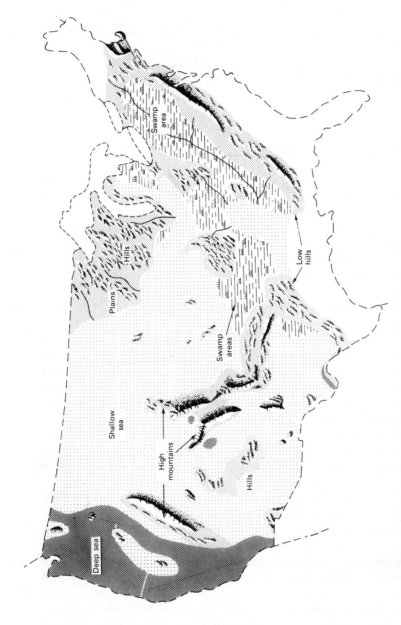

Figure 1-10. Generalized geographic map of the United States approximately 200 million years ago, indicating extensive swamp areas which subsequently formed into coal deposits. (Source: U.S. Department of the Interior, Geological Survey)

Effect on Design and Construction

Many large land areas have been formed with soils deposited primarily by one of the transportation methods described. One area may be underlain by glacially deposited soils, another by lake or marine deposits, another by river deposits.

However, it is not unusual to have stratification of soils that have been deposited by different methods. For instance, a glacial till may be covered with a glacial outwash or other alluvium. Soft lake deposits are known to exist over compact glacial till or river deposits of sands and gravels. Coarse soil alluvium may overlie soft lake deposits or marine clays. In areas that have been subjected to repeated glaciation, stratification of different types of glacially transported soils having different properties can exist. The result is that loose or soft soils may overlie compact or firm soils, and vice versa.

The type and condition of soil deposits underlying any proposed construction site must be an important consideration to the engineering and construction personnel concerned with the project, for it is the soil, or rock, that provides the support for the structure. Proper design for a structure includes investigation and evaluation of soil conditions underlying the proposed structure. However, knowledge of geology and the manner in which land forms and the soils in them have been created, as discussed throughout this chapter, can frequently serve to provide a preliminary evaluation of an area and the potential advantages or problems.

PROBLEMS

1. Name the three main types of rock and describe how they originated.

2. What factor of formation most influences the texture of igneous rocks?

3. What visual properties can frequently be used to distinguish between acidic rocks and basic rocks?

4. Describe the two typical processes that occur to transform soil sediments into sedimentary rock.

5. What processes occur to cause metamorphosis of rocks?

6. What type of bedrock formations would be thought to offer good foundation support for structures? What types would be more suspect of being a possible poor bedrock material?

7. What are loess soils and what is the potential danger to loss of stability in loess deposits?

8. Indicate the soil types to be expected in a glacial terminal moraine.

9. Indicate why sand or gravel deposits are frequently found along old river and stream locations.

10. What effect does the shape of a channel have upon stream flow and the related carried or deposited soil sediments?

11. How do natural levees originate? What is the major soil type expected in a natural levee?

12. What soil types would be expected in a river or stream delta?

13. Describe the process by which a glacial lake is formed.

14. How can lakes exist at elevations above the ground water table?

15. Why are naturally-filled-in lake locations often thought of as areas that offer poor support for building foundations?

16. How do eskers differ from drumlins, in regard to formation and soil types?

17. In glacially affected areas, what type of glacial formations represent possible good sources of sands and gravels for the construction industry?

18. Does an area covered by glacial till represent a location of advantage or disadvantage to the building and construction industry?

19. Do soils typical of beach deposits represent any advantage or disadvantage to the building and construction industry?

20. What is the potential danger to stability in areas where the land is formed from marine clays?

CHAPTER 2
Soil Composition: Terminology and Definitions

Soil deposits include the accumulated solid particles of soil or other materials plus the void spaces that exist between the particles. The void spaces are partially or completely filled with water or other liquid. Void spaces not occupied by fluid are filled with air or other gas. Since the volume occupied by a soil mass may generally be expected to include material in the three states of matter—solid, liquid, and gas—soil deposits are referred to as three-phase systems.

Significant engineering properties of a soil deposit, such as strength and compressibility, are directly related to or at least affected by basic factors such as how much volume or weight of a soil mass is solid particles or water or air. Information such as soil density (or weight per unit volume),[1] water content, void ratio, degree of saturation—terms defined in the following sections—is used in calculations to determine the bearing capacity for foundations, to estimate foundation settlement, and to determine the stability of earth slopes. In other words, such information helps to define the condition of a soil deposit for its suitability as a foundation or construction material. For this reason, an understanding of the terminology and definitions relating to soil composition is fundamental to the study of soil mechanics.

Soil Composition—Analytical Representation

A soil mass as it exists in nature is a more or less random accumulation of soil particles, water, and air space, as shown in Fig. 2-1(a).

---

[1]In soil mechanics, the term "density" is frequently used to indicate unit weight, i.e., pounds per cubic foot. For the metric system, unit weights are frequently expressed as grams per cubic centimeter, kilograms per cubic meter, or kilonewtons per cubic meter.

Soil particles

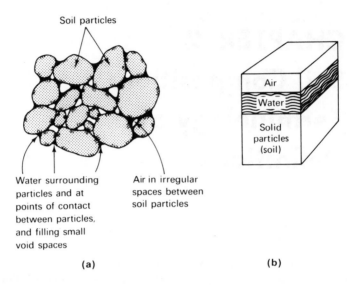

Water surrounding
particles and at
points of contact
between particles,
and filling small
void spaces

Air in irregular
spaces between
soil particles

(a)   **(b)**

Figure 2-1. *(a) Actual soil mass consisting of soil particles, water, and air; (b) block diagram representation of soil mass.*

For purposes of study and analysis it is convenient to represent this soil mass by a block diagram with part of the diagram representing the solid particles, part representing water or other liquid, and another part air or other gas, as shown in Fig. 2-1(b).

BASIC WEIGHT-VOLUME RELATIONSHIP / On the block diagram, the interrelationships of weights and volumes that make up the soil mass being analyzed can be shown. The relationships are summarized in Fig. 2-2.

As the diagram shows, the total *weight* of the soil volume is taken as the sum of the weight of solids plus water. In practical problems, all weighings are made in air, and the weight of air (in the voids) measured in air (the earth's atmosphere) is zero. If the gas is other than air, it

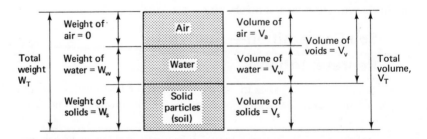

Figure 2-2. *Relationship between volumes and weights of a soil mass.*

may have a measurable weight, but it would normally be very small compared to the total weight of soil plus water and therefore can be neglected without causing serious error.

$$W_T = W_s + W_w \qquad (2\text{-}1)$$

The total *volume* of the soil mass includes the volume occupied by solids plus water (or liquid) plus air (or other gas). The total space occupied by water and air may collectively be indicated as the volume of voids.

$$V_T = V_s + V_w + V_a = V_s + V_v \qquad (2\text{-}2)$$

The relationship between weight and volume, for any material, is

$$W_x = V_x G_x \gamma_w \qquad (2\text{-}3)$$

where W_x = weight of the material (solid, liquid, or gas)
   V_x = volume occupied by the material
   G_x = specific gravity of the material, a dimensionless value
   γ_w = unit weight of water at the temperature to which the problem refers. (In most soils work, γ_w is usually taken as 62.4 pcf regardless of temperature. In experimental work, accurate values are used. A value of 1.0 gm/cc is also commonly used for γ_w in soils mechanics calculations.)[2]

For soil mechanics problems, then,

$$W_s = V_s G_s \gamma_w \qquad (2\text{-}4)$$

$$W_w = V_w G_w \gamma_w = V_w \gamma_w \qquad \text{since } G_w = 1 \qquad (2\text{-}5)$$

The specific gravity of most commonly occurring rock or soil materials is between 2.30 and 3.10. For many deposits, the specific gravity of soil solids lies within the range of 2.60 to 2.75.

---

[2]Actually, the gram is a unit of mass, not of weight as is the pound. A value of grams per cubic centimeter is in fact density (by scientific definition, density is mass per unit volume and is conventionally given the symbol ρ). Although unit weight and density are *not* identical, they are related by the expression $\gamma = \rho g$, where g is the acceleration of gravity. Consequently, in soil mechanics work the terms are generally taken to have synonymous meaning. Since comparative values of unit weight or density are normally sought to relate to other properties, this practice does not cause complications. Somewhat unfortunately, the use of γ_w being taken as equal to 1 gm/cm^3 is established in some technical fields. To a degree, the practice is carried through in this text so not to be at variance with other references in soil mechanics and to minimize confusion to students. But where applicable, the use of the proper term (density) is also indicated.

Basic Terms Relating to Soil
Composition and Condition

The unit weight of a soil γ is conventionally expressed as pounds per cubic foot or grams per cubic centimeter. Unit weights are reported as wet unit weight γ_{wet} or dry unit weight γ_{dry}:

Wet unit weight, $\gamma_{wet} = \dfrac{W_T}{V_T}$ (pcf or gm/cm^3)   (2-6)

Dry unit weight, $\gamma_{dry} = \dfrac{W_s}{V_T}$ (pcf or gm/cm^3)   (2-7)

From the definition, it can be seen that the wet unit weight includes the weight of water as well as soil particles in a soil volume. The dry unit weight is based upon only the weight of soil solids in the soil mass.

To relate a value given in grams per cubic centimeter (density) to pounds per cubic foot, multiply by 62.4.

By definition, water content w is the ratio of the weight of water in a soil mass to the weight of soil solids.

$$w = \frac{W_w}{W_s} \times 100\%$$ (2-8)

where w = water content expressed as a percentage
W_w = weight of water
W_s = weight of dry soil

(*Cautionary note:* Water content is *not* weight of water divided by the total weight W_T.)

The relationship of water content and weight of soil to total weight of a soil volume is shown below.

$$W_T = W_s + W_w \qquad \text{[from Eq. (2-1)]}$$

and since $W_w = wW_s \qquad \text{[from Eq. (2-8)]}$
$$W_T = W_s + wW_s$$
$$= W_s(1 + w)$$

and therefore

$$W_s = \frac{W_T}{1 + w}$$ (2-9)

(*Note:* In this equation w has to be expressed as a decimal.)

This equation enables the dry soil weight to be easily determined when the wet weight of a large soil sample is known and the water content is determined from a small representative portion taken from the sample. This procedure for determining dry soil weight is frequently used in laboratory and construction work.

Two terms, void ratio and porosity, express a relationship between the volumes of a soil mass occupied by solids and nonsolids. Void ratio e is

$$e = \frac{\text{vol. of voids}}{\text{vol. of solids}} = \frac{V_v}{V_s} \qquad (2\text{-}10)$$

Void ratio is expressed as a decimal. Porosity n is

$$n = \frac{\text{vol. of voids}}{\text{total vol.}} \times 100 \text{ percent} = \frac{V_v}{V_t} \times 100 \text{ percent} \qquad (2\text{-}11)$$

Porosity is conventionally expressed as a percentage. The relationship between void ratio and porosity is

$$e = \frac{n}{1 - n} \qquad (2\text{-}12a)$$

(*Note:* In this equation n is expressed as a decimal.)

$$n = \frac{e}{1 + e} \times 100 \text{ percent} \qquad (2\text{-}12b)$$

The term degree of saturation, S, indicates the portion of the void spaces in a soil mass that are filled with water. Degree of saturation is expressed as a percentage.

$$S \text{ percent} = \frac{V_w}{V_v} \times 100 \text{ percent} \qquad (2\text{-}13)$$

Full saturation, or 100 percent saturation, indicates that all voids are filled with water. A soil can remain 100 percent saturated even though its water content is changed if the soil experiences compression or expansion (since compression or expansion indicates a decrease or increase in void spaces).

With reference to a block diagram, as shown in Fig. 2-3, other useful weight-volume relationships can be developed.

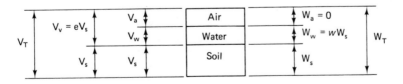

Figure 2-3. *Block diagram used to develop equations applicable to weight-volume relationship.*

Since $V_T = V_s + V_v$, and $e = \dfrac{V_v}{V_s}$,

$$V_T = V_s + eV_s = V_s(1 + e) \tag{2-14a}$$

or

$$V_s = \frac{V_T}{1 + e} \tag{2-14b}$$

Also, since $\quad \gamma_{\text{dry}} = \dfrac{W_s}{V_T}$ [from Eq. (2-7)]

$$\gamma_{\text{dry}} = \frac{V_s G_s \gamma_w}{V_T} = \frac{V_s G_s \gamma_w}{V_s(1 + e)} = \frac{G_s \gamma_w}{(1 + e)} \tag{2-15}$$

For a fully saturated soil, the unit weight becomes

$$\gamma_{\text{sat}} = \frac{W_T}{V_T} = \frac{W_s + W_w}{V_s(1 + e)} \quad \text{[from Eq. (2-6)]}$$

$$= \frac{V_s G_s \gamma_w + V_w \gamma_w}{V_s(1 + e)} = \frac{V_s G_s \gamma_w + eV_s \gamma_w}{V_s(1 + e)}$$

$$= \frac{(G_s + e)\gamma_w}{(1 + e)} \tag{2-16}$$

A very informative relationship can be obtained by proper substitution of terms into Eq. (2-5).

$$W_w = V_w G_w \gamma_w = V_w \gamma_w \quad \text{[from Eq. (2-5)]}$$

or

$$V_w = \frac{W_w}{\gamma_w}$$

Dividing both sides by V_s gives

$$\frac{V_w}{V_s} = \frac{W_w}{V_s \gamma_w}$$

Multiplying the left term by $\dfrac{V_v}{V_v}$ gives

$$\frac{V_w}{V_s} \times \frac{V_v}{V_v} = \frac{W_w}{V_s \gamma_w}$$

and

$$\frac{V_w}{V_v} \times \frac{V_v}{V_s} = \frac{wW_s}{V_s \gamma_w}$$

$$S \cdot e = w\left(\frac{W_s}{V_s \gamma_w}\right)$$

$$Se = wG_s \tag{2-17}$$

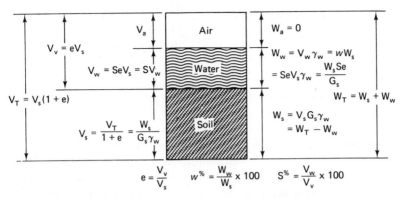

Figure 2-4. *Summary of weight-volume equations.*

For a given soil and density, this equation shows the relationship between void ratio and water content, and the limiting water content that can be obtained.

A summary of the weight-volume equations most frequently used in solving problems is presented in Fig. 2-4. In performing the analysis to determine weight or volume properties, it is extremely helpful to sketch the block diagram, and indicate on the diagram that information which is known (from given data or as it is developed). This procedure helps to guide the analyst through the proper steps or equations necessary for a complete solution and will eliminate unnecessary computations.

Illustration 2-1: A sample of soil obtained from a test pit is one cubic foot in volume and weighs 140 lb. The entire sample is dried in an oven and found to weigh 125 lb. Calculate the water content, wet unit weight, and dry unit weight.

$$\text{weight of water} = W_w = 140 - 125 = 15 \text{ lb}$$
$$\text{weight of dry soil} = W_s = 125 \text{ lb}$$
$$\text{total volume of sample} = V_T = 1.0 \text{ ft}^3$$

$$\text{wet unit weight} = \gamma_{\text{wet}} = \frac{W_T}{V_T} = \frac{140 \text{ lb}}{1.0 \text{ ft}^3} = 140 \text{ pcf}$$

$$\text{dry unit weight} = \gamma_{\text{dry}} = \frac{W_s}{V_T} = \frac{125 \text{ lb}}{1.0 \text{ ft}^3} = 125 \text{ pcf}$$

$$\text{water content} = w = \frac{W_w}{W_s} = \frac{15 \text{ lb}}{125 \text{ lb}} = 0.12 = 12 \text{ percent}$$

Illustration 2-2:   Calculate the dry unit weight, void ratio, water content, and degree of saturation for a sample of moist soil weighing 40 pounds and having a volume of 0.32 ft³. When dried in an oven, the soil weighed 36 lb. The specific gravity of the soil solids is 2.70.

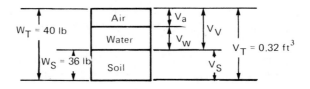

$$\text{dry unit weight} = \gamma_{\text{dry}} = \frac{W_s}{V_T} = \frac{36 \text{ lb}}{0.32 \text{ ft}^3} = 112.5 \text{ pcf}$$

$$\text{water content} = w = \frac{W_w}{W_s} = \frac{40 \text{ lb} - 36 \text{ lb}}{36 \text{ lb}}$$

$$= 0.11 = 11 \text{ percent}$$

$$\text{void ratio} = e = \frac{V_v}{V_s} = \frac{0.106}{0.214} = 0.495$$

$$\left[\text{where } V_s = \frac{W_s}{G_s \gamma_w} = \frac{36 \text{ lb}}{(2.70)(62.4 \text{ pcf})} = 0.214 \text{ ft}^3 \right.$$

$$\left. \text{and } V_v = V_T - V_s = 0.32 \text{ ft}^3 - 0.214 \text{ ft}^3 = 0.106 \text{ ft}^3 \right]$$

$$\text{degree of saturation} = S = \frac{w G_s}{e} = \frac{(0.11)(2.70)}{(0.495)}$$

$$= 0.60 = 60 \text{ percent}$$

Illustration 2-3:   A 150-cc sample of wet soil scales 250 g when 100 percent saturated. It is oven-dried and found to weigh 162 g. Calculate the dry unit weight (really density, since gram units are used), water content, void ratio, and G_s.

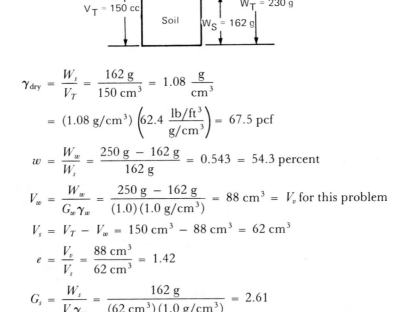

$$\gamma_{dry} = \frac{W_s}{V_T} = \frac{162 \text{ g}}{150 \text{ cm}^3} = 1.08 \frac{\text{g}}{\text{cm}^3}$$

$$= (1.08 \text{ g/cm}^3) \left(62.4 \frac{\text{lb/ft}^3}{\text{g/cm}^3}\right) = 67.5 \text{ pcf}$$

$$w = \frac{W_w}{W_s} = \frac{250 \text{ g} - 162 \text{ g}}{162 \text{ g}} = 0.543 = 54.3 \text{ percent}$$

$$V_w = \frac{W_w}{G_w \gamma_w} = \frac{250 \text{ g} - 162 \text{ g}}{(1.0)(1.0 \text{ g/cm}^3)} = 88 \text{ cm}^3 = V_v \text{ for this problem}$$

$$V_s = V_T - V_w = 150 \text{ cm}^3 - 88 \text{ cm}^3 = 62 \text{ cm}^3$$

$$e = \frac{V_v}{V_s} = \frac{88 \text{ cm}^3}{62 \text{ cm}^3} = 1.42$$

$$G_s = \frac{W_s}{V_s \gamma_w} = \frac{162 \text{ g}}{(62 \text{ cm}^3)(1.0 \text{ g/cm}^3)} = 2.61$$

Illustration 2-4: Laboratory test data on a sample of saturated soil show that the void ratio is 0.45 and the specific gravity of soil solids is 2.65. For these conditions, determine the wet unit weight of the soil and its water content.

This sample is saturated; thus all voids are filled with water.

$$e = \frac{V_v}{V_s} = 0.45$$

but V_v and V_s are not known. On the block diagram, assume that V_s is unity. Therefore

$$V_T = V_s + eV_s = 1.0 + 0.45 = 1.45$$

As a result, $W_s = V_s G_s \gamma_w$

$$= (1.0)(2.65)(62.4 \text{ pcf}) = 165 \text{ lb}$$

and $W_w = V_w \gamma_w = (0.45 \text{ ft}^3)(62.4 \text{ pcf}) = 28 \text{ lb}$

from which $W_T = W_s + W_w = 165 \text{ lb} + 28 \text{ lb} = 193 \text{ lb}$

$$\gamma_{\text{wet}} = \frac{W_T}{V_T} = \frac{195 \text{ lb}}{1.45 \text{ ft}^3} = 133 \text{ pcf}$$

$$w = \frac{W_w}{W_s} = \frac{28 \text{ lb}}{165 \text{ lb}} = 0.17 = 17 \text{ percent}$$

Submerged Soil

In many soil mechanics problems it is necessary to determine the effective weight of a soil when it is below the ground water table. For this condition, the soil solids are buoyed up by the water pressure, and the submerged soil weight becomes less than for the same soil above water. The effective soil weight then becomes the unit weight of the soil mass when it is weighed under water. The water in the voids has zero weight (when submerged, all voids can be assumed to be filled with water), and the weight of the soil solids is reduced by the weight of the water they displace. Therefore, a submerged weight (W_{sub}) equals the soil weight above water minus the weight of water displaced, or

$$W_{\text{sub}} = V_s G_s \gamma_w - V_s G_w \gamma_w = V_s \gamma_w (G_s - G_w)$$

$$W_{\text{sub}} = V_s \gamma_w (G_s - 1) \tag{2-18}$$

Since unit weight is total weight divided by total volume,

$$\gamma_{\text{sub soil}} = \frac{W_{\text{sub}}}{V_{\text{total}}} = \frac{V_s \gamma_w (G_s - 1)}{V_s (1 + e)}$$

$$\gamma_{\text{sub soil}} = \frac{(G_s - 1)}{(1 + e)} \gamma_w \tag{2-19}$$

Equation (2-19) indicates that an accurate determination of the submerged soil weight requires that the specific gravity of the soil solids and the void ratio be known. Unfortunately, in terms of time and expense, some testing or physical analysis is required to determine the specific gravity, which in turn is used to compute the void ratio. Also unfortunately, insofar as analytical studies for practical soil mechanics problems are concerned, void ratios and soil weights vary somewhat even in "uniform deposits." Because of this fact, the effort to make highly accurate determinations of submerged soil weights is rarely made when studies and designs are done. Instead, satisfactory estimates, which can be made from knowing a wet weight, are frequently utilized.

For most soils, and fortunately for ease of computation, the submerged weight is on the order of half the wet soil weight above the water table. The most notable exception to this rule is for soils containing significant decomposed vegetation or organic material.

$$\gamma_{\text{sub soil}} = \tfrac{1}{2}\gamma_{\text{wet soil}} \quad \text{(approximately)} \tag{2-20}$$

For many practical problems, errors resulting from applying this simplification are negligible. Where accuracy is required, Eq. (2-19) should be used. However, for the situation where the soil above the water table is 100 percent saturated, Eq. (2-19) for the submerged unit weight becomes (simple and exact)

$$\gamma_{\text{sub soil}} = \gamma_{\text{sat. soil}} - \gamma_w$$

$$\gamma_{\text{sub soil}} = \gamma_{\text{sat. soil}} - 62.4 \qquad \text{(in pcf)} \tag{2-21}$$

Illustration 2-5: An undisturbed, one-cubic-foot volume of soil obtained from a test pit is found to have a wet weight of 103.2 lb. The dry weight of the sample is 84.5 lb. What would be the effective unit weight of such a soil if it were submerged below the ground water table? The specific gravity of the soil is determined to be 2.70. By Eq. (2-20), the effective submerged weight is approximately

$$\gamma_{\text{sub}} = \tfrac{1}{2}\gamma_{\text{wet}} = \tfrac{1}{2}(103.2 \text{ pcf}) \simeq 52 \text{ pcf}$$

An accurate determination is as follows:

$$V_s = \frac{W_s}{G_s\gamma_w} = \frac{84.5 \text{ lb}}{(2.70)(62.4 \text{ pcf})} = 0.50 \text{ ft}^3$$

$$V_v = V_T - V_s = 1.0 \text{ ft}^3 - 0.50 \text{ ft}^3 = 0.50 \text{ ft}^3$$

$$e = \frac{V_v}{V_s} = \frac{0.50}{0.50} = 1.00$$

$$\gamma_{\text{sub}} = \left(\frac{G_s - 1}{1 + e}\right)\gamma_w \qquad \text{[from Eq. (2-19)]}$$

$$= \left(\frac{2.70 - 1}{1 + 1.0}\right)(62.4 \text{ pcf}) = 53.2 \text{ pcf}$$

or

submerged weight,

$$W_{\text{sub}} = V_s\gamma_w(G_s - G_w) = (0.50)(62.4 \text{ pcf})(1.70)$$

$$= 53.2 \text{ lb}$$

and for a volume of one cubic foot,

$$\gamma_{\text{sub}} = \frac{W_{\text{sub}}}{V_T} = \frac{53.2 \text{ lb}}{1.0 \text{ cf}} = 53.2 \text{ pcf}$$

Illustration 2-6: Assume that a one-cubic-foot volume of soil, similar to the soil from the preceding illustration is excavated from a location below the water table. The soil is now 100 percent saturated. What saturated weight would be expected?

$$\gamma_{sub} = \gamma_{sat.} - \gamma_w \tag{2-21}$$

or
$$\gamma_{sat.} = \gamma_{sub} + \gamma_w$$
$$= 53.2 \text{ pcf} + 62.4 \text{ pcf} = 115.6 \text{ pcf}$$

PROBLEMS

1. A sample of soil obtained from a construction site is found to have a wet weight of 27.5 lb. When completely dried, the soil weighs 24.3 lb. What is the water content of the soil sample?

2. A sample of soil obtained from a borrow pit has a wet weight of 35 lb. The total volume occupied by the sample when in the ground was 0.30 ft^3. A small portion of the sample is used to determine the water content. When wet, the sample weighs 150 g. After drying, it weighs 125 g.
    (a) What is the water content of the sample?
    (b) Determine the wet and the dry unit weights of the soil in the borrow pit.

3. The dry weight of soil particles in a soil sample is 250 g. When immersed in water, the soil particles displace 95 cm^3, and this is then the volume of soil solids. Using this data, determine the specific gravity of soil solids.

4. A 1-ft^3 sample of undisturbed soil is found to have a dry weight of 107 lb. If the specific gravity of soil solids is 2.70, what is the void ratio of the sample? What is the porosity?

5. A dry sand is placed in a container having a volume of $\frac{1}{4}$ ft^3. The dry soil weight is 27 lb. If the specific gravity of soil solids is 2.75, what is the void ratio of the sand in the container?

6. A dry sand is placed in a container having a volume of 0.30 ft^3. The dry weight of the sample is 31 lb. Water is carefully added to the container so as not to disturb the condition of the sand. When the container is filled, the combined weight of soil plus water is 38.2 lb. From these data, compute the void ratio of the soil in the container, and the specific gravity of the soil particles.

7. What will be the dry unit weight of a soil whose void ratio is 1.20, where the specific gravity of soil solids is 2.72?

8. A sand is densified by compaction at a construction site so that the

void ratio changes from 0.80 to 0.50. If the specific gravity of solids is 2.70, what is the increase in the dry unit weight of the sand?

9. Demonstrate that $e = \dfrac{n}{1 - n}$.

10. An undisturbed sample of clay is found to have a wet weight of 63 lb, a dry weight of 51 lb, and a total volume of 0.50 ft^3. If the specific gravity of soil solids is 2.65, determine the water content, void ratio, and degree of saturation.

11. A clay sample has a wet weight of 417 g and occupies a total volume of 276 cm^3. When oven-dried, the sample weighs 225 g. If the specific gravity of soil solids is 2.70, calculate the water content, void ratio, and degree of saturation.

12. Given the following data for an undisturbed soil sample

$$G_s = 2.69, \qquad e = 0.65, \qquad w = 10 \text{ percent}$$

determine wet unit weight, dry unit weight, and degree of saturation.

13. A saturated sample of undisturbed clay has a wet weight of 700 g. The sample has a total volume of 425 cm^3. When dry, the sample weighs 450 g. What is the specific gravity of the soil solids?

14. A sample of soil obtained from beneath the ground water table is found to have a water content of 20 percent. If it is assumed that the specific gravity of most soil particles is within the range between 2.60 and 2.75, what is the approximate void ratio of the soil sample?

15. In an undisturbed soil formation, it is known that the dry soil weight is 115 pcf. The specific gravity of the soil particles is 2.75.

   (a)   What is the saturated wet unit weight of the soil?
   (b)   What is the effective submerged weight of the soil?

16. An undisturbed soil sample has a wet unit weight of 120 pcf when the water content w is 15 percent. The specific gravity of the soil particles is 2.65.

   (a)   Approximate the effective submerged weight of this soil.
   (b)   Determine the effective submerged weight, using the exact formula.

CHAPTER 3
Soil Types and
Soil Structure

The term *soil*, as generally used, refers to the accumulation of particles of disintegrated rock, and, frequently, also man-made materials. Because of the experienced wide variation in characteristics and behavior, soil has been subdivided into categories based upon the materials' physical properties. In nature, soils are comprised of particles of varying size and shape. Size and, to some extent, shape are factors that have been found to be related to or to affect the material behavior of soil to some degree. Consequently, soil categories or types have been developed that are basically referenced to size. For the category that is below the size that can be discerned visually, an additional property, plasticity (or nonplasticity), is used as a criterion for indicating type.

Experiences and study have proved that a soil's important behavioral properties are not always controlled by particle size and plasticity. Soil structure and mineralogical composition, and the intereffect with water, may also have significant influence on the properties and behavior deemed important for design and construction. Under certain conditions, simple typing of the soil provides adequate information for design and construction, whereas certain other conditions require that detailed information about the soil's composition and structure be determined.

Major Soil Types

The major categories of soil are *gravel, sand, silt,* and *clay.* There is not unanimous agreement on the exact division between each of these major soil types, but gravel and sand *are* universally considered as coarse-grained soil, for the individual particles are large enough to be distinguished without magnification. The silts and clays are considered

35

as *fine-grained soil* because of their small-size particles, which, for the most part, are too small to be seen unaided.

The most commonly used divisions for classifying soils for engineering and construction purposes are shown in Table 3-1. On a comparative basis, the division sizes between gravel and sand (4.76 mm or 2.00 mm), and sand and silt-clay (0.074 mm or 0.05 mm) are actually quite close. As a result, lack of agreement on these division sizes normally does not cause serious problems.

TABLE 3-1 / SIZE RANGE FOR SOIL TYPES

Soil Type	Upper Size Limit	Lower Size Limit
Gravel	Varies between 3 in. up to about 8 in. (8 cm to 20 cm)	4.76 mm (about 0.20 in.) determined by #4 U. S. Standard Sieve) or 2.00 mm (#10 U. S. Standard Sieve)
Sand	4.76 mm or 2.00 mm	0.074 mm (#200 U. S. Standard Sieve) or 0.050 mm (#270 U. S. Standard Sieve)
Silt and Clay	0.074 mm or 0.05 mm	None

Particles larger than gravel are commonly referred to as cobbles or boulders. Again, no unanimous agreement exists on range of sizes. When gravel extends up to the eight-inch size (20 cm), anything larger would be termed a boulder. Where the three-inch (8 cm) size, or thereabouts, is taken as the upper size for gravel, the sizes between three inches and eight inches may be designated as cobbles, and anything larger than eight inches as boulders. However, it may also be expected to find that six inches or 12 inches (15 cm or 30 cm) is taken as the division between cobbles and boulders. As for sands and gravels, these discrepancies usually do not cause serious problems. Conventionally, when a construction project requires a particular material, it has become standard practice to indicate the soil or aggregate requirements on the basis of size, instead of or as well as classification.

From the above, it is apparent that particle size alone serves as the basis for classification of sands, gravels, cobbles, and boulders.

The classification of a fine-grained soil as either a silt or a clay is not on the basis of particle size but rather on the plasticity or nonplasticity of the material. Clay soil is plastic over a range of water content—that is, the soil can be remolded or deformed without causing cracking, breaking, or change in volume, and will retain the remolded shape. The clays are frequently "sticky." When dried, a clay soil possesses very high strength (resistance to crushing). A silt soil possesses little or no plasticity, and when dried has little or no strength. If a small sample of moist silt is shaken easily but rapidly in the palm of the hand, water will appear on the surface of the sample but disappear when shaking stops. This is referred to as *dilatancy*. When a sample of moist clay is similarly shaken, the surface will not become wetted.

The reason for the difference in behavior between clay and silt relates to the difference in mineralogical composition of the soil types and particle shape. Silt soils are very small particles of disintegrated rock, as are sands and gravels, and possess the same general shape and mineralogical composition as sands and gravels (which are nonplastic). The clay minerals, however, represent chemical changes that have resulted from decomposition and alteration of the original rock minerals. The effect is that their size and shape are significantly different from other types of soil particles. This is discussed further in a following section.

Naturally occurring soil deposits most generally include more than one soil type. When they are classified, all the soil types actually present should be indicated, but the major constituent soil type should dominate the description, while the soils of lesser percentage are used as modifying terms, e.g., a material that is mostly sand but includes silt would be classified as a silty sand, while a silt-clay mixture with mostly clay would be termed a silty clay.

Though a soil may be predominantly coarse-grained, the presence of silt or clay can have significant effect on the properties of the mixture. Where the amount of fine-grained material exceeds about a third of the total soil, the mixture behaves more like a fine-grained soil than a course-grained soil.

The condition also exists where small fragments of decomposed vegetation are mixed with the soil, particularly fine-grained soils. Organic material, mixed with the nonorganic soil, can have striking detrimental effects on the strength and compressibility properties of the material. The presence of organic material should be carefully watched for. A foul odor is characteristically though not always associated with such soils, as is a blackish or dark gray color. Soils in this category are designated as *organic* (i.e., *organic silt* or *organic clay*) in comparison to a nonorganic designation for soil free of decomposed vegetation.

Particle Shapes and Sizes

Particles in the sand, gravel, and boulder categories are considered as "bulky grain," indicating that particle dimensions are approximately equal, i.e., the dimension in a length, width and thickness direction would be of the same order of magnitude (that is, one dimension is no more than two or three times larger or smaller than another dimension). Individual particles are frequently very irregular in shape, depending somewhat on the rock they were derived from, their age, and exposure to weathering and transporting processes. Generally, a new particle is "angular" and rough-surfaced, and is then modified with time and exposure to become more smooth-surfaced and rounder. The various stages of transition—angular, subangular, rounded—are illustrated in Fig. 3-1. Generally, the angular particles possess better engineering properties, such as higher shear strength, then do weathered and smooth particles.

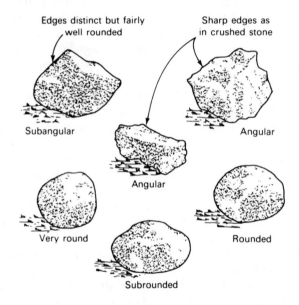

Figure 3-1.   *Shapes of granular soil particles.* (Ref. 1)

Particles in the silt category, though classified as "fines" along with the clays, are still angular or bulky in shape and of the same mineralogical composition as the coarse-grained soils. Because of the mineralogical composition, such particles rarely break down to less than $2\,\mu$ in size. (one μ = one micron = 0.001 mm).

The mineralogical composition of true clay is distinctly different from the mineral components of the other soil types, inviting the distinctions *clay-minerals* and *nonclay minerals*. Almost all clay minerals are crystalline minerals (having an orderly, sheet-like molecular arrangement) that are capable of developing cohesion (because of an attraction and bonding between soil sheets and water) and plasticity. Clay particles may be made of many sheets on top of one another. Clay particles are mostly found in sizes less than $2\,\mu$ or easily break down to this size. It is emphasized, however, that it is the mineral type and not the small size that is primarily responsible for the high cohesion and plasticity that clays possess. Where particles of nonclay minerals have been broken down to smaller than $2\,\mu$, the clay properties are not developed.

Because of the sedimentary origins of fine-grained soil deposits, and the overlap in sizes of the clay and nonclay minerals, it is unusual to find natural deposits of pure clay mineral soils. Very frequently, "clay deposits" are actually a mixture of clay minerals and the nonclay minerals. Because of this, the term *clay material* has been used to prevent confusion when one is designating a naturally occurring soil deposit consisting of fine-grained soils that have the general properties of cohesion and plasticity.

The clay minerals themselves can and do vary in their composition and, therefore, in their behavior properties. The building blocks, or constituent sheets, that combine to form most of the different types of clay mineral are the *silica tetrahedral* sheet (Fig. 3-2) and *alumina octahedral* sheet (Fig. 3-3). The silica tetrahedron consists of four oxygen ions and one silicon ion. The molecular arrangement is such that the four oxygens are spaced and located at what would be the corners and tip of a three-dimensional, three-sided pyramid, with the silicon located within the pyramid. Oxygen ions at the base are shared by adjacent tetrahedrons, thus combining and forming the sheet. The thickness of a silica sheet is 5×10^{-7} mm, or 5 Angstrom units. (One Angstrom unit = $1\,\text{Å} = 1 \times 10^{-7}$ mm.)

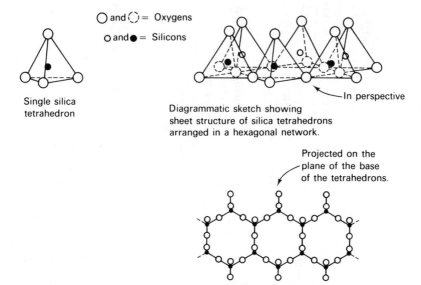

Single silica tetrahedron

○ and ◌ = Oxygens

o and ● = Silicons

Diagrammatic sketch showing sheet structure of silica tetrahedrons arranged in a hexagonal network.

—In perspective

Projected on the plane of the base of the tetrahedrons.

Figure 3-2. *Assembly of atoms forming the basic clay mineral sheet—silica tetrahedrol (Ref. 58)*

TABLE 3-2 / BASIC PROPERTIES OF SOME TYPICAL CLAYS

Clay Mineral	Composition	Layer Thickness	Shape of Mineral, General Properties, and Comments
Kaolinite	One silica, one alumina sheet. Very strongly bonded together.	7.5 Å	The most prevalent clay mineral. Very stable, little tendency for volume change when exposed to water. Kaolinite layers stack together to form relatively thick particles. Particles are plate-shaped. Form from crystalline rocks in humid climates.
Halloysite	One silica, one alumina sheet make up the layer. Has sheet of water molecules between layers. (Similar to kaolinite except for sheet of water.)	10 Å	Sheets of halloysite curl into tubes. Strength and plasticity are significantly affected by drying and removal of the water. After drying, the clay mineral will not reinstate a water layer if again exposed to water. Caution is required in identifying this mineral and in using remolded (and rewetted) samples in the laboratory testing to determine properties. Dried halloysite has characteristics of kaolinite. Rewetted samples appear stronger and less plastic than naturally wetted halloysite.
Illite	Alumina sheet sandwiched between two silica sheets to form the layer. Potassium provides the bond between layers.	10 Å	Irregular flake shape. Generally more plastic than kaolinite. Does not expand when exposed to water unless a deficiency in potassium exists. Illite clays seem most prevalent in marine deposits and soil derived from micaceous rock (shists, etc.).

Mineral	Structure	Spacing	Properties / Occurrence
Montmorillonite	Alumina sheet sandwiched between two silica sheets to form the layer. Iron or magnesium may replace the alumina in the aluminum sheet; aluminum may replace some silicons in the silica sheet (isomorphous substitution). Weak bond between layers.	9.5 Å	Irregular plate shapes or fiberous. Because of the weak bond between layers and the negative charge resulting because of isomorphous substitution, the clay readily adsorbs water between layers. This mineral has a great tendency for large volume change because of this property. Montmorillomite forms mostly from ferromagnesium rock in areas where high temperatures exist and rainfall is abundant, also from decomposition of volcanic ash.
Chlorite	Layers formed of alumina sheet sandwiched between two silica sheets, but layers are bonded together with an alumina sheet.		Irregular plate shapes. Nonexpanding. Formed from well-drained soils and micaceous rocks in humid areas.

41

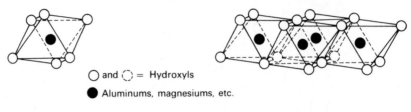

◯ and ◌ = Hydroxyls

⬤ Aluminums, magnesiums, etc.

Single octahedral unit

Diagrammatic sketch showing the sheet
structure of the octahedral units

Figure 3-3.   *Assembly of atoms forming the basic clay mineral sheet—
Alumina octahedron. (Ref. 58)*

The alumina octahedron consists of six oxygens and one aluminum. Three of these oxygens are in the top plane of the octahedron,
and three are in the bottom plane. The aluminum is within the oxygen
grouping. It is possible that the aluminum ion may be replaced with
magnesium, iron, or other neutral ions. To obtain a valence balance,
some of the oxygens may also carry a hydrogen iron (resulting in a
hydroxyl at some oxygen locations). The alumina sheet is also 5 ×
10^{-7} mm or 5 Å units thick.

It also happens that oxygens from the tip of a silica tetrahedron
can share in an alumina sheet, thus layering sheets. Different arrangements of sheets then can combine to form the different clay minerals.
The composition and typical properties of the more commonly occurring
clays are summarized in Table 3-2.

Though the thickness of a clay mineral sheet is limited, the dimensions in the length and width direction are not. As a result, the clay
minerals have a flat, plate-like shape (like an irregular sheet of paper),
where the length and width can be several tens or several hundreds
times the thickness.

Clay and Water

The surfaces of clay mineral particles have a net electrical charge
that is negative, while the edges have positive and negative charges.
This results from the molecular grouping and arrangement of ions.
However, the charges are not uniform, varying in intensity at different
locations on the particle. Because of the extremely small size of clay
particles and the very high ratio of particle surface to particle mass, the
forces of electrical charge have profound effect on the behavior of particles coming in association with other particles and water (or other
fluids) present in the soil. And because of the manner of sedimentary
deposition, clay deposits almost always exist in the presence of some
water.

The engineering behavior of coarse particles, because of a comparatively low ratio of particle surface to mass, are not significantly affected by surface electrical charges.

Clay mineral particles would, by themselves, tend to repel each other because of the net negative charges, unless edge-to-surface contact were made (positive to negative would attract). However, because of the net negative charge, the particles will attract cations (positive ions) such as potassium, sodium, calcium, and aluminum from moisture in the soil (ions present as a result of solutions from rock weathering), so as to obtain an electrically balanced or equilibrium condition.

Because of the net positive charge of the cations, they in turn can also attract negative charges. As a result of this phenomenon, water becomes bonded to the cations. Water, though neutral, has its oxygen and hydrogen atoms spaced in such a manner that the center of gravity of the positive and negative electrical charges do not coincide. The resulting molecule has a plus charge acting at one end and a negative charge acting at the opposite end, similar to a bar magnet. Water molecules are thus considered *polar molecules* (Fig. 3-4). The negative tips of water molecules are attracted and held to the cation, which in turn is held by the clay particle. The resulting effect is that significant water (significant with respect to the size and weight of the clay particle) becomes "bonded" to the clay. Water molecules are also held to the particle surface, where they become attracted directly to a location of negative charge.

Additional water molecules then also become attracted to the clay particle because of a chain-like arrangement of negative ends to positive ends of molecules, and by hydrogen bonding (the condition where hydrogen atoms in water are shared with hydrogen atoms in the clay).

The state or nature of the water immediately surrounding a clay particle is not clearly understood by soil scientists, but it is generally

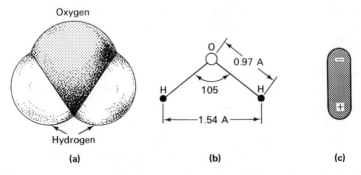

Figure 3-4. (a) Model of water molecule; (b) relative location of atoms in the water molecule; (c) polar representation of water molecule.

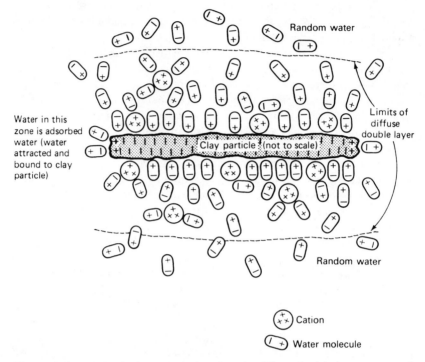

Figure 3-5. *Adsorbed water and cations in diffuse double layer surrounding clay particle.*

known to possess properties different from liquid water. It may be in a very dense and viscous state. It *is* certain that this water is very strongly attracted to the clay particle, however.

The attraction for cations necessary to balance the negative charge of the clay extends beyond the surface layer of molecules surrounding the particle. The further from the particle surface, though, the weaker the attraction becomes, and, therefore, the concentration of cations becomes lessened. The resulting effect is that water molecules are still attracted to the clay particles, indirectly, but the further from the particle, the weaker the attraction. At a distance beyond where cation and, therefore, water molecules are attracted to a clay particle, water in the soil is considered "loose" or "normal" pore water.

The distance from the clay particle surface to the limit of attraction is termed the *diffuse double layer* (Fig. 3-5). This term originates from the belief that immediately surrounding the particle a thin, very tightly held layer of water, perhaps 1×10^{-6} mm (10 Å) thick exists, and a second, more mobile, diffuse layer extends beyond this first layer to the limit of attraction. Molecular movement in the outer layer (or outer zone), and probably also in the tightly held layer, continually occurs, however. The water that is held in the diffuse double layer is frequently termed *adsorbed water* or *oriented water* to differentiate it from normal pore water, which is not oriented.

The plasticity that clay soils possess is attributed to the attracted and held water. And, as a restatement, water molecules are attracted because of their dipole structure. The unusual properties of plasticity possessed by clays result because of the unusual molecular structure and the common presence of water in soil deposits. Experiments performed with clay using nonpolar liquid in place of water result in "no plasticity" conditions, where the particles act similarly to a coarse-grained sandy soil.

Soil Structure

The particle arrangement of the equidimensional particles—i.e., gravel, sand, and silt—has been likened to arrangements that can be obtained by stacking marbles or oranges. For similarly sized spherical particles, a loose condition (condition with a high void ratio) is obtained from an arrangement as shown in Fig. 3-6(a). A dense condition (condition with a low void ratio) is obtained from an arrangement as in Fig. 3-6(b).

(a) Loose   (b) Dense

Figure 3-6. *Schematic diagram of grain arrangement for loose and dense granular soils.*

Actual soil deposits are made of accumulations of soil particles having at least some variation, but more frequently great variation, in the particle sizes. As a result, the soil structure is not quite like that presented in Fig. 3-6. Generally, the greater the range of particle sizes, the smaller the total volume of void spaces there will be. However, for a given soil deposit, a range of conditions between loose and dense is possible. Typical values for different types of soil mixtures are tabulated in Table 3-3.

In relating the volume of void spaces to properties desirable for building construction purposes, it is generally recognized that the smaller the void ratio (or the denser the material), the higher the strength and the lower the compressibility will be.

TABLE 3-3 / TYPICAL VOID RATIOS AND UNIT WEIGHTS
FOR COHESIONLESS SOILS

Soil Description	Range of Void Ratio		Condition (moisture)	Range of Unit Weight			
	e_{max} (loose)	e_{min} (dense)		γ_{min} pcf (loose)	γ_{max} pcf (dense)	$\gamma_{min} \frac{kN}{m^3}$ (loose)	$\gamma_{max} \frac{kN}{m^3}$ (dense)
Well-graded fine to coarse sand	0.70	0.35	Sat	125	140	19.5	22
			Dry	95	120	15	19
Uniform fine to medium sand	0.85	0.50	Sat	120	130	19	20.5
			Dry	85	110	14	17.5
Silty sand and gravel	0.80	0.25	Sat	115	145	18	22.5
			Dry	90	130	14	17
Micaceous sand with silt	1.25	0.75	Sat	110	125	17	19.5
			Dry	75	95	12	15

Coarse soil in an initially loose condition may be prone to quick volume reductions and loss of strength if subjected to shock or vibrations, unless there is some cementing at points of particle contact or "cohesive strength" provided by moisture menisci (discussed further in Chapter 10).

Experiences indicate that it is possible for sands or silts to be deposited in such a manner that an unusually loose or honeycomb structure results. Grains settling slowly in quiet waters, or a loosely dumped mass of moist soil, can develop a particle-to-particle contact that bridges over relatively large void spaces within the mass and can carry the weight of the overlying material. Possible particle arrangements are shown in Fig. 3-7.

The presence of flake-shaped particles, such as mica flakes, in a coarse soil has significant effect on the void ratio, density, and compressibility of a deposit. The flake-shaped particles are capable of bridg-

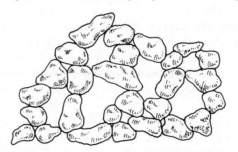

Figure 3-7. *Honeycomb structure in a granular soil.*

ing over open spaces so that relatively large void ratios develop. When subject to external loading, however, the flakes are incapable of providing great support, bending or breaking and rearranging under load.

For coarse-grained soils and silts, the mass of an individual particle is relatively great compared to the surface area. Therefore, the effect of gravity has the most influence over the arrangement of deposits of such soils. (The effect of electrical charges on the particle surface is negligible.) Conversely, the clay particles, because of their large surface-to-mass ratio, are more affected by the electrical forces acting on their surfaces than by gravity forces.

Clay deposits developed from clay particles that have settled out of suspension in a fresh-water or salt-water environment tend towards a *flocculated structure*, whereby the attraction and contact between many of the clay particles is through an edge-to-face arrangement. Clays settling out in a salt-water solution tend to a structure more flocculent than clays settling out in a fresh-water situation. Salt water (oceans are approximately a 3.5 percent saline solution) acts as an electrolyte in which the repulsion between particles is reduced. With respect to other particles, particle sedimentation then occurs with a random orientation, creating the flocculent structure. Sedimentation in a weaker electrolyte, such as fresh water (fresh surface or subsurface water is not "pure") produces a structure where some parallel orientation of settled particles occur, but the overall structure is flocculent; see Figs. 3-8(a) and 8(b). Clay particle sedimentation occurring in ponds and wetlands where organic decay is taking place will result in highly flocculent structures.

Clay deposits with flocculent structures will have high void ratios, low density, and quite probably high water contents. However, the structure is quite strong and resistant to external forces because of the

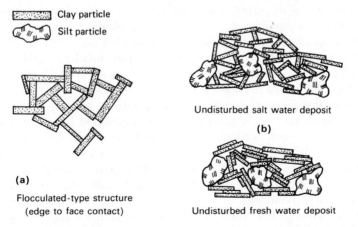

Figure 3-8. *Schematic diagrams of types of particle orientations.* (Ref. 82)

attraction between particles. However, if the environment surrounding the clays is changed, such as having the salts leached from the deposit (as has occurred where ocean deposits have raised to above sea level and fresh water subsequently has percolated through the soil, leaching out the salts with it), the attraction and strength between particles can be markedly decreased.

Clays that have been further transported after being deposited (such as from glacial action, or from man-made earth fills) are reworked or remolded by the transportation process. The particle structure that develops from remolding is a more parallel arrangement or orientation of particles than existed in the flocculent condition, as shown in Fig. 3-9. Such a particle arrangement is considered a *dispersed or oriented* structure.

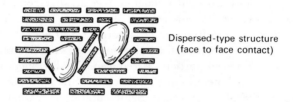

Dispersed-type structure
(face to face contact)

Figure 3-9.   *Schematic diagram of particle orientations.*

When used for construction, clays that were in a flocculent condition before use generally lose some strength as a result of remolding. Subsequent to remolding, and with the passage of time, however, the strength increases, though not back to the strength of the originally undisturbed clay. Reasons for this increase appear to be related to a time-dependent rebuilding from the remolded, dispersed structure towards a less dispersed, more flocculent structure. In the dispersed condition, the equilibrium of forces between particles is disturbed. With time the particles become sufficiently reoriented (only very small movements are necessary) to reacquire a structure in which the forces between particles are again in equilibrium (but not as originally structured); see Fig. 3-10. This phenomenon of strength loss-strength gain, with no changes in volume or water content, is termed *thixotropy.* Thixotropy has been defined as "a process of softening caused by remolding, followed by a time-dependent return to the (original) harder state." The degree of difference between the undisturbed strength and remolded strength, and the extent of strength gain after remolding, are affected by the type of clay minerals in the soil. Generally, the clay types that absorb large quantities of water, such as the montmorillonites, experience greater thixotropic effects than do the more stable clay types, such as kaolinite.

Shaded area represents
adsorbed water layer

Attraction > > repulsion
Water in high-energy
structure

(a) Structure immediately after
remolding or compaction

Attraction > repulsion

(b) Structure after thixotropic
hardening partially complete

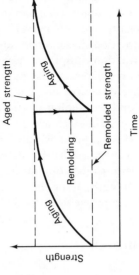

▨ Clay particle

▨ Silt particle

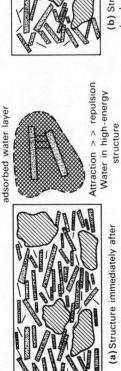

Attraction = repulsion
Water in low-energy
structure

(c) Final structure at end of
thixotropic hardening

(d)

*Figure 3-10. Schematic diagram of thixotropic structure change in a fine-
grained soil. (Ref. 100)*

49

For many construction situations, thixotropy is considered a beneficial phenomenon, since with the passing of time soil structures (dams, highway enbankments, etc.) and disturbed foundation soils get stronger and presumably safer. However, the phenomenon also causes its problems; construction sites may be quickly transformed into a mire of mud when construction equipment travels across the site, making handling of equipment and materials very difficult. Thixotropic influences have also affected piles driven in clay soils when the driving operation caused remolding and weakening of the clay surrounding the pile. The soil strength is sometimes recovered relatively quickly. This has been experienced where a pile has initially been driven part of its length, and attempts at continued driving after a one- or two-day wait have met with considerable resistance due to the increasing adhesion (directly related to a clay's strength) along the pile surface. This is one reason that piles embedded in cohesive soils should be fully driven whenever possible.

PROBLEMS

1. Clay is a soil material that possesses plasticity in the presence of water. What does the term "plastic" mean in relation to clay soils?

2. What is the essential reason for the difference in behavior of natural clays and other soil types such as silts and sands?

3. Comment on the difference between the shape and size of clay particles compared to other soil types such as silts and sands.

4. What are the "building blocks" of most clay minerals? Comment on the comparative length-width-thickness dimensions of a typical clay particle.

5. Referring to the attraction that typically exists between water and clay particles, what is adsorbed water?

6. Describe what is meant by the dipole nature of a water molecule. How is this related to adsorbed water and plasticity in a clay soil?

7. Why does the presence of water in a soil have so much greater effect on clays than on sand or gravel materials?

8. In a general way, the strength of a coarse-grained soil is related to the soil deposit's structure (or particle orientation) and void ratio (or density). What is this relationship?

9. (a) Briefly describe the difference in a flocculent structure and a dispersed structure in clay soils.
   (b) What type of structure is most likely expected in a clay deposit

that has resulted in a naturally occurring underwater environment?

(c) What type of structure would be expected where a clay had
been used for a compacted fill on a construction project?

10. With regard to the "thixotropy" phenomenon in clay soils, what is
it and what causes it?

CHAPTER 4
Index Properties and Classification Tests, and Soil Classification Systems

As an aid for the construction and engineering professions, soils have been divided into basic categories or classifications based upon certain physical characteristics. But because of the range of characteristics for the different soil variations that exist in nature, classification categories have been, of necessity, relatively broad in scope. All of a soil's properties are not checked to obtain a classification. Consequently, for proper evaluation of a soil's suitability for construction or foundation use, information about its properties in addition to classification is frequently necessary. Those properties which do help to define a soil's engineering qualities and which are used to assist in determining accurate classification are termed *index properties*. The tests necessary to determine index properties are *classification tests*. Index properties include those characteristics which can be determined relatively quickly and easily, and which will have bearing on items of engineering importance such as strength or load-supporting ability, tendency to settle or expand, and effect of water and freezing conditions.

Index Properties

Index properties refers to those properties of a soil that indicate the type and condition of the soil, and provide a relationship to structural properties such as the strength and the compressibility or tendency for swelling, and permeability.

Generally, for coarse-grained soils, properties of the particles and the relative state of compaction are most significant. For fine-grained soils, the consistency (firm or soft) and plasticity are particularly important. The index properties that provide the desired information for coarse-grained and fine-grained soils are summarized in Table 4-1.

TABLE 4-1 / INDEX PROPERTY AND RELATED CLASSIFICATION TEST

Soil Type	Index Property	Classification Test
Coarse-grained	Range of particle sizes and distribution of sizes.	Particle size distribution (mechanical analysis) by sieving or sedimentation test.
	Shape of particles	Visual.
	Presence of fine-grained particles.	From mechanical analysis (usually from use of a fine-mesh sieve).
	In-place density and relative state of compaction.	In-situ density determination, and relative density test.
	Classification.	From mechanical analysis, or visual identification based upon grain size.
Fine-grained	Consistency (strength and type of structure in the undisturbed state).	Field or laboratory evaluation of unconfined compressive strength or shear strength (cohesion).
	Change in consistency due to remolding.	Unconfined compressive strength or cohesion for the remolded soil.
	Water content.	Water content.
	Plasticity.	Atterberg limits (liquid limit and plastic limit).
	Classification.	From visual identification and Atterberg limits.
	Presence and type of clay.	Indirectly from determination of plasticity and change in consistency, and/or directly from a clay mineral analysis.

It should be recognized that in studies and analyses for construction projects it frequently is not necessary to determine all the index properties for the soil. Properties to be determined relate to the information that is needed and how such information eventually is to be used. For example, a clay mineral analysis requires very specialized equipment and is not performed in studies for foundation designs, unless the soils are unusual.

For organic soils, it is important to know of the presence and at least the approximate amount of organic material because of its influence on compressibility and strength.

For all soils, the description should include the color. Color may have bearing on the mineralogical composition, but it is also extremely useful for determining homogeneity of a soil deposit, and is an aid for identification and correlation during field construction.

Classification Tests

PARTICLE SIZE DISTRIBUTION (MECHANICAL ANALYSIS) / This classification test determines the range of sizes of particles in the soil, and the percentage of particles in each of the sizes between the maximum and minimum.

Two methods are in common use for obtaining the necessary information. Sieving is generally used for coarse-grained soils, while a sedimentation procedure is used for analyzing fine-grained soils. Sieving is a most direct method for determining particle sizes, but there are practical lower limits to sieve openings that can be used for soils. This lower limit is approximately at the smallest size attributed to sand particles. Information on sieves in common use is shown in Table 4-2.

TABLE 4-2 / COMMON SIEVE TYPES AND MESH OPENINGS

Sieve Size Designation	U. S. Standard		Tyler Standard		British Standard	
	Inches	Millimeters	Inches	Millimeters	Inches	Millimeters
#4	0.187	4.76	0.185	4.70	—	—
#8	0.0937	2.38	0.093	2.362	0.081	2.057
#10	0.0661	1.68	0.065	1.651	0.0661	1.676
#20	0.0331	0.84	0.0328	0.833	—	—
#40	0.0106	0.42	—	—	—	—
#60	0.0098	0.25	0.0097	0.246	0.0099	0.251
#100	0.0059	0.149	0.0058	0.147	0.0060	0.152
#200	0.0029	0.074	0.0029	0.074	0.0030	0.076
#270	0.0021	0.053	0.0021	0.053	—	—
#400	0.0015	0.037	0.0015	0.038	—	—

In the sieve analysis, a series of sieves having different-size openings are stacked with the larger sizes over the smaller (see Fig. 4-1). The soil sample being tested is dried, clumps are broken up if necessary, and the sample is passed through the series of sieves by shaking. Larger particles are caught on the upper sieves, while the smaller particles filter through to be caught on one of the smaller underlying sieves. The weight of material retained on each sieve is converted to a percentage of the total sample. The resulting data are conventionally presented as a grain- or particle-size distribution curve plotted on semi-log coordinates, where the sieve size opening is on a horizontal *logarithmic* scale, and the

Figure 4-1.   *Set of sieves being assembled. Sieves shown are U.S. Standard Sieves, typically used in laboratories.*

percentage (by weight) of the size smaller than a particular sieve opening is on a vertical *arithmetic* scale. Results may be presented in tabular form also. A detailed procedure for performing the sieve analysis is presented in the *ASTM* (American Society for Testing and Materials) *Testing Manual*, under ASTM Test Designation D-422. A typical presentation is shown in Fig. 4-2. Note the "reversed" scale of the logarithmic scale.

Most soil grains are not of an equal dimension in all directions. Hence, the size of a sieve opening will not represent the largest or the smallest dimension of a particle, but some intermediate dimension. As an illustration, assume a brick-shaped particle whose length, width, and

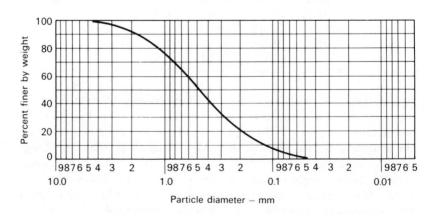

Figure 4-2.   *Particle-size (or grain-size) distribution curve.*

thickness are different. The dimension that controls whether or not the particle passes through the sieve opening is the intermediate dimension, if the particle is aligned so that the greatest dimension is perpendicular to the sieve opening.

Sieve tests can be performed in a laboratory or in the field (at the area being explored, such as a proposed borrow pit or construction site).

The appearance of the particle-size distribution plot depends on the range and amounts of the various sizes of particles in the soil sample. These, in turn, have been affected by the soil's origin or the method of deposition. Well-graded soils (a distribution of particles over a relatively large range of sizes) produce a longish straight curve; see Fig. 4-3(a). A uniform soil (soil having most of the particles of approximately similar size) plots as shown in Fig. 4-3(b). A gap-graded soil (an absence of intermediate sizes) plots as in Fig. 4-3(c).

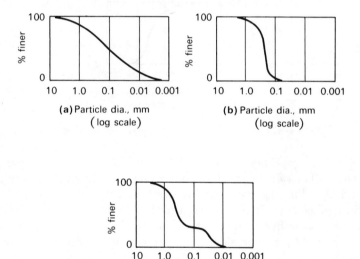

Figure 4-3. *General types of particle-size distribution curves.*

The grain-size plot can provide an indication of a soil's history. A residual deposit has its particle sizes constantly changing with time as the particles continue to break down, and typically produces grain-size curves, as shown in Fig. 4-4. The curves shown in Figs. 4-5(a) and 4-5(b) represent glacial and glacial-alluvial deposits. River deposits may be well-graded, uniform or gap-graded, depending on the water velocity, the volume of suspended solids, and the river area where deposition occurred.

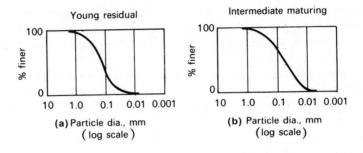

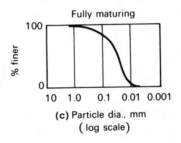

Figure 4-4.   *Typical particle-size curves for residual soils.*

Certain properties of *clean sands* have been related to particle diameters. The *effective size* of a sand is taken as the particle size corresponding to the 10 percent passing size from the grain-size curve, and is indicated as D_{10}. It is this size that is related to permeability and capillarity. Another relation, the ratio D_{60}/D_{10}, is termed the *uniformity coefficient* C_u, and provides a comparative indication of the range of particle sizes in the soil (see Fig. 4-6). Sand having a wide range of

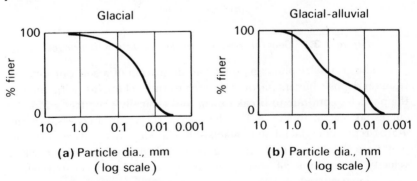

Figure 4-5.   *Typical particle-size curves for transported soil.*

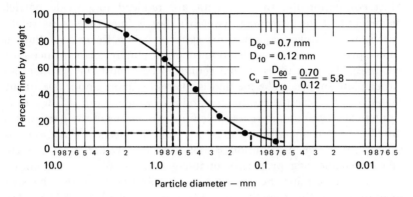

Figure 4-6. *Determining uniformity coefficient from particle-size curve.*

particle sizes is considered well-graded and has C_u values greater than 10. A uniform soil has C_u values less than about 5.

The procedure commonly used for obtaining particle-size distribution information for silts and clays is the sedimentation method. In this method, the soil is placed into a solution with distilled water, and the soil particles are permitted to settle out of the solution. As settling occurs, the average specific gravity of the solution decreases. Readings of specific gravity made at different time intervals provide an indication of the weight of soil remaining in solution, and also information on the sizes of particles that have settled out of the solution (see Fig. 4-7).

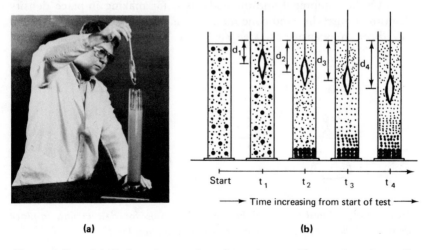

Figure 4-7. *(a) Hydrometer used to determine specific gravity of a soil-water solution being inserted into the solution; (b) diagram showing change in hydrometer depth as soil particles settle out of solution.*

Most conventionally, the test data are reduced to provide particle diameters and the percentage (weight) that is finer than a particular size by using Stokes equation for spheres falling freely in a fluid of known properties. This application is not absolutely correct, since most fine-grained soil particles are not round; in fact, most clay particles are flat or plate-shaped. The method is, therefore, more applicable to silts than to clays. Nevertheless, the method is in wide use, for it is felt to be a practical way to obtain reasonable approximations of the particle-size distribution for fine-grained soils. Resulting effects are not serious, since particle-size distribution is not used for evaluation of the significant engineering properties of fine-grained soils. Details on the method to determine the particle-size distribution by using the sedimentation test (hydrometer method) are described by ASTM Test Designation D-422.

IN-PLACE DENSITY / The in-place density refers to the volumetric weight, usually pounds per cubic foot or kilograms per cubic meter, of a soil in the undisturbed (or in-situ) condition.[1] Generally, for the coarse-grained soils, the greater the density, the better the shear strength and the lesser the tendency for compression (settlement). In-place density determinations are made of borrow pit soils so to estimate the volume of shrinkage or swell that will occur as the soil is transported or compacted in place at a fill location. Where compacted earth fills are being constructed, it is standard practice to make in-place density determinations of the soil after it is placed to determine if the compaction effort has been adequate or if more compaction is required.

Of the equipment and methods used for making in-place density determinations, the sand-cone method and the rubber balloon method have been in extensive use (see Fig. 4-8.) With these methods, a small

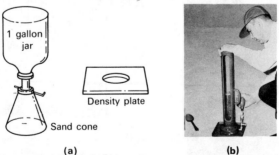

(a)   (b)

Figure 4-8. Sand-cone and balloon apparatus for determining in-place density. (Balloon apparatus photo courtesy of Soiltest Inc.)

---

[1]In-place density is also given in $\dfrac{kN}{m^3}$, where 1 kg = 9.81 N.

test area is selected and a volume of compacted soil is dug up and weighed. The sand cone equipment or the balloon equipment is used to determine the volume of the dug hole. Knowing soil weight and corresponding volume permits the density to be calculated.

A modern development for making in-place density determinations involves the use of nuclear equipment (Fig. 4-9). Through controlled use of a nuclear material, gamma rays (photons) are emitted into the tested soil. These photons collide with electrons in the soil materials, some being scattered and some being absorbed. The quantity of photons reaching a detection device (part of the test equipment) relates to the soil density. To determine water content, a neutron-emitting material and detector are used. Compared to the other methods of making in-place density determinations, a significant advantage of the nuclear method is the rapid speed with which results are obtained.

Figure 4-9.   *Nuclear moisture-density meter in use. (Courtesy of Soiltest, Inc.)*

RELATIVE DENSITY / For a granular soil, the shear strength and resistance to compression are related to the density (or unit weight) of the soil; higher shear strength and more resistance to compression are developed by the soil when it is in a dense or compact condition (high density) than when it is in a loose condition (low density). In a dense condition, the void ratio is low; in a loose condition, the void ratio is high. To evaluate the relative condition of a granular soil, the in-place void ratio can be determined and compared to the void ratio when the soil is in its densest condition and when it is in the loosest condition; see Fig. 4-10. This comparison is the relative density D_R. Relative density is expressed as a percentage. High values indicate a dense or compact material; low values represent a loose material.

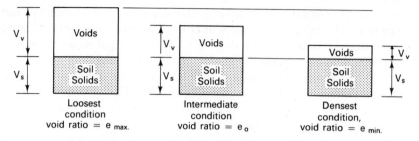

Figure 4-10.   *Relative conditions of a granular soil.*

$$D_R = \frac{e_{max} - e_0}{e_{max} - e_{min}} \times 100 \text{ percent} \qquad (4\text{-}1)$$

where e_{max} = void ratio of the soil in its loosest condition

e_{min} = void ratio of the soil in its densest condition

e_0 = void ratio of the soil in the natural condition or condition in question

The maximum density (or minimum void ratio) is determined in the laboratory by compacting the soil in thin layers in a container of known volume and subsequently weighing the soil. The compaction is achieved by applying a vibratory and compressive force simultaneously. The compressive force needs to be sufficient to compact the soil without breaking the individual particles of soil. Because of the irregular size and shape of granular particles, it is not possible to obtain a zero volume of void spaces. Practically, there will always be voids in a soil mass.

The minimum density (or maximum void ratio) can be determined in the laboratory by carefully letting the soil slowly flow into the test container through a funnel. When this task is carefully performed, the soil will be deposited and remain in a loose condition, from whence the loose density and void ratio can be calculated.

In terms of dry unit weights (which are more convenient to work with), relative density is

$$D_R = \frac{\dfrac{1}{\gamma_{min}} - \dfrac{1}{\gamma_0}}{\dfrac{1}{\gamma_{min}} - \dfrac{1}{\gamma_{max}}} \times 100 \text{ percent} \qquad (4\text{-}2)$$

where γ_{min} = dry unit weight (or density) in the loosest condition

γ_{max} = dry unit weight in the densest condition

γ_0 = dry unit weight in the condition in question

Typical relative density values are presented in Table 4-3.

TABLE 4-3 / REPRESENTATIVE VALUES OF RELATIVE DENSITY

Descriptive Condition	Relative Density, %	Typical Range of Unit Weight	
		PCF	$\dfrac{kN}{m^3}$
Loose	Less than 35	Less than 90	Less than 14
Medium dense	35 to 65	90 to 110	14 to 17
Dense	65 to 85	110 to 130	17 to 20
Very dense	Greater than 85	Above 130	Above 20

Illustration 4-1: An undisturbed sample of fine sand is tested in the laboratory and found to have a dry weight of 8 lb, a total volume of 0.07 ft^3 and a specific gravity G_s of 2.70. Other laboratory tests were performed to determine the maximum and minimum density for the sand. At the maximum density, it is determined that the void ratio is 0.35; at the minimum density the void ratio is 0.95. Determine the relative density of the undisturbed sample.

Void ratio of undisturbed sample e_0:

$$V_T = 0.07 \text{ ft}^3$$

$$V_s = \frac{W_s}{G_s \gamma_w} = \frac{8 \text{ lb}}{(2.70)(62.4 \text{ pcf})} = 0.0474 \text{ ft}^3$$

$$V_v = V_T - V_s = 0.070 - 0.0474 = 0.0226 \text{ ft}^3$$

$$e_0 = \frac{V_v}{V_s} = \frac{0.0226 \text{ ft}^3}{0.0474 \text{ ft}^3} = 0.476$$

$$D_R = \frac{e_{max} - e_0}{e_{max} - e_{min}} = \frac{0.95 - 0.476}{0.95 - 0.35}$$

$$= 0.79 = 79 \text{ percent}$$

$$\text{Unit weight of soil} = \frac{W_s}{V_T} = \frac{8.0 \text{ lb}}{0.07 \text{ ft}^3} \simeq 115 \text{ pcf}$$

In its natural condition, the soil is probably dense. (see Table 4-3).

WATER CONTENT / For coarse- and fine-grained soils, water content can have a significant effect on the soils' behavioral properties when used for construction purposes. By definition and as previously described, water content (w) is the ratio of the weight of water in a soil to the dry weight of the material. As a result, in laboratory and field work the weight of water in the test sample has to be determined.

Conventionally this is done by drying the original wet sample, recording the wet and dry weights, and performing the required calculation.

One device, the Speedy Moisture Tester (Fig. 4-11), bases its operation on the reaction that occurs between a carbide reagent and soil moisture to determine a soil's water content. The wet soil sample is placed in a sealed container with calcium carbide, and the pressure generated by the vaporized moisture is then related to water content. This method provides a rapid procedure for determining moisture content and is finding increased usage in laboratory and field work.[2]

Figure 4-11. *Speedy moisture tester: Speedy Kit including tester chamber, scales, and reagent for vaporizing soil moisture.*

CONSISTENCY OF CLAYS / Consistency refers to the texture and firmness of a soil and is often directly related to the strength. Consistency is conventionally described as *soft, medium stiff (or medium firm), stiff (or firm),* or *hard*. These terms unfortunately are relative and have different meaning to different observers. For standardization, it is reasonable and practical to relate consistency to strength. With clays, shear strength is discussed in terms of cohesion and unconfined compressive strength. The unconfined compressive strength is obtained by imposing an axial load to the ends of an unsupported cylindrical sample of clay and determining the load to shear the cylinder. Unconfined compression tests can be performed in the laboratory or in the field. The unconfined compressive strength, is, under practical conditions,

---

[2]Somewhat unfortunately, the Speedy moisture reading is expressed as a percentage of the soil's wet weight. Conversion to moisture content by dry weight can be performed using the relationship

$$w = \frac{w_{sp}}{1 - w_{sp}} \times 100 \text{ percent}$$

where w_{sp} is the moisture content obtained with the Speedy device, expressed as a decimal.

twice the cohesion (or shear strength) of a clay soil. The cohesion can be determined in a laboratory strength test. For a consistency classification in the field or laboratory, special soil testing equipment such as a vane-shear device or pocket penetrometer provides means of making quick and easy determinations (Fig. 4-12). A tabulation of strength values for various consistency terms is shown in Table 4-4.

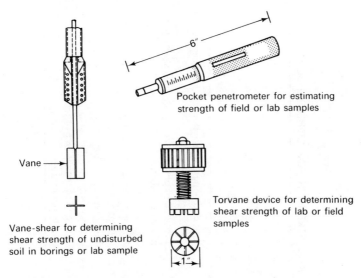

Pocket penetrometer for estimating strength of field or lab samples

Vane

Torvane device for determining shear strength of lab or field samples

Vane-shear for determining shear strength of undisturbed soil in borings or lab sample

Figure 4-12. Testing equipment for making strength or consistency determinations on cohesive soils.

It is well established that the strength of a clay soil is related to its structure. If the original structure is altered because of changes in particle arrangement (from reworking or remolding) or chemical changes, the strength of the altered clay is less than the original strength. *Sensitivity* is the term that provides an indication of remolded strength related to original strength. When remolded, the strength of a clay is affected by the water content. At lower water contents, strength is generally greater. However, sensitivity should be based upon comparison of remolded to undisturbed soil strength at an identical water content.

Sensitivity S_t =

$$\frac{\text{unconfined compressive strength, undisturbed clay}}{\text{unconfined compressive strength, remolded clay}} \quad (4\text{-}3)$$

For most clays, sensitivities range between 2 and 4. Clays considered

TABLE 4-4 / CONSISTENCY AND STRENGTH FOR COHESIVE SOILS

Consistency	Shear Strength T/SF or Kg/cm^2 (kN/m^2)	Unconfined Compressive Strength T/SF or Kg/cm^2 (kN/m^2)	Feel or Touch
Soft	less than 0.25 (<24)	less than 0.5 (<48)	Blunt end of pencil-size item makes deep penetration easily.
Medium (medium stiff or medium firm)	0.25 to 0.50 (24 to 48)	0.50 to 1.0 (48 to 96)	Blunt end of pencil-size object makes $\frac{1}{2}$-in. penetration with moderate effort.
Stiff (firm)	0.50 to 1.0 (48 to 96)	1.0 to 2.0 (96 to 190)	Blunt end of pencil-size object can make moderate penetration (about $\frac{1}{4}$ in.).
Very stiff (very firm)	1.0 to 2.0 (96 to 190)	2.0 to 4.0 (190 to 380)	Blunt end of pencil-size object makes slight indentation; fingernail easily penetrates.
Hard	greater than 2.0 (>190)	greater than 4.0 (>380)	Blunt end of pencil-size object makes no indentation; fingernail barely penetrates.

sensitive have S_t values between 4 and 8. Clays classified as extrasensitive have values between 8 and 16. Clays with sensitivity values greater than 16 are classified as *quick clays*. Such clays are very unstable. Normally, clays with a high degree of sensitivity possess a very flocculent structure in the undisturbed condition.

CONSISTENCY IN THE REMOLDED STATE AND PLASTICITY / In the remolded state, the consistency of a clay soil varies in proportion to the water content. At high water content, the soil-water mixture possesses the properties of a liquid; at lesser water contents, the volume of the mixture is decreased and the material exhibits the properties of a plastic; at still lesser water contents, the mixture behaves as a semi-solid and finally as a solid.

The water content indicating the division between the liquid and plastic state has been designated the *liquid limit*. The division between the plastic and semi-solid state is the *plastic limit*. The water content at the division between the semi-solid and the solid state is the *shrinkage limit*.

At water contents above the shrinkage limit, the total volume of the soil-water mixture changes in proportion to change in water content. Below the shrinkage limit, there is little or no change in volume as water content varies (Fig. 4-13).

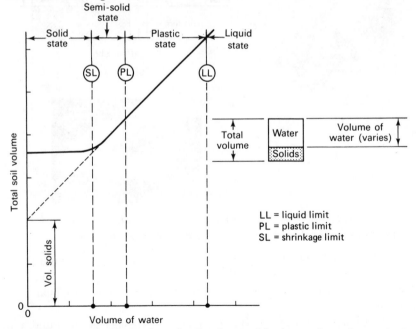

Figure 4-13. *Variation of total soil volume with change in water content for a fine-grained soil.*

Definition of the various states of consistency and the establishment of criteria to determine these various states were first formally proposed by A. Atterberg, a Swedish soil scientist, in the early part of the twentieth century.   Initially intended for agricultural use, the method has been adapted for engineering use in classifying soils.   Although the limit values (water contents) have little direct meaning insofar as engineering properties of soils are concerned, correlations between liquid or plastic limits and engineering properties have been established over the years which aid in evaluating a soil for use as structural fill (dams, embankments, land fills) for highway construction and for building support.

The liquid limit is taken as the water content at which the soil "flows" (like a viscous liquid).   Special equipment and procedures, defined in ASTM Test Designation D-423,* are required for determining the liquid limit (see Fig. 4-14).   The plastic limit is the water content where the soil can just be rolled into a thread $\frac{1}{8}$ inch diameter, as described in ASTM Test Designation D-424.*

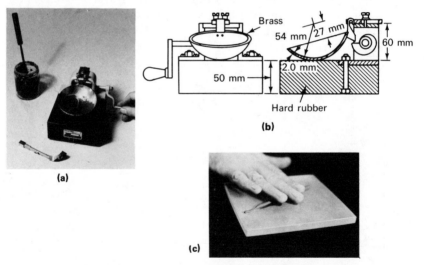

Figure 4-14.   *(a) Liquid limit trial being performed; (b) details of liquid limit device; (c) soil thread being rolled to determine plastic limit.*

Various designations have been used to indicate the liquid and plastic limit.   Using *LL* and *PL* for liquid limit and plastic limit, respectively, is an easily understood terminology that has little chance of misinterpretation.

The *plasticity index* (PI) is the numerical difference of the liquid and plastic limit, and indicates the range of water content through which

*Refer to Appendix I for details on test procedure.

the soil remains plastic. For proper evaluation of a soil's plasticity properties, it has been found desirable to use both the liquid limit and plasticity index values. Engineering soil classification systems use these values as a basis for classifying the fine-grained soils.

For fine-grained soils, determining the natural water content (the water content of a soil in an undisturbed condition in the ground) and relating it to the plastic and liquid limits can provide an indication of the soil's consistency and/or sensitivity potential. One such relationship is the *liquidity index LI,*

$$LI = \frac{w - PL}{LL - PL} = \frac{w - PL}{PI} \qquad (4-4)$$

where w is the natural water content of the soil. A value less than one indicates that the natural water content is less than the liquid limit. A very low value for the LI, or a value near zero, indicates that the water content is near the plastic limit, where experience has shown that the sensitivity will be low and the cohesive strength relatively high (a stiff or hard consistency). As the natural water content approaches or exceeds the liquid limit, the sensitivity increases. The undisturbed strength still very much depends on the undisturbed structure. Negative values of the LI are possible and normally indicate a desiccated (dried) hard soil.

PRESENCE OF CLAY MINERALS / The presence of even small amounts of certain clay minerals in a soil mass can have significant effect on the properties of the soil. Identifying the type and amount of clay mineral may be necessary in order to predict the soil's behavior or to develop methods for minimizing detrimental effects.

The identification of clay minerals requires special techniques and equipment and trained personnel. Many different techniques are available. Some are useful for identifying only a particular type of clay, whereas other methods are suitable for identifying several types of minerals. These techniques include microscopic examinations, X-ray diffraction, differential thermal analysis, infrared absorption, optical property determination, and electron micrography. Description of these techniques and equipment is beyond the scope of this text, and further reference should be made to publications dealing with clay mineralogy. Even with the techniques available to today's soil scientists, the accurate determination of some clay minerals is not possible. Generally, however, qualitative if not quantitative identifications can be made that are adequate for many engineering problems.

A somewhat indirect method of obtaining information on the type and effect of clay mineral in a soil is to relate plasticity to the quantity of clay-size particles. It is known that for a given amount of clay mineral the plasticity resulting in a soil will vary for the different types of clays.

One relationship is *activity,* defined as

$$\text{Activity} = \frac{\text{plasticity index}}{\text{percentage of clay sizes}} \tag{4-5}$$

For this analysis, the percentage of clay sizes is that portion of the soil, by weight, consisting of particles sizes below 0.002 mm. Such information is available from the conventional hydrometer analysis used to determine a particle-size distribution. Activity, therefore, can be determined from standard laboratory tests. Clay materials with kaolinite, a stable clay mineral, will have low activity, whereas those soils with montmorillonite, known to be a type subject to large volume changes depending on available water, will have a high activity value. Limited study has also indicated the possibility of a relation between activity and shear strength. A relative activity classification is as follows:

Activity	Classification
less than 0.75	inactive clays
0.75 to 1.25	normal clays
greater than 1.25	active clays

OTHER PROPERTIES / Where a soil is to be used to support building foundations or other types of buried structures (such as pipelines, tunnels, and storage vaults), it is good practice to have pH and soluble sulfate determinations made, so as to obtain information on the corrosion potential to buried metal (piles or piping) or the deteriorating effect on concrete foundations.

Classification Systems

Soil classification systems have been devised primarily to facilitate the transfer of information between interested parties. In the engineering and construction field, the broad general properties of concern relate to the performance or usefulness for supporting structures and the handling or working qualities of a soil. Because of the wide variation in properties and behavior of soils, classification systems generally group together in relatively broad categories the soils that have similar features or properties that are considered to be of importance. As a result, a classification system is not necessarily an identification system in which all pertinent engineering properties of a material are determined. Because of this, soil classification should not be used as the sole basis for design or construction planning.

Historically, the most widely used method of classifying soils has been through visual identification, the size of soil grains and the plasticity of the soil being used as the basis for indicating the soil type.

To a great degree, the refined and more recently developed classification systems that rely on laboratory-determined properties for accurate identification still use these two criteria as the basis for indicating soil type.

The desirable requirements for a satisfactory engineering soil classification system include:

1. There should be a limited number of different groupings, so that the system is easy to remember and use. Groupings should be on the basis of only a few similar properties and generally similar behavioral characteristics.

2. The properties and behavioral characteristics should have meaning for the engineering and construction profession. Generally, the properties should, at least crudely, relate to a soil's handling characteristics, shear strength, volume change characteristics, and permeability.

3. Descriptions used for each grouping should be in terms that are easily understood and are in common use for indicating the soil type and its properties. Symbols or shorthand used to describe the grouping should easily relate to the soil type. Coded symbols are not desirable.

4. Classification into any grouping should be possible on the basis of visual identification (generally limited to differentiating between particle sizes, coarse- and fine-grained soils, and plasticity) without special tests or equipment being necessary.

The Unified Soil Classification System satisfies the above requirements of a classification system and is the system that is coming into prevalent use in the engineering and construction industry. This system is shown in Fig. 4-15. Classifications are on the basis of coarse- and fine-grained soils, and retain the four common groupings of soil—gravel, sand, silt, and clay. The symbols are easily associated with the classification, being simply the first letter of the soil type (except for silt, which has the designation M, from "mo," the Swedish word for silt). Refinement in grouping is based upon a coarse soil's being well or poorly graded, and a fine-grained soil being of a high or low plasticity. What at first appears to be a large number of groupings are in fact very logical categories. Experience has proven that previously untrained personnel very quickly learn the system and use it with accuracy. The Unified System includes the use of a plasticity chart for aiding classification of fine-grained soils.

Major division				Group Symbols	Typical Names			Classification Criteria

Coarse-grained soils — More than 50% retained on No. 200 sieve

Gravels — 50% or more of coarse fraction retained on No. 4 sieve

- **Clean gravels**
  - **GW** — Well-graded gravels and gravel-sand mixtures, little or no fines
    - $C_u = D_{60}/D_{10}$ Greater than 4
    - $C_z = \dfrac{(D_{30})^2}{D_{10} \times D_{60}}$ Between 1 and 3
  - **GP** — Poorly graded gravels and gravel-sand mixtures, little or no fines
    - Not meeting both criteria for GW
- **Gravels with fines**
  - **GM** — Silty gravels, gravel-sand-silt mixtures
    - Atterberg limits plot below "A" line or plasticity index less than 4
  - **GC** — Clayey gravels, gravel-sand-clay mixtures
    - Atterberg limits plot above "A" line and plasticity index greater than 7

Sands — More than 50% of coarse fraction passes No. 4 sieve

- **Clean sands**
  - **SW** — Well-graded sands and gravelly sands, little or no fines
    - $C_u = D_{60}/D_{10}$ Greater than 6
    - $C_z = \dfrac{(D_{30})^2}{D_{10} \times D_{60}}$ Between 1 and 3
  - **SP** — Poorly graded sands and gravelly sands, little or no fines
    - Not meeting both criteria for SW
- **Sands with fines**
  - **SM** — Silty sands, sand-silt mixtures
    - Atterberg limits plot below "A" line or plasticity index less than 4
  - **SC** — Clayey sands, sand-clay mixtures
    - Atterberg limits plot above "A" line and plasticity index greater than 7

Classification on basis of percentage of fines: Less than 5% Pass No. 200 sieve — GW, GP, SW, SP. More than 12% Pass No. 200 sieve — GM, GC, SM, SC. 5% to 12% Pass No. 200 sieve — Borderline classification requiring use of dual symbols

Fine-grained soils — 50% or more passes No. 200 sieve

Silts and Clays — Liquid limit 50% or less

- **ML** — Inorganic silts, very fine sands, rock flour, silty or clayey fine sands
- **CL** — Inorganic clays of low to medium plasticity, gravelly clays, sandy clays, silty clays, lean clays
- **OL** — Organic silts and organic silty clays of low plasticity

Silts and Clays — Liquid limit greater than 50%

- **MH** — Inorganic silts, micaceous or diatomaceous fine sands or silts, elastic silts
- **CH** — Inorganic clays of high plasticity, fat clays
- **OH** — Organic clays of medium to high plasticity

Check plasticity chart

Highly organic soils

- **Pt** — Peat, muck and other highly organic soils
  - Fibrous organic matter; will char, burn, or glow

(a)

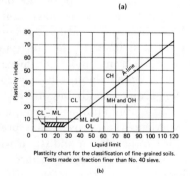

Plasticity chart for the classification of fine-grained soils.
Tests made on fraction finer than No. 40 sieve.

(b)

Figure 4-15. *Unified Soil Classification System (ASTM Designation D-2487).*

American Association of State Highway Officials Soil Classification System (AASHO Designation M-145)

General Classification*	Granular Materials (35 per cent or less passing No. 200)							Silt-Clay Materials (more than 35 per cent passing No. 200)				
	A-1		A-3	A-2				A-4	A-5	A-6	A-7	
Group Classification	A-1-a	A-1-b		A-2-4	A-2-5	A-2-6	A-2-7				A-7-5	A-7-6
Sieve analysis per cent passing:												
No. 10	50 max											
No. 40	30 max	50 max	51 min									
No. 200	15 max	25 max	10 max	35 max	35 max	35 max	35 max	36 min	36 min	36 min	36 min	36 min
Characteristics of fraction passing No. 40:												
Liquid limit				40 max	41 min	40 max	41 min	40 max	41 min	40 max	41 min	41 min
Plasticity index	6 max		N.P.[b]	10 max	10 max	11 min	11 min	10 max	10 max	11 min	11 min	11 min
Usual types of significant constituent materials	Stone fragments—gravel and sand		Fine sand	Silty or clayey gravel and sand				Silty soils		Clayey soils		
General rating as subgrade	Excellent to good							Fair to poor				

[a] Classification procedure: With required test data in mind, proceed from left to right in chart; correct group will be found by process by elimination. The first group from the left consistent with the test data is the correct classification. The A-7 group is subdivided into A-7-5 or A-7-6 depending on the plastic limit. For PL < 30, the classification is A-7-6; for PL > 30, A-7-5.

[b] N.P. denotes nonplastic.

Figure 4-16. *American Association of State Highway Officials Soil Classification System (AASHO designation M-145).*

Another system that is familiar to the engineering and construction industry is the AASHO (American Association of State Highway Officials) Classification System (Fig. 4-16), which has been in wide use for the highway construction field. The criteria for the groupings are logical, but shortcomings include the requirement for laboratory testing in order to determine a classification, and the difficulty in using code designations and remembering the requirement for each of the designations. To obtain an indication of the relationship between the Unified and AASHO systems, a comparison is presented in Table 4-5.

TABLE 4-5 / APPROXIMATE EQUIVALENT GROUPS OF AASHO AND UNIFIED SOIL CLASSIFICATION SYSTEMS

AASHO	Unified
A-1-a	GW, GP, GM
A-1-b	SW, SM
A-2-4	GM, SM
A-2-5	GM, SM
A-2-6	GC, SC
A-2-7	GC, SC
A-3	SP
A-4	ML, OL
A-5	MH
A-6	CL
A-7-5	CL, OL
A-7-6	CH, OH

PROBLEMS

1. Indicate the difference between index properties of a soil and classification tests.

2. The following information is obtained from a sieve analysis to determine the range of particle sizes in a granular soil sample.

Sieve Size	Sieve Opening (mm)	Percent Finer By Weight
#4	4.76	96
#10	2.00	80
#20	0.84	52
#40	0.42	38
#60	0.25	25
#100	0.149	12
#200	0.074	5

Present the information as a grain size curve on semilog co-ordinates of percent finer versus particle diameters. From the plot, determine the uniformity coefficient Cu.

3. (a) An in-place density determination is made of a sand in a borrow pit using a balloon type density apparatus. A sample dug from a test hole is found to weigh 8.5 lb. The volume of the test hole is 0.065 ft^3. From these data, compute the wet unit weight of the soil.

   (b) This soil is found to have a water content of 15 percent. Compute the dry unit weight of the soil and, by referral to Table 4-3, indicate the probable condition of the natural material.

4. A sand at a borrow pit is determined to have an in-place dry unit weight of 115 pcf. Laboratory tests performed to determine the maximum and minimum unit weights give values of 122 pcf and 102 pcf, respectively. From these data, what is the relative density of the natural soil?

5. An undisturbed sample of sand has a dry weight of 4.20 lb and occupies a volume of 0.038 ft^3. The soil solids have a specific gravity of 2.75. Laboratory tests performed to determine the maximum and minimum densities indicate void ratios of 0.42 at the maximum density and 0.92 at the minimum density. Com-pute the relative density of this material.

6. A fine-grained soil is found to have a liquid limit of 70 percent and a plastic limit of 38 percent.

   (a) What do the percentages for these limit values represent?
   (b) What is the plasticity index for this soil?
   (c) On the plasticity chart for the Unified Soil Classification System, what soil type is this?

7. A silt-clay soil has a plastic limit of 25 and a plasticity index of 30.

   (a) If the natural water content of the soil is 35 percent, what is the liquidity index?
   (b) What soil type is this, according to the Unified Soil Classi-fication System?

8. A fine-grained soil is found to have a liquid limit of 90 percent and a plasticity index of 51. The natural water content is 28 per-cent.

   (a) Determine the liquidity index and indicate the probable consistency of the natural soil.
   (b) Classify this soil according to the Unified Soil classification System.

9. A 100-cc clay sample has a natural water content of 30 percent. It is found that the shrinkage limit occurs when the water content is 19 percent. If the specific gravity of soil solids is 2.70, what will the volume of the sample be when the water content is 15 percent?

10. A clay soil is found to have a liquid limit of 75 percent, a plastic limit of 45 percent, and a shrinkage limit of 25 percent. If a sample of this soil has a total volume of 30 cc at the liquid limit and a volume of 16.7 cc at the shrinkage limit, what is the specific gravity of the soil solids?

11. Limit tests performed on a clay indicate a liquid limit of 67 and a plastic limit of 32. From a hydrometer analysis to determine particle sizes, it is found that 40 percent of the sample consists of particles smaller than 0.002 mm. From this information, indicate the activity classification for this clay, and the probable type of clay mineral.

CHAPTER 5
Movement of Water Through Soil: Basic Principles
Permeability and Capillarity

Natural soil deposits, as they exist at most locations, include water. Under many conditions, the soil water is not immobile but does move or is capable of moving through the ground. Moving water affects the properties and behavior of soil and can influence both construction operations and the performance of completed construction. Since ground water conditions are frequently encountered on construction projects, those in the construction profession have found it necessary to understand the manner in which movement of water through soil can occur and the effects that can result. The discussion on permeability and capillary action in soil in this chapter relates to type and manner of water movement. The discussion on drainage, seepage, and frost heave in the following chapter relates to the practical effects of water movement.

Permeability

Soil, being a particulate material, has many pore or void spaces existing between the solid grains because of the irregular shape of the individual particles. Thus, soil deposits are porous.

In a mass of particles that are rounded and roughly equidimensional in shape such as the gravels, sands, and silts, or are platey or flake-like, such as clays, the pore spaces are interconnected. Fluids (and gases) can travel or flow through the pore spaces in the soil, and the material is considered a *permeable* material. It should be realized that flow is occurring through the void spaces between particles, and not through the particles (see Fig. 5-1).

FACTORS AFFECTING FLOW / The actual path taken by a fluid particle as it flows through void spaces from one point towards

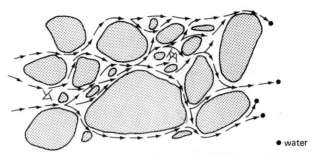

● water

Figure 5-1.   *Schematic diagram indicating manner in which water flows through soil.*

another is a tortuous and erratic one in most soils because of the random arrangement of the soil grains. It is highly probable that the direction of flow and velocity of flow vary considerably.

The factors that can affect the flow of a fluid through soil are known, but the *influence* of all factors has not been clearly established. These factors include:

(a)   The pressure difference existing between the two points where flow is occurring.

(b)   The density and viscosity of the fluid.

(c)   The size, shape, and number of pore openings.

(d)   The mineralogical, electrochemical, or other pertinent properties of the fluid and the soil particles, which affect the attraction between the two materials.

The effects of items (c) and (d) on flow are the most difficult to evaluate, partially because of the tremendous variation that occurs in natural deposits (in even homogeneous soils) and partially because of insufficient knowledge.

In many engineering and construction problems, the concern is primarily with the *quantity* of fluid, usually water but not always, that is flowing through or out of a soil mass. The seepage *velocity* is frequently sufficiently low that no problems result because of this factor, or it may not even require consideration. An exception to this is where the velocity is great enough to cause movement of the soil particles, or erosion. This problem is discussed in Chapter 6 on flow nets. For the situation where the *quantity* of flow is to be determined, an *average discharge velocity* is assumed. This is a fictitious velocity compared to the actual velocity of flow. The discharge velocity is simply the volume of fluid flow per unit of time divided by the total area (soil plus voids) measured normal to the direction of flow. If an average *seepage velocity* (average actual velocity of flow) is desired, it can be obtained by dividing the average discharge velocity by n, the porosity of the soil (recall that porosity n equals V_v / V_T).

Experimental studies have shown that fluid flow is affected by the shape and dimensions of the channel through which flow is occurring. To relate these effects to flow velocity, the terms *hydraulic radius* and *shape factor* have been developed. Hydraulic radius R_H provides a relation between the cross-sectional area of flow and the channel walls that are in contact with the fluid;

$$R_H = \frac{\text{area of flow}}{\text{wetted perimeter}} \qquad (5\text{-}1)$$

The coefficient that reflects the shape factor is C_S. For a circular tube flowing full (Fig. 5-2), the hydraulic radius is one-half the tube diameter and $C_S = \frac{1}{2}$.

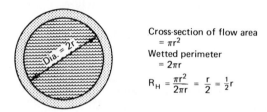

Cross-section of flow area
$= \pi r^2$
Wetted perimeter
$= 2\pi r$
$R_H = \frac{\pi r^2}{2\pi r} = \frac{r}{2} = \frac{1}{2}r$

Figure 5-2. *Hydraulic radius for pipe flowing full.*

To indicate how *permeable* a porous material will be to *any* flowing fluid, the term K, permeability, is used. Mathematically,

$$K = C_S R_H^2 n \qquad (5\text{-}2)$$

where n = porosity of the material.

It should be understood this term applies *only to the material* through which the flow could occur. It reflects the effect of the size, shape, and number of flow channels, and is completely independent of any fluid properties.

To indicate the ease or difficulty that a particular *fluid* will encounter when flowing through a permeable material, the properties of the fluid are incorporated with K, the properties of the permeable material, to provide a *coefficient of permeability* k (the lowercase letter is used for this term), where

$$k = K\frac{\gamma}{\eta} \qquad (5\text{-}3)$$

where
γ = unit weight of the fluid (e.g., pcf).
η = viscosity of the fluid.

This relation permits a coefficient of permeability to be determined for any fluid whose properties are known.

DARCY'S LAW FOR FLOW / In the mid-eighteenth century, H. Darcy performed experiments to study the flow of water through sands. With an arrangement represented by Fig. 5-3, it was found that the quantity of water flowing through the soil in a given period was proportional to the soil area normal to the direction of flow and the difference in levels indicated in the piezometers (open standpipes), and inversely proportional to the length of soil between piezometers through which flow took place. Mathematically,

$$\frac{Q}{t} \propto \frac{(\Delta h) \times (A)}{L} = \text{(a constant)} \times \frac{(\Delta h) \times (A)}{L}$$

where　Q = volume of water flowing through the soil in a time t.

　　　　t = time period for the volume Q to flow.

　　$\Delta h = h_1 - h_2$.

　　　A = cross-sectional area of the soil sample.

　　　L = length of soil through which flow occurs.

The factors A and L relate to the volume of the soil mass but not to the properties of the soil. The value of Δh relates to the pressure acting to force the water to flow through the soil. The constant of proportionality, a factor that indicates if the volume of flow is to be great or small, relates to the ease or difficulty with which the water moves through the soil. This constant of proportionality is k, Darcy's *coefficient of permeability*. This is the same coefficient of permeability indicated in Eq. (5-3), but attributed to Darcy who first established it. Consequently,

$$\frac{Q}{t} = k \frac{(\Delta h)(A)}{L} \tag{5-4}$$

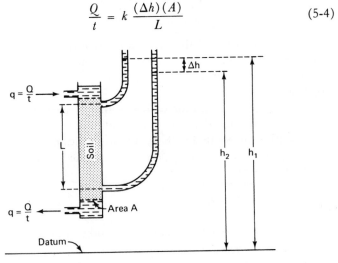

Figure 5-3. *Darcy's sand filtration experiment.*

which is Darcy's law. The ratio of $\Delta h/L$ is termed the hydraulic gradient i, and therefore

$$\frac{Q}{t} = q = kiA \qquad (5\text{-}5)$$

where q = volume of flow per unit time.

The units of k are length per unit time; i is dimensionless (length divided by length); area is in square units.

For steady flow, the volume of flow q passing a point is equal to the product of flow velocity v and the cross-sectional area through which flow occurs, i.e.,

$$q = Av \qquad (5\text{-}6)$$

where units of v are length or distance per unit of time.

From Eq. (5-5) and (5-6), an expression for velocity of flow is obtained.

$$q = kiA,$$

and $\qquad\qquad\qquad\qquad v = ki \qquad (5\text{-}7)$

This is a theoretical average velocity, and will be lower than an actual velocity. Its determination is of use in practical problems, however, where only order of magnitude is required.

LAMINAR AND TURBULENT FLOW / The movement of a fluid through a channel or pore space can be described as laminar or turbulent flow, depending on the path followed by the flowing water particles. Laminar (layered) flow indicates that adjacent paths of water particles are parallel, even when changing direction, and the paths never cross. This is an orderly flow with no mixing. Turbulent flow indicates a disorderly random path for moving water particles, with lines of movement crossing and frequently moving at an angle with or contrary to the general direction of flow. A high degree of mixing occurs.

Velocity has direct bearing on whether a flow is laminar or turbulent.

Darcy's law for fluid flow applies provided that the flow is laminar. In soils, the velocity of flow is affected by the size of the void opening as well as the hydraulic gradient i. Studies show that for soils in the coarse sand and finer range, and frequently for small gravel, laminar flow occurs provided that the hydraulic gradient is 5 or less. In practical soil mechanics work, Darcy's law thus has a wide range of application.

EFFECT OF SOIL TYPE / The volume of water that can flow through a soil mass is related more to the size of the void openings than to the number or total volume of voids. This is shown by observing that the values of k for coarse soils are greater than for fine-grained soils (even though void ratios are frequently greater for the fine-grained soils) along with the knowledge that voids in a soil mass are approximately of the same size as the particles themselves.

This phenomenon of higher permeability for coarse-grained soil can be explained, at least in part, by the manner in which water flows through a conduit. The fluid flow measured at increments of distance extending between the walls of the conduit indicates that the velocity varies from a very low value adjacent to the wall of the conduit (or against the soil particle) to a maximum at the center of the conduit, as indicated in Fig. 5-4. This variation in flow is caused by the friction developed at the conduit wall and the viscous friction developed in the moving fluid. For fine-grained soil, where void spaces are very small, all

Figure 5-4.   *Variation of flow velocity across the cross section of a tube.*

lines of flow are physically close to the "wall" of the conduit, and therefore only low-velocity flows occur. In clays, flow in already small "flow channels" is further hampered because some of the water in the voids is held, or adsorbed, to the clay particles, reducing the flow area and further restricting flow.

TABLE 5-1 / TYPICAL RANGES OF PERMEABILITY FOR DIFFERENT SOIL TYPES

Soil Type	Relative Degree of Permeability	k, Coeff. of Permeability (cm/sec)	Drainage Properties
Clean gravel	High	1 to 10	Good
Clean sand, sand and gravel mixtures	Medium	1 to 10^{-3}	Good
Fine sands, silts	Low	10^{-3} to 10^{-5}	Fair through poor
Sand-silt-clay mixtures, glacial tills	Very low	10^{-4} to 10^{-7}	Poor through practically impervious
Homogeneous clays	Very low to practically impermeable	Less than 10^{-7}	Practically impervious

Note: To convert cm/sec to ft/min, multiply cm/sec by 2; i.e., 1 cm/sec = 2 ft/min; also ft/day = cm/sec × 3 × 10^3.

Typical ranges of permeability for different soil types and resulting drainage characteristics are listed in Table 5-1.

EMPIRICAL RELATIONSHIPS / Considerable information on the flow of fluids through porous media has been obtained from studies of flow through tubes and conduits. From such information, attempts have been made to relate permeability to a soil's grain size. Practically, such a relationship appears more possible for sands and silts than for clays, because of the particle size, shape, and overall soil structure. One of the more widely known relationships is

$$k = 100 D_{10}{}^2 \tag{5-8}$$

where k is given in centimeters per second and D_{10} is the 10 percent particle size, expressed in centimeters, from the grain-size distribution analysis (from the curve resulting from plotting percent finer by weight vs. particle diameters). This relationship was developed from the work of Hazen (1911) on sands. This expression applies *only* to uniform sands in a relatively loose condition.

PERMEABILITY TESTS / Much of the available information from studies of flow through uniform porous media is not directly applicable to soils because of the variation of the size and shape of void spaces in a soil mass. For soil, it has been found to be more practical and accurate to evaluate flow directly through use of laboratory or field tests on the soil in question. Experience has shown that for a given soil a relationship exists between permeability and void ratio. Generally, a semilog plot (e plotted on an arithmetic scale, k plotted on a logarithmic scale) produces approximately a straight line for most soils. Thus, permeability tests can be performed on a soil at two or three widely different void ratios, and the results then be plotted. Permeabil-

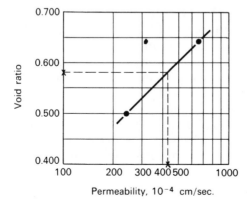

Figure 5-5. *Void ratio vs. permeability plotted on semilog coordinates.*

ity at intermediate void ratios would then be determined by interpolation (see Fig. 5-5).

The permeability of a soil is significantly affected by its in-place structure. A loose granular (coarse-grained) soil would have a higher void ratio than a dense soil, and therefore would permit greater flow. Clays are very significantly affected by structure. Even at similar void ratios, a clay with an undisturbed flocculated structure will possess larger void openings than the same clay having a remolded dispersed (or oriented) structure, with the result that the permeability is many times greater for the flocculated soil (see Fig. 5-6).

Undisturbed flocculent structure   Remolded or dispersed structure

Figure 5-6. *Cohesive soil with flocculent structure will have higher permeability than soil with dispersed structure.*

Extremely important is the significance of stratification. In layered soils, the permeability measured for vertical flow, across layers, can be greatly different from that for horizontal flow (or parallel to the layering). Fine-grained deposits, such as clay or alternating layers of silt and clay, have a permeability in the direction parallel to bedding of the layers many times the permeability of vertical, or cross bedding, flow. Thus, the use of undisturbed test samples, where the sample has retained its original structure and is tested so that flow through the sample is in the correct direction (to correspond to horizontal or vertical flow, as will actually occur in the field), is important if reliable results are to be obtained. This also points out an important consideration regarding use of test data: data from remolded samples may not apply to field conditions if the natural soil remains undisturbed, or test data from undisturbed samples may not apply if, in the field, the soil is to be disturbed and rehandled (as in placing a compacted fill).

Conditions other than the size and number of voids will affect the quantity of flow through a soil deposit. Trapped air or gases prevent flow, while seams, cracks, fissures, and cavities that exist in a soil deposit increase the opportunity for fluid movement. Field investigations need to provide information on the presence of these conditions if reliable measures of flow through the soil mass at a construction site are to be obtained. For this reason, *field permeability tests* are felt to provide more accurate information than laboratory tests. Field testing has its disadvantages, however. Costs and time involved are usually greater than for a laboratory test, and the field test provides only information on conditions in the limited proximity of the test location.

In practice, determining permeability from field or laboratory tests or from indirect analytical methods is for *order of magnitude* use only. This is realistic when it is realized that subsurface conditions and soil properties most probably vary over even short horizontal distances, that soil conditions and properties are generally not known in all areas of influence at a site, and that frequently external factors causing or affecting flow are not accurately known during planning and design.

LABORATORY PERMEABILITY TESTS / Two of the more conventional laboratory permeability tests are the constant head test and the falling head test. Schematic diagrams showing each of these methods, and the mathematics to calculate permeability, are shown in Figs. 5-7 and 5-8.

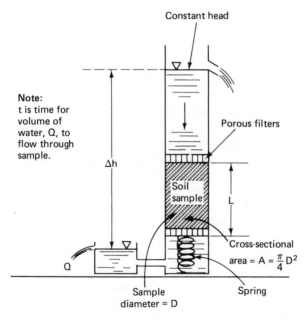

Constant head test: $\dfrac{Q}{t} = kiA$

$$k = \frac{Q}{t} \cdot \frac{1}{iA} = \frac{Q}{t} \cdot \frac{L}{A(\Delta h)}$$

Figure 5-7. *Constant-head permeameter.*

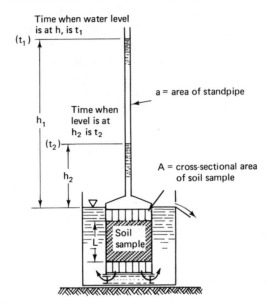

$$\text{Falling head test: } k = \frac{L}{(t_2 - t_1)} \cdot \frac{a}{A} \cdot \ln \frac{h_1}{h_2}$$

$$\text{or} \qquad k = \frac{(2.303)L}{(t_2 - t_1)} \cdot \frac{a}{A} \log_{10} \frac{h_1}{h_2}$$

Figure 5-8.   Falling-head permeameter.

The constant-head permeability apparatus (permeameter) is in wide use for testing the coarse-grained soils, such as sands, where the volume of flow through the soil will be relatively large. For fine-grained soils, such as silt and clay, the falling-head permeameter is generally used. In the constant-head test, permeability is computed on the basis of fluid that passes through the soil sample. In the falling-head test, permeability is computed on the basis of fluid flowing into the sample. The reason for the distinction is simple. In a fine-grained soil, a very limited volume of fluid will flow through the sample. With the constant-head test, time is required to accumulate the fluid volume necessary to perform computations. Extreme care would be required to prevent leaks in the apparatus and evaporation of discharged water. With the falling-head method, the duration of the test is shortened, and there is no concern about the volume of discharge. Care is required to prevent evaporation of water in the inlet tube, however.

Darcy's coefficient of permeability is the factor for a condition of steady flow through a soil. In performing laboratory permeability tests it is essential that volumes be measured only after steady flow has been occurring for some period. It is important to assure that no air or other gases are trapped within the soil to interfere with flow. A vacuum may be required to remove trapped air. In general, the constant-head test is easier to perform, and requires less skill and experience than the falling-head test. Care is required during testing of fine granular soils (such as in the fine sand range) to prevent the particles from being carried along with the discharging water. Details for performing permeability tests are presented in the *ASTM Procedures for Testing Soils*.

Illustration 5-1: A constant-head permeability test is performed on a sample of granular soil. The test setup is as indicated in in Fig. 5-7. The length of soil sample is 15 cm and the cross-sectional area is 10 cm². If a volume of 24 cm³ passes through the soil sample in a 3-minute period, when Δh is 30 cm, compute the coefficient of permeability.

$$k = \left(\frac{Q}{t}\right)\left(\frac{L}{A \, \Delta h}\right)$$

where
$$Q = 24 \text{ cm}^3$$
$$t = 3 \text{ minutes}$$
$$L = 15 \text{ cm}$$
$$A = 10 \text{ cm}^2$$
$$\Delta h = 30 \text{ cm}$$

$$k = \left(\frac{24 \text{ cm}^3}{3 \text{ min}}\right)\left(\frac{15 \text{ cm}}{10 \text{ cm}^2 \times 30 \text{ cm}}\right)$$

$$= 0.4 \frac{\text{cm}}{\text{min}} = 0.006 \frac{\text{cm}}{\text{sec}}$$

Illustration 5-2: A falling-head permeability test is performed on a silty soil. The test setup is as shown in Fig. 5-8. For the test data summarized below, what is the coefficient of permeability for this sample?

Sample length = 8 cm

Cross-sectional area of sample = 10 cm²

Area of standpipe = 1.5 cm²

Height of water in standpipe at start of test period
$h_1 = 100$ cm

Height of water in standpipe at end of test period
$h_2 = 90$ cm

Time for change from h_1 to h_2 = 60 minutes

TABLE 5 – 2/METHODS FOR PEFORMING FIELD PERMEABILITY TESTS
(Ref. 40)

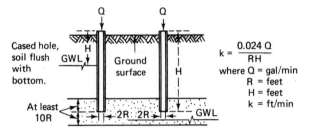

Cased hole, soil flush with bottom.

$$k = \frac{0.024\,Q}{RH}$$

where Q = gal/min
R = feet
H = feet
k = ft/min

Used for permeability determinations when water is above or below bottom of casing. Q is quantity of water to keep casing filled

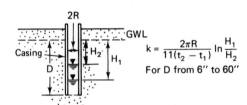

Cased hole, soil flush with bottom.

$$k = \frac{2\pi R}{11(t_2 - t_1)} \ln \frac{H_1}{H_2}$$

For D from 6" to 60"

Used for permeability determination at shallow depths below the water table. May yield unreliable results in falling head test with silting of bottom of hole.

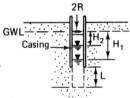

Cased hole, uncased or perforated extension of length L."

$$k = \frac{R^2}{2L(t_2 - t_1)}$$

$$\ln\left(\frac{L}{R}\right)\ln\left(\frac{H_1}{H_2}\right)$$

For $\frac{L}{R} > 8$

Used for permeability determinations at greater depths below water table.

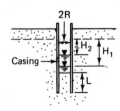

Cased hole, column of soil inside casing to height L."

$$k = \frac{2\pi R + 11L}{11(t_2 - t_1)} \ln\left(\frac{H_1}{H_2}\right)$$

Principal use is for permeability in vertical direction in anisotropic soils.

88

$$k = \frac{(2.303)L}{t_2 - t_1} \left(\frac{a}{A}\right) \log \frac{h_1}{h_2}$$

$$= \frac{(2.303)(8 \text{ cm})}{60 \text{ min}} \left(\frac{1.5 \text{ cm}^2}{10 \text{ cm}^2}\right) \log \frac{100 \text{ cm}}{90 \text{ cm}}$$

$$= 0.00212 \frac{\text{cm}}{\text{min}} = 3.5 \times 10^{-5} \frac{\text{cm}}{\text{sec}}$$

Where consolidation tests are performed on soil samples (Chapter 10), the permeability test can be adapted to determine the rate of flow through the consolidation sample, and the coefficient of permeability computed from these data. Or the coefficient can be determined from the consolidation data obtained to calculate the rate of consolidation of the soil (described in Chapter 10).

FIELD PERMEABILITY TESTS / Field permeability tests have the advantage of testing undistorted soil in its natural location with respect to the ground surface, water table, and other factors that could influence the rate of flow. Various methods for determining permeability are available, depending, among other things, on the soil's being above or below the ground water table. The methods described herein are of the type where a cased boring is made into the soil that is to be tested. The casing and related equipment necessary are of the type normally utilized by soil-boring contractors. Whenever possible, it has been found expedient to perform field permeability tests during the investigative stage of planning a project, at the time that the subsurface investigation (soil borings) is being made.

Essentially, the field permeability test involves obtaining a record of the time that it takes for a volume of water to flow out of, or into, the boring casing. A schematic presentation of different conditions and the related equations for calculating the coefficient of permeability is presented in Table 5-2.

Capillarity

The *groundwater table* (or *phreatic surface*) is the level to which underground water will rise in an observation well, pits, or other open excavations into the earth. All voids or pores in soil located below the groundwater table would be filled with water (except possibly for small isolated pockets of trapped air or gases). In addition, however, soil voids for a certain height above the water table will also be completely filled with water (full saturation). Even above this zone of full saturation, a condition of partial saturation will exist.

Any water in soil located above the water table is referred to as *soil moisture*. The phenomenon in which water rises above the groundwater table against the pull of gravity but is in contact with the water table as its source is referred to as *capillary rise*. The water associated with capillary rise is *capillary moisture*.

WATER IN CAPILLARY TUBES / The basic principles of capillary rise in soils can be related to the rise of water in glass capillary tubes (tubes with very small diameters) under laboratory conditions. When the end of a vertical capillary tube is put in contact with a source of water, the water rises up in the tube and remains there. The rise is attributed to the attraction between the water and the glass and to a *surface tension* which develops at the air-water interface at the top of the water column in the capillary tube. This surface tension can be thought of as an infinitely thin but tough film, such as a stretched membrane. (The surface tension phenomenon is one of the reasons that small insects can "walk" on water.) The water is "pulled up" in the capillary tube, to a height regulated by the diameter of the tube, the magnitude of the surface tension, and the density of the water.

The attraction between the water and capillary tube affects the shape of the air-water interface at the top of the column of water. For water and glass, the shape is concave downward; i.e., the water surface is lower at the center of the column than at the walls of the tube. The resulting curved liquid surface is termed the *meniscus* (Fig. 5-9).

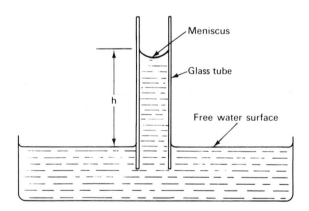

Figure 5-9 *Capillary rise and water meniscus in a glass tube.*

The column of water in the capillary tube has risen above the surface of the water supply and against the pull of gravity. For a condition of equilibrium, the effect of the downward pull of gravity on the capillary column of water has to be resisted by the ability of the surface film to adhere to the wall of the capillary tube and hold the column of water. This wall adhesion must equal the surface tension of the fluid. If T_s is the value of surface tension, expressed in units of force per unit length, the vertical loading that can be supported is

$$\text{(tube circumference)} \times (T_s) \times (\cos \alpha)$$

where α is the angle formed between a tangent to the meniscus and the capillary wall. With water and glass, the meniscus at the wall of the capillary tube is tangent to the wall surface, and the angle α is zero degrees. Therefore, $\cos \alpha$ is one, and the column weight that is capable of being supported because of surface tension is

$$(2\pi r)(T_s)$$

where r is the radius of the capillary tube.

The weight of the column of water in the capillary tube is

$$(\pi r^2) \times (h) \times (\rho) \times (g)$$

where h = height of the column of water

ρ = density of water, mass per volume, taken as 1 gm/cm^3 or 1.95 slugs/ft^3.

g = acceleration of gravity, 980 cm/sec^2 or 32.2 ft/sec^2.

If h_c is the maximum height of capillary rise that can occur for prevailing physical conditions (such as for a particular temperature), equilibrium requires that

$$(2\pi r)(T_s) = (\pi r^2 h_c)(\rho)(g) = \pi r^2 h_c \frac{(\text{weight})^1}{(\text{volume})}$$

$$= \pi r^2 h_c \gamma_w$$

where γ_w = unit weight of water, taken as
62.4 pcf or 980 dynes/cm^3.

and the maximum height of capillary rise is

$$h_c = \frac{(2\pi r)(T_s)}{(\pi r^2)(\rho)(g)} = \frac{2T_s}{(r)(\rho)(g)} =$$

$$= \frac{2T_s}{r(\gamma_w)} = \frac{4T_s}{d(\gamma_w)} \qquad (5\text{-}9)$$

where d = the diameter of the capillary tube.

The value of T_s for water varies according to temperature. At normal room temperatures, T_s is close to 0.005 lb/ft or 73 dynes/cm. In applying the development of capillary rise in tubes to capillary rise in

---

[1]Since weight equals mass times acceleration of gravity ($W = mg$), the product of ρg equals $\dfrac{(\text{mass})}{(\text{volume})} \times (g)$, or $\dfrac{\text{weight}}{\text{volume}}$. Also see footnote, p. 23.

soils, these values for T_s are sufficiently accurate for many practical problems. Thus, the equation for capillary rise can be expressed as

$$h_c = \frac{0.31}{d} \text{ cm (approximately)} \qquad (5\text{-}10)$$

provided that d is in centimeters.

Illustration 5-3:   Compute the height of capillary rise for water in a tube having a diameter of 0.005 cm.
In metric units:

$$h_c = \frac{4\,T_s}{(d)\,(\rho)\,(g)} = \frac{(4)\left(73\,\dfrac{\text{dynes}}{\text{cm}}\right)}{(0.005\text{ cm})\left(1\,\dfrac{\text{gm}}{\text{cm}^3}\right)\left(980\,\dfrac{\text{cm}}{\text{sec}^2}\right)}$$

$$\cong 60 \text{ cm} \cong 2 \text{ feet}$$

In British units:

$$0.005\text{ cm} = \frac{0.005\text{ cm}}{\left(2.54\,\dfrac{\text{cm}}{\text{in.}}\right)\left(12\,\dfrac{\text{in.}}{\text{ft}}\right)} = 1.64 \times 10^{-4}\text{ ft}$$

$$h_c = \frac{4\,T_s}{d\gamma_w} = \frac{(4)\,(0.005\text{ lb/ft})}{(1.64 \times 10^{-4}\text{ ft})\,(62.4\text{ pcf})} \cong 2 \text{ ft}$$

There *are* situations, however, where the temperature effects should be considered. Generally, as temperature increases, the value of T_s decreases, indicating a lessening height of capillary rise under warm conditions or an increasing height of capillary rise for conditions of falling temperatures. At freezing, T_s for water is about 76 dynes/cm.

As the column of water stands in the capillary tube, "hanging" by the surface tension at the meniscus, the *weight* of the column is transferred to the walls of the capillary tube, creating a compressive force in the walls. The effect of this on soil is discussed in a later section, "Effects of Surface Tension."

The height of capillary rise is not affected by a slope or inclination in the direction of the capillary tube, or by variations in the shape and size of the tube at levels below the meniscus (Fig. 5-10). But for water migrating up a capillary tube, a large opening can prevent further movement up an otherwise small-diameter tube. The determining factor is the relation between the size of the opening (tube diameter) and the particular height of its occurrence above the water supply.

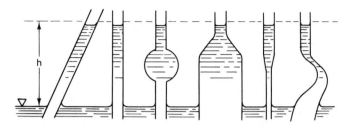

Figure 5-10. *Capillary heights of capillary tubes of various shapes are the same if their menisci diameters are the same.*

In the capillary tube, at the level equal to the free surface of the water supply, the hydrostatic pressure in the water is zero. Hydrostatic pressures increase below the free water surface according to the relationship

$$p_w = \gamma_w Z \qquad (5\text{-}11)$$

where Z = the depth below the water surface.

p_w = the fluid pressure at the depth Z.

Conversely, hydrostatic pressure measured in the capillary column *above* the free water surface is negative, according to the relationship

$$p_w = -\gamma_w h \qquad (5\text{-}12)$$

where h = the distance measured upwards,
from the free water surface.

The negative sign indicates that water pressures in the capillary tubes are at less than atmospheric pressure. As a result, water in the capillary columns is said to be in *tension*. The maximum negative pressure, or capillary tension, exists at the maximum height of rise h_c.

Capillary rise is not limited to tube, or enclosed, shapes. If two vertical glass plates are placed so that they touch along one end and, in plan, form a "V," a wedge of water will rise up in the "V" because of the capillary phenomena (Fig. 5-11). The height of rise relates to the attraction between the water and plates and the physical properties of the water, as in tubes, but also to the angle formed by the "V." The significance of this type of capillary rise is discussed in the following section, "Capillary Rise in Soil."

CAPILLARY RISE IN SOIL / In soils, the shapes of void spaces between solid particles are unlike those in capillary tubes. The voids are of irregular and varying shape and size, and interconnect in all directions, not only the vertical. These properties make the accurate prediction of the height of capillary rise in soil almost impossible. How-

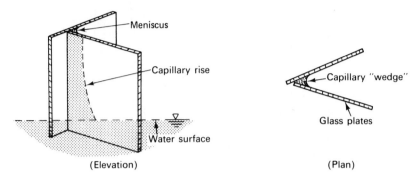

Figure 5-11.   *Capillary rise in corner formed by glass plates.*

ever, the features of capillary rise in tubes are applicable to soils insofar as they facilitate an understanding of factors affecting capillarity, and help to establish an order of magnitude for capillary rise in the different types of soils.

Illustration 5-4:   Limited laboratory studies indicate that for a certain silt soil, the effective pore size for height of capillary rise is $\frac{1}{5}$ of D_{10}, where D_{10} is the 10 percent particle size from the grain-size distribution curve. If the D_{10} size for such a soil is 0.02 mm, estimate the height of capillary rise.

$$d = \text{effective capillary diameter} = \tfrac{1}{5}D_{10} = \tfrac{1}{5}(0.02 \text{ mm})$$

$$= 0.004 \text{ mm} = 0.0004 \text{ cm}$$

$$h_c = \frac{0.31}{d} \qquad\qquad\qquad\qquad \text{[from Eq. (5-10)]}$$

$$= \frac{0.31}{0.0004 \text{ cm}} = \frac{0.31}{4 \times 10^{-4} \text{ cm}}$$

$$= 7.75 \times 10^2 \text{ cm} = 775 \text{ cm} \simeq 25 \text{ ft}$$

Review of Eq. (5-8) indicates that even relatively large voids will be filled with capillary water if the soil is close to the source of water, i.e., the groundwater table. As the distance from the water table increases, only the smaller voids would be expected to be filled with capillary water. The larger voids represent interference to upward capillary flow and would not be filled. Consequently, the soil closest to the water table but above it is fully saturated as a result of capillarity. Above this zone of full saturation lies a zone of partial saturation due to capillarity (see Fig. 5-12.) The height to which capillary water rises is · termed *capillary fringe.*

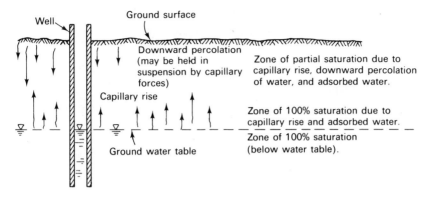

Figure 5-12. *Distribution of soil moisture in a soil profile.*

In the upper zone of the capillary fringe, where partial saturation exists, capillary movement occurs through the capillary tube phenomenon but may also in the wedges of the capillary "V" formed where soil particles are in contact (similar to the "V" formed by vertical plates discussed in the preceding section); see Fig. 5-13.

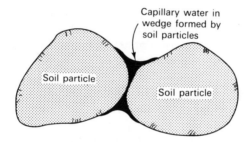

Figure 5-13. *Wedge of capillary water at point of contact between particles.*

Since void spaces in soil are of the same order of magnitude as the particle sizes, it follows that the height of capillary rise would be greater in fine-grained soils than in coarse-grained soils. For orders of magnitude, relative values of capillary rise for different type soils are presented in Table 5-3.

Temperature plays an important practical role in the capillary rise in soil. The height of rise is greater at lower temperatures than at high temperatures. Demonstrations of this fact have frequently been observed in construction performed in a cool season following a warm season (i.e., the fall season). Cool temperatures cause an increase in migration of water toward the surface. Frequently the additional moisture increases the difficulty in handling or working with the soil,

TABLE 5-3 / REPRESENTATIVE HEIGHTS
OF CAPILLARY RISE IN SOILS

Soil Type	Approximate Capillary Heights	
	cm	ft
Small gravel	2–10	0.1–0.4
Coarse sand	15	0.5
Fine sand	30–100	1–3
Silt	100–1000	3–30
Clay	1000–3000	30–90

or creates a muddy surface, which makes the moving of equipment a problem. *Horizontal migration* of water can also occur near a heated building. Water may move from beneath a structure in cold seasons if the presence of the structure affords some protection from cold temperatures. If the building itself contributes to a lowering of the ground temperature, as in a poorly insulated cold-storage warehouse, frozen foods processing plant, or ice skating rink, capillary flow will be induced towards the building.

Attempts at establishing a relationship between height of capillary rise for a given soil and some aspect of the soil's grain size or void ratio have been made. The general applicability and accuracy of proposed relationships appear to be limited to providing estimates of order of magnitude.

TIME RATE OF CAPILLARY RISE / The time necessary for the expected maximum height of capillary rise to occur requires consideration in some practical problems, such as where a structural fill has been placed for highways, buildings, or other purposes. On the basis of typical void sizes, clay and fine silt soils will have significant heights of capillary rise. However, the time period required for the rise to occur may be so great that other influences, such as evaporation and change in groundwater level, also have to be considered.

The term indicating the *rate* of capillary rise is *capillary conductivity* or *capillary permeability*, k_{cap}. Factors known to affect a soil's capillary conductivity are void sizes, moisture content, and temperature in the soil. Generally, capillary conductivity is greater at higher moisture contents and lower temperatures. Absolute values of capillary conductivity are not available. Though not identical to Darcy's coefficient of permeability, the *relative rates* of capillary conductivity can be thought of in terms of the comparative values for permeability, i.e., rapid for coarse soils, low and very low for silts and clays.

SUSPENDED CAPILLARIES / Water attempting to percolate downward through a soil (such as percolating surface water from rain or melting snow) or pore water resulting from a formerly higher water table can be held suspended in the soil voids because of the same surface tension phenomenon responsible for capillary rise. For this, the column of water would have a meniscus at both ends of the suspended column. Each meniscus would be in tension. The maximum length of a column would be controlled by the same factors affecting height of capillary rise.

ELIMINATION OF CAPILLARY WATER IN SOIL / For capillary rise to occur, the existence of an air-water interface is required. Capillary water will not exist in soil at a level below the groundwater table. It follows that a condition of capillarity will cease where submergence of a soil zone occurs.

Capillary water can be removed from a soil by evaporation. As a result, capillary water can represent a very mobile category of water where the evaporation is continually replaced by new capillary water.

EFFECTS OF SURFACE TENSION / At the location of the meniscus the surface tension imposes a compressive force onto the soil grains in contact with the meniscus that is equal in magnitude to the weight of the water in the capillary column, as indicated in the earlier discussion for capillary tubes. This effect applies to both a meniscus resulting from capillary rise and for pore water suspended above a capillary zone. The compressive force is imposed throughout the soil in contact with the "held column" of water, and serves to compress or shrink the soil.

In clays, this phenomenon results where the groundwater table drops subsequent to the time the clay deposits were formed. Such "drying" of an area causes sufficient internal compressive forces within the clay mass that the clay becomes firm and strong. This is referred to as drying by dessication, and the clays are referred to as *dessicated clays*. A land area formed by clay deposits that were originally soft and weak may have a firm crust or upper layer overlying the still soft and weak deeper clays because of dessication. The thickness and strength of a dessicated zone has frequently been found sufficiently adequate that road and light building structures could be satisfactorily supported by it.

Typical magnitudes of compressive stress that can result in fine-grained soils are tabulated in Table 5-4.

TABLE 5-4 / REPRESENTATIVE VALUES OF COMPRESSIVE STRESS RESULTING FROM CAPILLARY FORCES

Soil Type	Compressive Stress		
	psf	Kg/cm^2	kN/m^2
Silt	200 to 2000	0.1 to 1.0	10 to 100
Clay	2000 to 6000	1.0 to 3.0	100 to 300

In partially saturated granular (coarse-grained) soils, the surface tension phenomenon also contributes to the strength of the mass. At partial saturation, all voids are not filled with water. The available water collects in the interstices adjacent to where soil particles touch, forming a wedge of moisture but leaving the center portion of the void filled with air. Thus, an air-water interface, or meniscus, is formed. The surface tension in the meniscus imposes a compressive force onto the soil particles, increasing the friction between particles and thus the shearing strength. The strength gain in granular soils due to partial saturation and the surface tension phenomenon is termed *apparent cohesion*. The strength gain can be quite significant, as the firm condition of sandy beach surfaces, which easily support the weight of vehicles, indicates. The menisci and surface tension, along with the apparent cohesion, will disappear when the soil is fully saturated.

PROBLEMS

1. Compute the value of the hydraulic radius for a circular pipe flowing half-full.

2. Calculate the value for the hydraulic radius for flow in an open channel where the bottom of the channel is 3 ft wide, the side slopes are 1 on 1 (slope of one foot horizontal to one foot vertical), and the depth of flow is 2 ft.

3. A thick fluid having a high viscosity and a thin fluid having a low viscosity are to be passed through a permeable material. Indicate the condition for which the resulting coefficient of permeability will be greatest and briefly explain why. For which condition (thick or thin fluid) is the quantity of flow through the material expected to be greatest?

4. The coefficient of permeability is generally greater for coarse soils (sands and gravels) than for fine-grained soils (silts and clays). What effect does particle size have on permeability?

5. What effect does the presence of adsorbed water in clay have on the coefficient of permeability for this type of soil?

6. In the flow of water through soil, what are the conditions necessary in order for Darcy's Law to apply?

7. Estimate the coefficient of permeability for a uniform sand where a sieve analysis indicates that the D_{10} size is 0.15 mm.

8. Why might the permeability in fine-grained soil deposits be expected to be greater for horizontal flow than for vertical flow?

9. Briefly give reasons why the coefficient of permeability in an undisturbed clay deposit possessing a flocculent structure would be ex-

pected to be greater than if the same clay had a dispersed (re-molded) structure.

10. A constant-head permeability test is performed in a laboratory where the soil sample is 25 cm in length and 6 cm² in cross section. The height of water at the inflow end is maintained at 2 ft, and at 6 in. at the outlet end. The quantity of water flowing through the sample is 200 cm³ in 2 minutes.

(a) Make a sketch of the described conditions.
(b) What is the coefficient of permeability in centimeters per minute?

11. A constant-head permeability test is performed where the hydraulic gradient is 0.75. The cross-sectional area of the sample is 0.25 ft². The quantity of water flowing through the sample is measured to be 0.004 ft³/min.

(a) What is the coefficient of permeability in feet per minute?
(b) What is the coefficient of permeability in centimeters per second?

12. How much water will flow through a soil mass in a 5-minute period when the sample length is 15 cm, the cross section is 2 cm by 2 cm, and a constant head of 2 feet is maintained? The soil has a k value of 1×10^{-2} cm/sec.

13. A falling-head permeability test is performed on a fine-grained soil. The soil sample has a length of 12 cm and a cross-sectional area of 6 cm². The water in the standpipe flowing into the soil is 60 cm above the top of the sample at the start of the test. It falls 5 cm in 30 minutes. The standpipe has a cross section of 2 cm².

(a) Make a sketch of the described conditions.
(b) What is the coefficient of permeability in centimeters per second?
(c) What is the coefficient of permeability in feet per minute?

14. A field permeability test is performed by measuring the quantity of water necessary to keep a boring casing (pipe) filled. The distance from the top of the casing to the bottom (in the ground) is 10 ft. The groundwater table is below the bottom of the casing. The casing has an inside diameter of 6 in. In a 10-minute period, one gallon of water was used to keep the casing filled. What is the coefficient of permeability for the soil at the bottom of the casing?

15. A field permeability test indicates that the coefficient of permeability for a certain soil is 2×10^{-5} cm/sec. Is this a relatively high or low coefficient of permeability? What type of soil would this probably be?

16. To what height would water rise in a glass capillary tube that is 0.01 mm in diameter?

17. What is the water pressure just under the meniscus in a capillary tube where the water has risen to a height of 6 ft?

18. A glass capillary tube is 0.001 mm in diameter.

    (a) What is the theoretical maximum height of capillary rise for a tube of this size?

    (b) What compressive pressure results in the capillary water just under the meniscus?

19. In a silt soil, the D_{10} size is 0.01 mm. If the effective pore size for estimating capillary rise is taken as $\frac{1}{5}$ of D_{10}, approximately what height of capillary rise will occur?

20. Why is it expected that the maximum height of capillary rise is greater for fine-grained soils than for coarse-grained soils?

21. How is the height of maximum capillary rise of water in soil affected by temperature?

22. Explain how capillarity is related to the dried and firm condition frequently observed to exist in the surface zone of fine-grained soil deposits.

23. Indicate the ways that capillary water and the effects of capillarity can be removed from a soil.

CHAPTER 6
Movement of Water Through Soil: Practical Effects

Flow Nets and Seepage, Drainage, Frost Heave

The handling of mobile and stationary underground water during construction operations, and making provisions so that the effects of its presence will not interfere with the function of completed structures are of vital concern to the construction profession. The following discussion on flow nets and seepage, drainage, and frost heave relates to the practical aspects of controlling groundwater during and after construction.

Flow Nets and Seepage

FLOW OF SUBSURFACE WATER / The flow of water beneath the ground surface through all soils except coarse gravel and larger materials occurs as laminar flow; that is, the path of flow will follow a regular pattern, with adjacent paths of water particles all flowing parallel. For this condition, Darcy's Law for water traveling through soils can be applied to determine the rate and quantity of flow, and the seepage forces that result from this flow. In its most direct form, Darcy's Law is

$$q = kiA \qquad (6\text{-}1)$$

(as developed in Chapter 5, Eq. 5-5).

where q is the quantity of flow in a unit time period (or rate of flow), k is the coefficient of permeability for the soil, A is the cross-sectional area of the soil through which flow is occurring (normal to the direction of flow), and i is the hydraulic gradient (the difference in the energy head of water between two points divided by the distance between the same two points), all as previously defined in Chapter 5.

When underground water is flowing over a relatively long distance and within a soil zone having well-defined boundaries, such as that shown in Fig. 6-1, the quantity of flow can be determined by using the above expression directly.

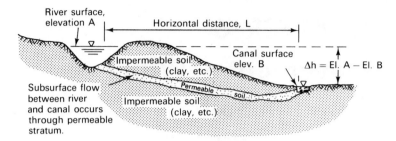

Figure 6-1.   *Instance in which direct application of Darcy's Law can be used to determine underground flow.*

Illustration 6-1:   A river and a canal run parallel to each other but at different elevations, as indicated by Fig. 6-1. If the difference in the water surface elevations is 16 ft, the horizontal distance is 400 ft, and the thickness of the permeable stratum is 6 ft, compute the seepage loss between river and canal, per mile of river-canal length. Permeability $k = 1.0$ ft/day.

$$q = kiA = k \frac{\Delta h}{L} A = (1.0 \text{ ft/day}) \frac{16 \text{ ft}}{400 \text{ ft}} (6 \text{ ft} \times 5280 \text{ ft/mi})$$

$$= 1,270 \text{ ft}^3/\text{day/mile of length}$$

THE NEED FOR FLOW NETS AND FLOW NET THEORY / Where the zones of flow or directions of flow are irregular, where water enters and escapes from a permeable zone of soil by travelling a short distance, or where the flow boundaries are not well defined (a boundary being the separation between where flow does and does not occur), it may be necessary to use flow nets to evaluate flow.   Flow nets are a pictorial method of studying the path that moving water follows.   Darcy's Law can be applied to flow nets to evaluate the effects of flow.

For a condition of laminar flow, the path that the water follows can be represented by *flow* lines.   In moving between two points, water tends to travel the shortest distance.   If changes in direction occur, the changes take place along smooth curved paths.   A series of flow lines to represent flow through a soil mass would be parallel except where a change in the size of an area through which flow occurs takes place. Figure 6-2 shows flow lines for water seeping through a soil in a simple laboratory model.

Flow of water occurs between two points because of a difference in energy, (usually expressed as head of energy or as pressure).   Since it is the relative energy difference between two points, energy head can be

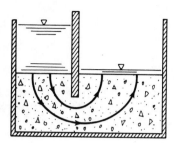

Figure 6-2. *Laboratory seepage tank showing flow lines indicating path of water flowing through the soil.*

referenced to an arbitrary datum for convenience. The total head of any point is that resulting from the sum of velocity head, potential head, and pressure head.[1] For water flowing through soils, velocity head is neglected, since it is small compared to potential and pressure head. In Fig. 6-3(a) piezometers (open standpipes) inserted at points A and B would have the water level rise to the elevation shown. This figure shows the pressure and potential energy heads.

At certain points on different flow lines, the total energy head will be the same as seen in Fig. 6-3(a). Lines connecting points of equal total energy head can be drawn and are termed *equipotential lines;* see Fig. 6-3(b). Equipotential lines must cross flow lines at right angles, since they represent pressure normal to the direction of flow. The flow lines and equipotential lines together form the flow net and are used to determine the quantities and other effects of flow through soils.

When seepage analyses are made, flow nets can be drawn with as many flow lines as desired. The number of equipotential lines will be determined by the number of flow lines selected. A general recommendation is to use the fewest flow lines that still permit reasonable depiction of the path along the boundaries and within the soil mass. For many problems, three or four flow channels (a channel being the path between adjacent flow lines) are sufficient.

DEVELOPMENT OF THE FLOW NET—ISOTROPIC SOIL[2] / Figure 6-4 shows a condition where flow occurs through a soil mass. Flow occurs because of the difference in energy head or pressure caused by the unequal heights of water.

---

[1]From the Bernoulli equation for steady irrotational (or laminar) flow, total head, $H = v^2/2g + p/\gamma + Z$, where v = velocity of flow, g is the acceleration of gravity, p is pressure, γ is the unit weight of the fluid flowing and Z is the vertical distance between the point where the pressure p is determined and a reference elevation or datum.

[2]When applied to fluid flow, the term *isotropic* infers that soil permeability is equal in all directions.

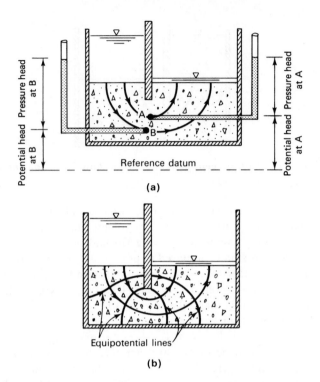

Figure 6-3. *Points of similar energy head indicated on flow lines, and drawing of the flow net by addition of equipotential lines.*

1. To begin the flow net, establish the boundaries of the soil mass through which the flow occurs and conditions at the boundaries. In Fig. 6-4, lines a-b and c-e-f-d are drawn to represent flow lines at the boundary. Line c-a is an equipotential line with total head equal to h_t. Line b-d is an equipo-

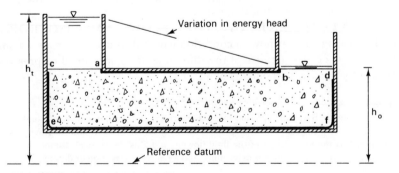

Figure 6-4. *Laboratory seepage tank showing soil and water conditions, and boundaries for flow.*

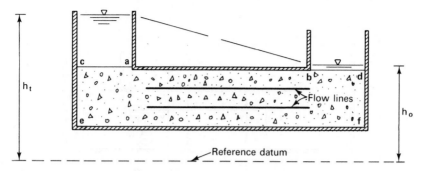

Figure 6-5.   *Intermediate flow lines between flow boundaries.*

tential line with a total head equal to h_0. These four lines
establish the boundaries for flow through the soil in this problem. Note the line that indicates the variation in head between the ends of the soil mass.

2.   Next sketch in flow lines parallel to those flow lines along the
boundaries. Start by spacing the flow lines an equal distance
apart in the sections where it is known that the flow lines will
be parallel to each other (where the direction of flow is in a
straight line and not curved); see Fig. 6-5.

3.   After the flow lines have been selected, locate a point of equal
total energy head on each flow line. A line connecting these
points will be an equipotential line; see Fig. 6-6. Draw additional equipotential lines, selecting a spacing so that the distances y_1, y_2, y_3 permit the proportion

$$\frac{y_1}{l_1} = \frac{y_2}{l_2} = \frac{y_3}{l_2}$$

to be maintained. Note the corresponding changes in total
head at each equipotential line, (Fig. 6-7).

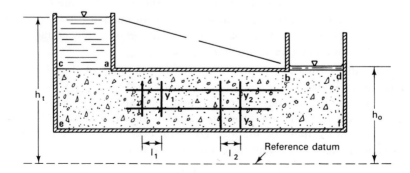

Figure 6-6.   *Equipotential lines to establish the flow net.*

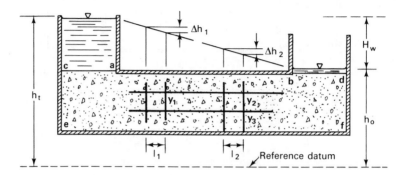

Figure 6-7. *The developing flow net indicating relationship of pressure head differences between equipotential lines.*

4.   Apply Darcy's Law, $q = kiA$. Since a volume (of flow) is to be determined, width as well as a length and height of soil mass must be considered. For convenience, assume a thickness (perpendicular to the page) of unity. Let the flow past y_1 be q_1, the flow past y_2 be q_2, and the flow past y_3 be q_3. Referring to Fig. 6-7, and noting that i (the hydraulic gradient) is the difference in head divided by the distance between the points where the head is measured, or $\Delta h/\Delta L$, (where l_1, l_2, are ΔL), obtain

$$q_1 = \frac{\Delta h_1}{l_1}(y_1)(1)(k)$$

$$q_2 = \frac{\Delta h_2}{l_2}(y_2)(1)(k)$$

$$q_3 = \frac{\Delta h_2}{l_2}(y_3)(1)(k)$$

Since y_1 and y_2 are bounded by the same flow lines, q_1 and q_2 must be equal, because flow is continuous. Also, since the ratios of the sides of each block are equal, y_3 is equal to y_2 and therefore q_2 equals q_3.

If all the blocks are proportional, the quantity of flow between flow lines is equal, and it follows that Δh_1 equals Δh_2.

In flow net solutions, the blocks are usually drawn as squares, although actually only the proportionality of distances between the two sides of a block must be maintained. However, by using squares the ratio of the sides is one, and the expression for flow becomes

$$q_1 = k\,\Delta h$$

where Δh is the total head divided by the number of equipotential line pressure drops N_d, or

$$\Delta h = \frac{H_w}{N_d}$$

Total flow is the summation of the flow in each flow channel or $q = q_1 + q_1 + q_3 + \ldots$, which is also q_1 times the number of flow channels, or $q_1 N_f$.

Therefore,

$$q = q_1 N_f = k \, \Delta h N_f = k \, \frac{H_w}{N_d} \, N_f$$

or $\qquad\qquad q = k H_w \left(\frac{N_f}{N_d}\right) \qquad$ (for isotropic soil) $\qquad\qquad$ (6-2)

In constructing flow nets for most problems, it is improbable that the figures resulting from the assumed flow lines and equipotential lines will all be squares. The requirement for this condition is that, for each block, the distance across the center of the block between flow lines must equal the distance between the two equipotential lines, and that the equipotential lines cross the flow lines at right angles. Figure 6-8 traces the steps used in drawing a flow net for a practical problem, and shows the computations to determine seepage.

BOUNDARIES FOR THE FLOW NET / Fig. 6-9 provides illustrations of typical flow net problems. These illustrations can serve as a guide to establish the boundaries of flow and the general pattern of a flow net for many types of problems. For problems like that shown in Fig. 6-8, the upper boundary is the flow line following along the base of the structure. The lower boundary is a flow line which, for application to the flow net, is a line following along the surface of the impermeable stratum. The locations and directions of other flow lines within these boundaries are selected by using the methods described earlier.

Earth dams represent the condition in which the upper limit of flow is not defined by a natural boundary.[3] In some cases, such as homogeneous earth dams situated on an impervious stratum (see Fig. 6-10), seepage will outlet on the dam's downstream face. The flow line representing the upper limit of seepage moving through the dam is approximately parabolic in shape. The point where the seepage breaks

---

[3]The method for studying seepage through earth dams is presented in Arthur Casagrandes paper "Seepage Through Dams," which appeared in the *Journal of the New England Water Works Association*, June 1937. This paper has been reprinted in *Contributions to Soil Mechanics, 1925–1940*, Boston Society of Civil Engineers, Boston, Mass.

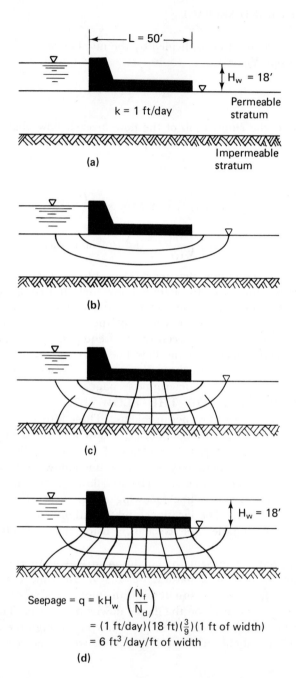

Figure 6-8. *Steps in drawing a flow net: (a) scale drawing of conditions; (b) trial flow lines; (c) trial equipotential lines; (d) final flow net and related seepage computation.*

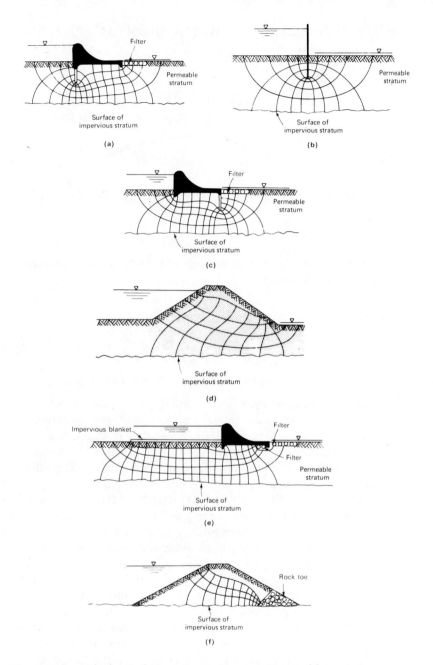

Figure 6-9. Flow nets for representative seepage problems: (a) masonry dam with sheetpile cutoff at heel; (b) sheetpile; (c) masonry dam with sheetpile cutoff and filter at toe; (d) earth dam; (e) masonry dam with upstream impervious blanket and downstream filter; (f) earth dam with downstream rock toe filter.

109

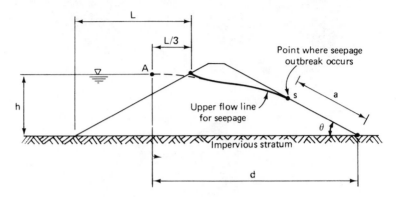

Figure 6-10.   *Seepage through an earth dam; definition of terms necessary to locate seepage outbreak.*

out onto the downstream face can be established from

$$a = \frac{d}{\cos \theta} - \sqrt{\frac{d^2}{\cos^2 \theta} - \frac{h^2}{\sin^2 \theta}}$$ (6-3)

(for slope angles, θ, less than 30 deg.)

where values of a, d, and h are defined on Fig. 6-10. The method used to establish the shape and position of the upper line of seepage through the dam is outlined in Fig. 6-11. After the upper flow is established, the remainder of the flow net can be sketched. The quantity of seepage can be estimated using the methods discussed previously. The effect of seepage forces is discussed in a later section of the chapter.

FLOW NETS FOR NON-ISOTROPIC SOIL / The method of flow net analysis discussed above applies for the frequently found condition in which the soil permeability in the horizontal and vertical directions is similar. However, in stratified soil deposits, the horizontal and vertical coefficients of permeability may differ; usually the horizontal permeability is greater than the vertical. In such instances, the methods for drawing the flow net need to be modified. Use of a *transformed section* is an easily applied method which accounts for the different rates of permeability. Vertical dimensions are selected in accord with the scale desired for the drawing. Horizontal dimensions, however, are modified by multiplying all horizontal lengths by the factor $\sqrt{k_v/k_h}$, where k_v and k_h are the vertical and horizontal coefficients of permeability, respectively. A distorted diagram, with shortened horizontal dimensions, results, as illustrated in Fig. 6-12. The conventional flow net is

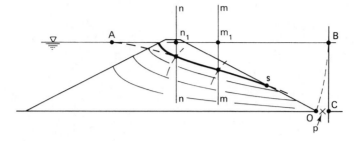

1. Establish point A and swing an arc OB having the radius AO (point O will be the focus of the parabola);

2. Draw vertical line BC (this will be the directrix of the parabola);

3. Locate point p midway between O and C;

4. Draw the vertical line m-m in the general location indicated, and determine the horizontal distance from line BC to line m-m;

5. Using point O as a center, swing an arc with radius $m_1 B$ to intersect line m-m. This intersection establishes a point on the parabolic flow line;

6. Draw the vertical line n-n in the general location indicated, and determine the distance $n_1 B$;

7. Using point O as a center, swing an arc with radius $n_1 B$ to intersect line n-n. This intersection establishes another point on the parabolic flow line;

8. Continuing the procedure just outlined, establish as many additional points as necessary to sketch the total length of flow line between A and s;

9. Modify the beginning section of the flow line freehand. The freehand line must intersect the upstream face of the dam at a right angle;

10. The parabolic flow line is assumed to follow the downstream face of the dam below point s;

11. Other flow lines are sketched in, assuming a shape generally similar to the upper seepage line.

Figure 6-11. *Procedure for locating the upper seepage line for flow through an earth dam.*

then drawn on the transformed section using the procedures presented previously. For flow through the non-isotropic soil, the seepage equation becomes

$$ q = H_w \left(\frac{N_f}{N_d} \right) \sqrt{k_v k_h}. \qquad \text{(for non-isotropic soils)} \qquad (6\text{-}4) $$

The values for N_f and N_d in this equation are taken directly from the transformed section.

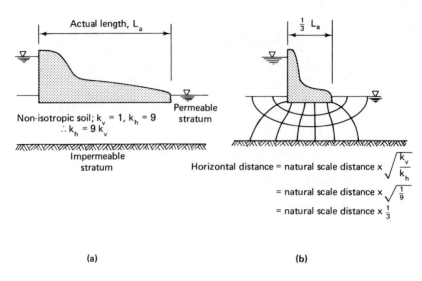

Figure 6-12. *Method of drawing the transformed section and flow net for non-isotropic soil conditions: (a) actual section (to scale); (b) transformed section (distorted scale).*

UPLIFT FORCES / For the problem of seepage beneath a structure such as a dam, the uplift force acting on the base of the structure because of this seepage can be evaluated from the flow net. The pressure acting at the upstream and downstream end of the structure (and at selected in-between points if necessary) is determined and a pressure diagram is made. The total uplift force (per unit of width perpendicular to the drawing) is then the area of the pressure diagram; see Fig. 6-13.

OTHER SEEPAGE FORCES / Water seeping beneath a structure and escaping by flowing upward at the downstream end imposes an upward force on the soil. If the upward force is sufficiently great, the soil will be carried away by the escaping water. There follows an illustration of how the flow net can be used to determine if the pressure of the escaping water is great enough to erode the soil. The submerged unit weight of soil is

$$\gamma_{sub} = \left(\frac{G_s - 1}{1 + e}\right)\gamma_w \tag{2-18}$$

If the value of G_s (specific gravity of soil) is about 2.7 and the void ratio is about 0.70 (both reasonable values), the submerged weight of

Note:
In this figure, the h values on the equipotential lines indicate
the height above the ground surface to which the water level
in a piezometer would rise.
Pressure head drop across adjacent equipotential lines =

$$\frac{H_w}{N_d} = \frac{18 \text{ ft}}{9 \text{ drops}} = 2 \text{ ft/drop.}$$

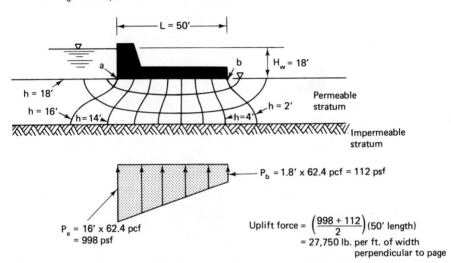

Figure 6-13. *Procedure used to compute uplift pressure acting on base of
a masonry dam.*

the soil would be about equal to the weight of water. The hydraulic
gradient between the last two equipotential lines is the difference in
pressure head across these two points divided by the distance between
these equipotential lines, or $\Delta h/l$. The pressure difference between the
two equipotential lines is Δh times the unit weight of water, γ_w. If the
hydraulic gradient is unity, the upward force due to the moving water
is equal to the unit weight of water. There then exists an upward pressure of 62.4 psf acting on a soil where gravity effects a downward pressure of 62.4 psf. This is a fully buoyant condition in the soil. A slightly
greater uplift pressure would carry away soil particles. For this reason,
an escape hydraulic gradient of about unity or greater is usually considered an indication that erosion may occur. To prevent erosion, the
pattern of flow could be modified to reduce the pressure of the escape
gradient (increasing the length of flow either by making the structure
larger or by embedding sheeting beneath the structure are two alterna-
tives), or a coarse material of carefully graded particle sizes that is not
susceptible to erosion by the escaping water could be placed beneath
the tip of the structure (a soil filter).

Illustration 6-2:   A masonry dam having a sheetpiling cutoff at the upstream end is located at a reservoir site, as indicated by the sketch. Draw a flow net for the subsurface flow and compute the seepage. Also calculate the uplift force acting on the base, and the escape gradient of the water at the downstream tip of the dam.

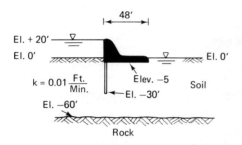

Note:
In this figure, the h values on the equipotential lines indicate the height above the ground surface to which the water level in a piezometer would rise.

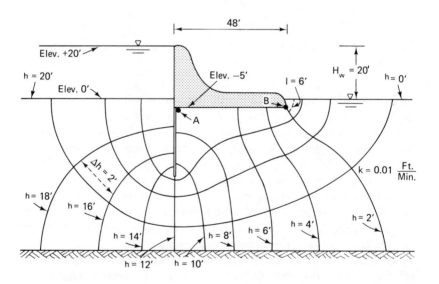

Number flow channels, $N_f = 4$.

Number pressure drops, $N_d = 10$

(note feet of head acting at each equipotential line).

$$\Delta h = \frac{h}{N_d} = \frac{20.0 \text{ ft}}{10} = 2.0 \frac{\text{ft}}{\text{drop}}$$

Seepage

$$q = kH_w \; \frac{N_f}{N_d} \; \text{(width)}$$

$$= \left(0.01 \; \frac{\text{ft}}{\text{min}}\right)(20 \text{ ft})\left(\frac{4}{10}\right)(1.0 \text{ ft wide})$$

$$= 0.8 \; \frac{\text{ft}^3}{\text{min}} \; \text{per foot of width}$$

Uplift Force on Base

$$p_A = (5 \text{ ft} + 7 \text{ ft})\gamma_w$$

$$p_B = (5 \text{ ft} + 2 \text{ ft})\gamma_w$$

$$\text{Uplift} = \left(\frac{p_A + p_B}{2}\right)(L) = \frac{(12 \text{ ft} + 7 \text{ ft})(62.4 \text{ lb/ft}^3)(48 \text{ ft})}{2}$$

$$= 29{,}200 \text{ lb per foot of width}$$

Escape Gradient
at downstream tip

Δh between last two equipotential lines $= 2.0$ ft

$l = 6$ ft

$$\imath = \frac{\Delta h}{l} = \frac{2}{6} = 0.33$$

PRACTICAL CONSIDERATIONS / The correct drawing of a flow net can be a tedious operation, particularly in the case of non-uniform soil conditions. Fortunately, even crudely drawn flow nets can provide reliable information on quantities of seepage and uplift pressures. Proper evaluation of an escape gradient requires a correct and carefully drawn flow net, however. For flow nets, as well as for all seepage studies, it should be expected that there will be a variation in the coefficient of permeability for the soil in the seepage zone under study. For practical problems, answers obtained from a seepage analysis should be considered as approximate, and are not to be believed as being precise.

QUICKSAND / The dreaded quicksand condition occurs at locations where a sand or cohesionless silt deposit is subjected to the seepage force caused by an upward flow of groundwater. This condition can occur in depressions of areas where the water table is high (close to the ground surface) or where artesian conditions exist. The upward gradient of the water is sufficient to hold the soil particles in suspension, in effect creating a material with the properties of a heavy liquid. A no-support condition exists even though the soil gives the appearance of being firm ground.

The suction or pull attributed to quicksand is in reality gravity exerting its normal effect but in a heavy liquid environment whose viscous properties also exert "drag" on a thrashing body. A person falling into a quicksand area can float as in water and quietly swim to solid ground.

Since the condition is caused by the forces of seepage and not some mystical property of the soil, elimination of the seepage pressure will return the soil to a normal condition capable of providing support.

Drainage

Among the more commonly occurring problems in the construction industry is the necessity for handling subsurface water encountered during the construction sequence and the handling of subsurface water after construction so that the completed facility is not damaged or its usefulness impaired.

During construction, lowering of the groundwater table and removal of water from working areas is desirable from the standpoint of better working conditions for men and equipment. In some cases dewatering may be a necessity to ensure that proper construction can be obtained. Conditions at some locations require that the structure be protected from erosive effects of flowing groundwater to prevent the loss of foundation stability. Additionally, usable interior portions of structures located below groundwater elevation normally are required to be free from serious seepage or leakage.

CONDITIONS REQUIRING DRAINAGE / In planning excavations for construction projects that extend below the water table in soils where the permeability is greater than about 1×10^{-5} ft/min (or 5×10^{-6} cm/sec), it is generally anticipated that at least some continuous drainage or dewatering procedures will be required if the work area is to be kept "dry" for construction. Excavations in the more impermeable soil types may remain "dry" after an initial dewatering, especially if the excavation is to remain open for only a limited time. (For all soil conditions, the actual amount of seepage expected would be affected by the depth below the water table and by the period of time that the excavation is to be open, as well as by the soil properties.)

DEWATERING SHALLOW EXCAVATIONS / For construction of shallow foundations and for other excavations of limited depth made in coarse soil, open drainage or interceptor ditches can be an expedient and relatively inexpensive method for lowering the groundwater table a slight distance. The interceptor ditch has to penetrate deeper than the elevation of the work area because of the pattern that the underground water surface takes (termed *drawdown curve*) in the area surrounding the interceptor ditch. Water collecting in such ditches normally has to be pumped out of the ditch for disposal. Since gravity flow is relied upon to bring the water to the ditch, the continued inflow is dependent on the water level in the ditch's being kept low. With this method, it is common to construct small pits in the ditch, termed *sump pits*, for locating the necessary pumps (sump pumps).

Where shallow foundations are to be installed at just about the water table elevation, it has frequently been found possible to obtain sufficient lowering of the water level to permit working in the dry by locating sump pits within, or immediately adjacent to, the foundation excavation.

The drawing down of the water table can also be accomplished by constructing a series of sump pits (Fig. 6-14), or, if greater depth is required, some type of drainage wells around the construction area and pumping the water from these pits or wells.

Subsurface water that flows in an upward direction into an excavation area that is being dewatered imparts a seepage force that tends to loosen the soil, reducing the soil strength. The change in strength should be considered in designing excavation bracing and foundations.

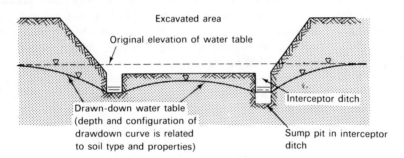

Figure 6-14. Shallow drawdown of water table by use of interceptor ditches and sump pit in excavation.

DEWATERING INTERMEDIATE DEPTHS / Where excavations in coarse-grained and silty soils are to extend more than a few feet below groundwater level, open ditches or pits are not practical, and more advanced methods are used. Discussion of some methods follows.

Well Points / For dewatering to intermediate depths (to about 30 feet but more if sufficient area is available for installing the necessary equipment) well-point systems are normally used. Basically, a well

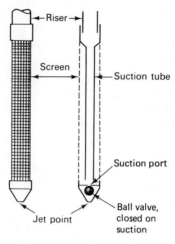

Figure 6-15.   A well-point assembly.

point is a closed-end pipe or tube, having perforations along its lower end, that is installed to the desired depth below the water table (see Fig. 6-15). Groundwater entering the well point through the perforations is pumped to the surface through a riser pipe connected to the well point (Fig. 6-16). For construction dewatering, the well-point perforations are provided with a protective screen or filter to prevent soil particles from clogging the perforations. Well points are conventionally installed in drilled holes or, most usually, by jetting. In jetting, water is pumped through (into) the riser and discharges through special openings at the tip of the well point, an action that displaces the soil below the tip of the point. This procedure is continued until the desired penetration is achieved.

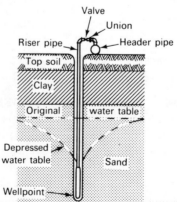

Figure 6-16.   Installed well point.

To dewater an area, a series of well points is installed around the perimeter of the area. The groundwater level within the perimeter will be lowered when the well-point system is put in operation. The spacing of the well points varies according to the soil type and depth of dewatering. Spacings conventionally vary between three and 10 feet (1 to 3 m). Each well-point riser is connected by hose or pipe to a horizontal header or collector pipe, which is connected to a pump. Usually, many risers are connected to a common header. Well-point systems are usually rented from specialty contractors who also install the system. With most contractors, the riser pipes are about two inches (5 cm) in diameter, and headers are 6 to 12 inches (15 to 30 cm), depending on the flow to be handled.

Normally, the screen provided with the well point is sufficient to keep medium sands and coarser materials from clogging the perforations. Where finer soils are to be dewatered, it is necessary also to provide a sand filter around the well point. The sand filter has the added benefit of acting to increase the size of the well point, permitting easier and greater flow into the point. To install a filter, it is necessary to create a hole having a larger diameter than the well point, to install the well point and riser, and then to place the sand filter and backfill around the riser. Large-diameter holes can be formed by providing attachments to the well point so that it can still be jetted to the desired depth, or by installing a casing by jetting or other method, after which the well point and surrounding sand filter are installed and the casing is pulled. For normal requirements, the sand filter material consists of a uniform graded medium to coarse sand.

With the type of pumping equipment conventionally used for well points, the depth of dewatering that can be achieved by a single line of well points located around the perimeter of an excavation is about 18 to 20 feet (6 to 7 m). This is due to the limit on the practical lifting, or suction, capacity of the pumping equipment. Lowering the

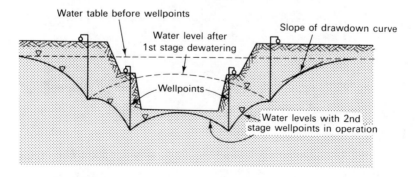

Figure 6-17. *Multistage well-point operation.*

water table through a greater distance may require the use of a two-(or more) -stage (multistage) installation. Where a two-stage installation is required, the well points for the first stage of drawdown are located near the extreme perimeter limits of the area that can be excavated, and are put into operation. Well points for the second stage are subsequently located *within* the area that has been excavated, near to the bottom elevation that has been dewatered by the first stage. The second stage well points then lower the water table to the additional depth necessary to complete the excavation dry (see Figs. 6-17 and 6-18).

TABLE 6-1 / APPROXIMATE SLOPE OF DRAWDOWN CURVE DUE TO A SINGLE ROW OF WELL POINTS

Soil Being Dewatered	Slope of Drawdown Curve, %*
Course sand	1 to 3
Medium sand	2 to 5
Fine sand	5 to 20
Silt-clay	20 to 35

*Refer to Fig. 6-17.

Vacuum Dewatering / Where well points are in use, flow to the well point is by gravity. In the coarser soils (soils with less than 25 percent of particles smaller than 0.05 mm) gravity flow is normally adequate to achieve dewatering. For silts, gravity flow is restricted because of capillary forces that tend to hold the pore water. However, it has

Figure 6-18. *Installed two-stage well-point system in operation.* (Courtesy of Moretrench American Corporation)

been found that by applying a vacuum to the piping system, satisfactory dewatering of silty soils can be achieved. For maximum efficiency, the vacuum dewatering system requires that the well point and riser be surrounded to within a few feet of the ground surface with filter sand and that the top few feet be sealed or capped with an impervious soil or other material. By having the pumps maintain a vacuum pressure, the hydraulic gradient for flow to the well points is increased. With this system, closer well-point spacings are required than for the conventional system.

Where suction pumps are used to draw the collected groundwater from the well points, as is common, the practical maximum height of lift is about 15 to 18 (5 to 6 m). Where an excavation is to extend more than this distance below the groundwater table, it will be necessary to dewater the area in two or more stages. With multistage dewatering, the well points for the first stage are installed on a perimeter line outside of the actual excavation area required for the construction and put in operation. Excavation proceeds within the perimeter formed by the well points to the depth where groundwater is encountered. Well points for the second stage are next installed within this excavation to lower further the water table for the actual construction area.

Though the principle on which well-point dewatering is based is a simple one, care and skill are required during installation and operation to obtain the intended results. Proper assembly of the piping system is particularly critical, because if air leaks develop at connections, the efficiency of the pump suction is reduced.

Electro-osmosis / Sand and the coarser silt soils usually can be dewatered by using gravity draining or vacuum-assisted well points. In fine silts, clay, or coarse-grained-fine-grained mixtures, the effective permeability may be too low to obtain a successful dewatering by using well-point methods. Drainage of such low-permeability soils may be achieved by electro-osmosis, provided that no more than 25 percent of the soil particles are smaller than 0.002 mm.

The basic principle of electro-osmosis is as follows: if a flow of direct current electricity is induced through a saturated soil between a positive and negative electrode (anode and cathode, respectively), pore water will migrate towards the negative electrode. If a well or well point is made the cathode, collected water drained from the soil can be removed by pumping as with a conventional well-point system.

Movement of water towards an electrically negative terminal occurs because of the attraction of the cathode for positive ions (cations) that are present in groundwater. Cations in pore water are the result of dissolved minerals going into solution with the groundwater. The cations concentrate around the negatively charged surface of clay particles to satisfy the electrical charge on the particle.   These cations, in turn, attract the negative "end" of "dipole" water molecules. As the cations are drawn to the cathode, water molecules held to the cation follow.

In a field installation it is considered most beneficial to locate the anode close to the excavation perimeter and place the cathode further back, so that the direction of drainage is away from the excavation. This arrangement eliminates the danger that seepage forces caused by the draining water could act to cause the exposed slopes of the excavation to slide or slough-in towards the excavation (construction area).

Because of techniques involved, the required special equipment, and the high electric consumption, drainage by electro-osmosis is expensive compared to gravity drainage methods. Consequently, the method is used only when other methods cannot be applied.

DEEP DRAINAGE / If excavations are to extend deep below the groundwater table, or to penetrate through a deep permeable stratum, well points may not be applicable because of the limits to which well points can raise water. For this situation, deep wells and deep well pumps (jet or venturi pumps) can be utilized. Deep wells are frequently relatively large-diameter (on the order of two feet or 60 cm) drilled holes. A perforated protective casing is installed, and the deep well pump is then placed inside the casing near the bottom. Coarse filter material is placed between the outside of the casing and the walls of the drilled hole.

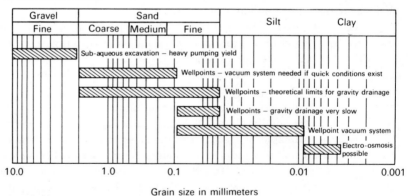

Grain size in millimeters

Figure 6-19.   *Summary-Relation between method for dewatering soil and range of particle sizes. (Ref. 7).*

CONSOLIDATION DRAINAGE / *Consolidation* refers to the process that occurs in clay deposits when the development of compressive stresses in the soil mass from external loading causes the soil to undergo a reduction in volume (due to a reduction in void spaces) that is simultaneously accompanied by water in the soil pores' being expelled to permit the decrease in voids. Thus, subsurface drainage of fine-grained soils by consolidation can be achieved by imposing a surface load onto an area. In general, a reduction in water content results in an increase in the shear strength of the clay. Unfortunately, the rate of drainage is quite slow in clay soils because of the low coefficient of permeability.

Sand drains represent one method that has been used with success to accelerate the rate at which consolidation drainage occurs in fine-grained materials. The method is based upon reducing the distance that pore water has to travel to escape from the consolidating soil. Sand drains are vertical columns of freely draining sand installed so as to penetrate the soil strata to be consolidated. In clays, the permeability in the horizontal direction is frequently many times the vertical permeability because of the stratified manner in which clay deposits have been formed. The presence of relatively closely spaced sand drains thus reduces the distance that pore water must travel to escape and also permits an easier (horizontal) drainage path. In draining from the clay, water flowing to the sand drains is under pressure as a result of the external loading that is applied to cause the consolidation (usually an earth fill to act as a surcharge). Therefore, the water can flow upward as well as downward in the sand drain. Further details on the mechanics of the drainage that occurs in sand drains is presented in Chapter 9.

DRAINAGE AFTER CONSTRUCTION / Preventing groundwater from seeping into a structure or through a structure after it has been built may be necessary in order to obtain proper use of the structure or to protect it from damage.

In some cases, the dewatering methods utilized during construction can be kept in use to protect the structure. Normally, however, this procedure is not followed; the equipment or methods may interfere with the usefulness of the finished project, may be too expensive for long-term use, or may be in some other way impractical.

When it is known that a usable part of a structure will be located below the groundwater table, it is desirable to build the facility utilizing waterproof design and construction techniques. One of the desirable features for a subaqueous structure is to have all seams and joints provided with water stops, and/or to have the structure built with no seams or joints wherever possible. For example, in small conventional buildings, concrete basement walls and floors can be poured monolithically.

FOUNDATION DRAINS / Where groundwater will be flowing in the vicinity of the structure, provision can be made so that the water will be quickly carried away from the building area, and at worse only a limited height of groundwater buildup against the exterior will occur. With adequate provisions, the large hydrostatic pressures that tend to force seepage entry thus will not develop. A method for achieving this control when the depth below the water table is not too great is with foundation drains, conventionally placed around the building exterior at footing level and adjacent to the footing (see Fig. 6-20). Such drains should not be lower than the bottom of the footing. Where there is concern over the ability of the exterior drain to handle the expected groundwater, the foundation drains may be located on the interior side of the foundation as well. With a high water table, it may also be desirable to place interceptor drains at some short distance from the building and at an elevation higher than the footings so that the water table is lowered in stages.

The installed drain normally consists of pipe provided with perforations or installed with open joints so that groundwater can enter into the pipe. Gravel and/or sand filter material must surround the drain pipe so that soil particles tending to be carried along by the inflowing water are prevented from clogging the drain pipe openings or causing erosion. Criteria for filter material are presented in subsequent paragraphs.

An outlet for collected water is a necessity; preferably disposal will be by gravity flow to a storm drain system or other drainage facility such as a ditch, dry well, or pool located at an area of lower elevation than the building. If disposal by gravity flow is not possible, drainage water will have to be directed to a sump pit or other collector and pumped to a disposal.

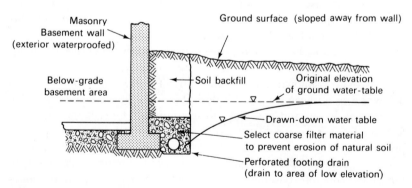

Figure 6-20. *Footing drain tile installation for disposing of groundwater against a basement wall.*

BLANKET DRAINS / Where a basement is to be located below groundwater level, a blanket or layer of filter material can be placed beneath the floor slab to provide a highly permeable drainage path for removal of groundwater acting against the bottom of the slab (see Fig. 6-21). Providing an escape path serves to reduce uplift pressures and the possibility for seepage to occur through the floor. The blanket connects to a sump pit where collected water is pumped out, or to drainage pipe where disposal occurs by gravity flow. The filter blanket normally consists of a sand layer placed over the natural subgrade soils, and a coarser, small gravel or crushed rock layer through which most of the horizontal flow is intended to take place.

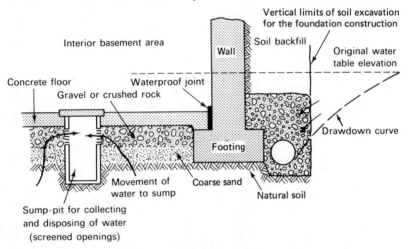

Figure 6-21.   *Details of a blanket drain beneath a concrete floor.*

INTERCEPTOR DRAINS / In paved highways and airfield runways, the design frequently calls for interceptor trench drains located parallel to the shoulder, as shown in Fig. 6-22. The purpose of the drains is to lower the groundwater table to a level beneath the pavement and to permit easy lateral drainage (escape) for water finding its way into the coarse base material (water resulting from upward migration from capillarity and from infiltration through cracks, joints or voids in the pavement) provided beneath the pavement.

The intent of the drainage facility is to keep the base and subgrade soils dry so that they will maintain high strength and stability. With coarser soils, the intent is probably achieved quite successfully. In fine-grained soils, the degree of achievement is questionable. Most probably, a groundwater table in such soils is not significantly lowered because of the material's low permeability. However, the drainage ditches do serve to prevent excess pore water pressures from building

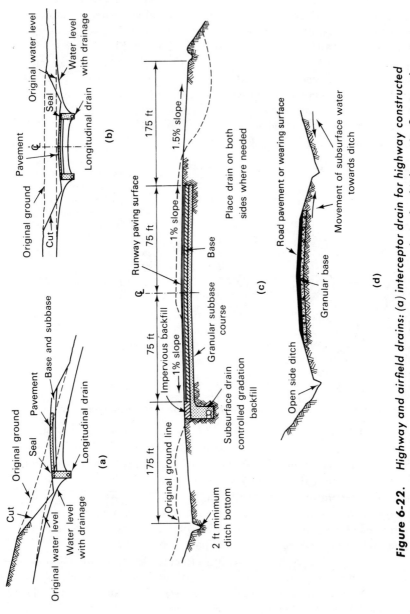

Figure 6-22. *Highway and airfield drains: (a) interceptor drain for highway constructed in a sidehill (after Cedergren, Ref. 24); (b) interceptor drain for highway in flat terrain (Ref. 24); (c) typical airport runway interceptor drains (Civil Aeronautics Administration); (d) typical open-shoulder ditch for roads.*

up in the subgrade soils (excess pore water pressures have the effect of weakening the soil). The drains also provide a means for disposal of surface and near surface water, and will intercept underground flow trying to enter the roadway from the side areas.

Similarly, paved and unpaved roads will benefit somewhat from even open drainage ditches located adjacent to shoulder areas; see Fig. 6-22(d). Such ditches handle surface water, provide a lateral drainage outlet for soil near the level of the road surface that helps prevent the development of excess pore water pressures, intercept surface and near surface water flowing toward the roadway area from the sides, and provide storage for plowed snow and a flow channel for melting snow so that it does not run onto the roadway.

FLOW THROUGH A STRUCTURE / In some types of structures, such as earth dams and dikes, it is known that water will be flowing through the structure after it is put in use. The design of the structure, therefore, considers the effect that such flow will have on the stability of the structure. Flow is not prevented, but rather is permitted and directed so as not to create a stability problem. For instance, in a homogeneous earth dam or dike, flow is typically as indicated in Fig. 6-23. If the velocity of the water flowing through the structure is too great, erosion and instability of the downstream side will result. The placement of a toe drain or of an underdrain blanket that consists of coarse materials helps direct the flow to those areas that will not erode or lose stability because of the flowing water (Fig. 6-24).

There are basically similar situations encountered in building construction, where localized underground flow occurs through seams or a stratum of pervious soil (frequently referred to as underground springs) and presents a potential seepage problem for the structure. Providing some type of drainage conduits for the flow to pass around or through the structure is frequently technically more desirable and less expensive than attempting to cut off or stop the flow. The main precaution to be observed is that the flow conduit be designed so that seepage forces do not affect the building, and soil erosion does not occur.

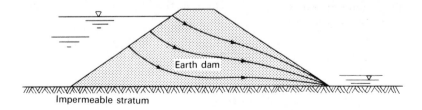

Figure 6-23.   *Pattern of flow through a homogeneous earth dam.*

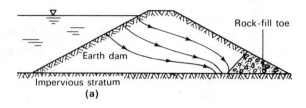

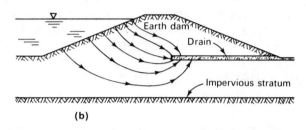

Figure 6-24.   *Flow through an earth dam modified by toe drain or under-drain: (a) earth dam with rock fill toe; (b) earth dam with downstream drain.*

FILTER DESIGN / Water flowing towards a drainage structure must be able to enter the drain easily and quickly but must not be permitted to carry with it small soil particles that could eventually clog up the drain, or cause erosion or cavitation in the areas surrounding the drain. As a result, where a drainage structure is open to the soil, it is necessary to provide a *drainage filter*.

To permit quick easy flow into the drain facility, and to "draw" water to the drain, it is necessary to have the permeability of the filter material considerably greater than the permeability of the surrounding natural soil. To satisfy this criterion, a ratio between particle sizes for the filter material and natural soil has been established, viz.,

$$\frac{D_{15} \text{ of filter material}}{D_{15} \text{ of natural soil}} \text{ should be 4 or greater.}$$

To prevent particles of the natural soils from migrating into the filter material and eventually clogging it, the following ratios of particle sizes should be used as criteria:

$$\frac{D_{15} \text{ of filter material}}{D_{85} \text{ of natural soil}} \text{ is less than 5.}$$

$$\frac{D_{50} \text{ of filter material}}{D_{50} \text{ of natural soil}} \text{ is less than 25.}$$

$$\frac{D_{15} \text{ of filter material}}{D_{15} \text{ of natural soil}} \text{ is less than 20.}$$

In these criteria the D_{15}, D_{50}, and D_{85} values refer to the particle dimensions obtained from a particle-size distribution curve (e.g., the D_{15} refers to the particle size for which 15 percent of the soil, by weight, is smaller). Generally speaking, these criteria require that natural and filter materials have parallel but transposed particle-size distribution curves.

To ensure that filter particles will not enter and clog the drain structure, where holes or slots are provided for the entry of water, the following criteria should be followed:

For circular holes, the ratio

$$\frac{D_{85} \text{ of filter material}}{\text{Diameter of opening}} \text{ is greater than 1.2.}$$

For slotted openings, the ratio

$$\frac{D_{85} \text{ of filter material}}{\text{Width of slot}} \text{ is greater than 1.4.}$$

For proper performance, the filter material must surround the drain, under as well as over. The filter material should also be compacted so that it does not settle. To help minimize any tendency for fine particles to migrate through the filter, and to drain as much water as possible, the holes or slots in the drain structure should be placed facing downward, or as near to downward as possible, so that the flow into the drain facility is *not* downward but instead is laterally or slightly upward.

If the natural soil is fine-grained, it may be necessary to provide a filter that consists of two or more layers, each having a different gradation. The material closest to the drain would be coarsest, with the outer material filter that is next to the natural soil being less coarse. For the situation in which two or more filter materials are used, the criteria presented earlier for filter and natural soil would also apply between any two adjacent filter materials.

LAND DRAINAGE / One of the earliest methods for reclaiming flat expanses of marginal wetland areas for use consisted of constructing a network of deep drainage ditches to lower the water table and obtain a stable surface suitable for agriculture and for supporting light structures. With a network of drainage ditches, drainage is by gravity, and best results are obtained in soils with relatively high coefficients of permeability. In clays or fine silts, where permeability is low and capillary forces holding the pore water are high, drainage from gravity effects will be limited. Water can be replaced by normal precipitation as quickly as drainage occurs. However, where the subsurface soils include strata or seams of coarse soils, as frequently exist in coastal and lake areas, predominently fine-grained soil deposits have been drained. The drainage ditch network, generally laid out in some form of grid pattern, requires outletting to areas of elevation lower than the draining area.

EFFECTS OF DRAINAGE / Water flowing towards a drain also causes seepage forces that act on the soil particles. These forces are capable of moving soil particles, particularly small particles. Unless the migration of particles by a properly functioning filter is prevented, the small particles will be washed into the drain, eventually eroding soil from the area surrounding the drain. This occurrence could cause the drainage structure to collapse because of inadequate support from the surrounding soil. This process is referred to as *internal erosion.*

If an open drainage system (i.e., open ditches) causes seepage forces to act towards an open excavation, there is the possibility that the forces will precipitate a sloughing-in of the soil embankment.

Normally, removing water from a soil (particularly a fine-grained soil) will increase the shear strength of the soil, in effect making the soil mass stronger. However, lowering of a water table also results in an increased effective vertical pressure acting on the soil mass, because the soil weight changes from a submerged weight to a saturated but un-buoyed weight. The usual result of an additional loading to the area within the zone of drawdown is settlement. The magnitude of the settlement depends on the amount of water table lowering and the strength and compressibility of the soil. If an undeveloped (no buildings, etc.) area is dewatered and the surface settles, usually no difficulties develop. However, if developed areas are in the zone of drawdown, settlement of buildings and utilities in the area could occur.

If the settlement of structures adjacent to a construction site being dewatered is anticipated, effects can be minimized by the process of *recharging.* Recharging consists of pumping water back into the ground, usually with well points, in the area between where the endangered structures are located and where the dewatering takes place. Careful monitering of groundwater levels is required for proper overall results.

Frost Heave in Soils

When exposed to a freezing climate, soil temperatures will depress to below freezing with sufficient passage of time. Temperatures will vary in the soil mass, being lowest where contact with the freezing source is made (usually this would be the ground surface in contact with freezing air temperatures, but the source of a freezing temperature could also be from the floor of a frozen storage warehouse or ice rink) and gradually increasing with distance from the freezing source to reach the stable, above-freezing temperature that exists underground.

When freezing temperatures develop in a soil mass, most of the pore water in the soil also is subject to freezing. As water crystallizes, its volume expands approximately nine percent. In considering normal void ratios and the degree of saturation for soils in general, an expansion of the soil mass as a result of freezing might be expected to be on the order or three or four percent of the original volume. This may mean a vertical expansion on the order of one or two inches (2.5 to 5 cm) for an area with a climate that exists in the northerly half of the United States. However, experience indicates that in some soil types the volume of expansion resulting during freezing periods, termed *frost heave*, is frequently considerably greater than could be expected from expansion of normal pore water. Frost heaves exceeding one foot (30 cm) are not uncommon.

Investigation of heaved soils indicate that pore water indeed has become frozen. However, much of the frozen water has been segregated in discontinuous layers, or lenses, throughout the frozen soil, and the volume of the water in the ice lenses is considerably greater than the volume of the original pore water. Further, the magnitude of ground surface heave has been found to equal the combined thickness of ice lenses existing in the soil at the heave location (see Fig. 6-25).

Studies conducted at field sites and in the laboratory to investigate the frost heave phenomena have been able to define conditions and occurrences responsible for the ice lens formation, which is the basic cause of heaving.

As freezing temperatures develop, water in the *larger* voids of the frozen soil zone also becomes frozen. Such frozen water is potentially the start of an ice lens. Water remaining unfrozen in capillary size (very small) voids is subsequently attracted to such beginning ice lens and enlarges the volume of the lens. The process of ice lens growth continues as long as freezing temperatures prevail and free water can be attracted to the ice lens.

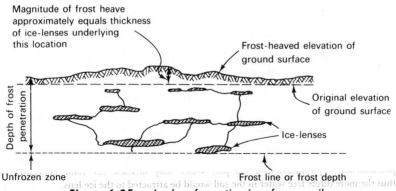

Figure 6-25.   *Ice-lens formation in a frozen soil.*

The development of ice lens formations significant enough to cause large heaves to occur requires that a source of water be sufficiently close to the freezing zone of soil that it can be drawn into this zone. In the normal frost heave occurrence the source of water is the ground-water table. Upward movement from a water table to the freezing zone occurs through capillary rise. Water in fine capillary openings can continue to migrate through even a zone of subfreezing temperatures because of the depressed freezing point for water in small capillaries and water adsorbed to the surface of soil particles. In a "frozen" silty soil, it is known that there will exist some frozen water and some still liquid water with temperatures as low as, and possibly lower, than 25°F (−4°C).

All factors causing moisture migration through a frozen soil are not fully understood, but current theories attribute movement to a thermal gradient where capillary flow is in the direction of higher toward lower temperatures, and to osmotic flow.[4]

Height of capillary rise is quite limited in clean, coarse-grained soils, and unless a source of water is close to the zone of freezing soil, frost heave problems are normally not expected in such soils.

Theoretical height of capillary rise is greatest in the fine-grained silt and clay soils. Practically speaking, however, and under typical seasonal conditions of limited cold periods, ice-lens growth is affected by

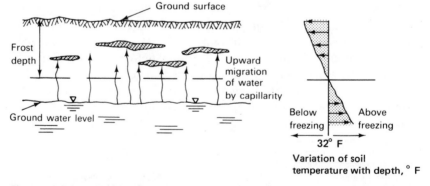

Figure 6-26. *Ice-lens formation by capillary movement of ground-water into zone of freezing.*

[4]Osmotic flow relates to the movement between liquid solutions having different ion concentrations. Simply stated, there is the tendency for the more dilute solution to flow into and mix with the highly concentrated solution. Water in soil normally has some concentration of cations because of dissolved minerals. Applied to the ice-lens phenomenon, crystallization of soil pore water may increase the ionic concentration; thus the more dilute free water in the soil would be attracted to the ice lens.

the *rate of capillary rise* in soil. In clay, this rate is so slow that under normal conditions sufficient water can not be obtained to cause ice-lens growth. Silts, however, have a relatively high rate of capillary rise, and water can migrate easily to the zone where ice lenses are forming. Silts, therefore, because of their potential for high capillary rise (distance) coincident with the potential for rapid rise are conducive to ice-lens formation and frost heave. As a result, silts are termed *frost-susceptible* soils. Silt mixtures, such as silty sands and silty clays, are similarly regarded as frost susceptible. Because of the practical limitations on ice-lenses growth within them, coarse-grained and clay soils are normally considered as *non-frost-susceptible* soils.

In summary, several simultaneously occurring conditions are required for the notorious frost-heave problem to occur:

(a)   Presence of a frost-susceptible soil.
(b)   Presence of freezing temperatures.
(c)   Presence of a supply or source of water to help form and feed the ice lenses.

The rate at which freezing temperatures penetrate the frost-susceptible soil has an effect on the total ice-lens formation and, consequently, on the frost heave of the ground surface. The maximum effects result when there is a gradual decrease in temperature as opposed to a rapid or flash decrease. The longer and the colder the freezing period, the greater will be the depth of frost penetration, and the greater the frost heave will be.

Damages from frost heave result to structures supported on the soils being heaved. Commonly, this includes roadways and pavements, and building foundations and exposed ground level building slabs. The heaving force can be considerable; it has the ability to lift pavements, slabs, and building foundations. Because the magnitude of heave is rarely uniform over even short horizontal distances, cracking of pavements and slabs frequently occurs. In buildings, walls or floors may crack or distort so that doors, windows, and equipment supported on the walls may not operate properly. If heave is great enough, utility pipelines passing through or over the walls may be ruptured because of the movement.

Damage is not restricted to that resulting from the heaving. Much of the ice-lens volume represents new water introduced to the soil, and when the soil thaws from the surface downward the free water can not drain through the still frozen underlying soil. The effect is that the melting ice lenses significantly increase the water content of the soil, and loss of soil strength occurs. Loss of soil bearing and settlement follow. For building foundations not located below frost depth, the cycle of heaving in winter and settling in spring is potentially an annual occurrence that is capable of causing progressive damage.

Similar effects can occur in pavements and unprotected ground-level building slabs. In roadway surfaces, the effects are compounded by the repeated loadings applied to an area by vehicle movement. Under the action of traffic, the roadway surface will be broken up because of loss of underlying soil support. This is generally referred to as the spring breakup. Further, the soft soil will be displaced from the road base, causing holes and depressions, which are referred to as potholes or chuck holes.

To prevent the effects of frost-heave damage to buildings, it is common practice to construct foundations to a depth at least equal to depth of frost penetration for the area. Such procedures might be thought to be necessary only in a frost-susceptible situation. If non-frost-susceptible soils exist at a building site or if there is no source of water, locating foundations below frost depth could be unnecessary unless required for other reasons. However, in natural soil deposits there should also be some suspicion that even "non-frost-susceptible" soils will have some properties that could make them susceptible. For example, many naturally occurring coarse-grained soils include some fines, or include seams or strata of finer, more susceptible material. Clay soils frequently have considerable silt content, or the clay deposit may include fissures and hair line cracks that can act as capillary tubes for the migration of water.

It should also be recognized that during the life of a structure, the depth of the groundwater table is subject to change.

In considering poor performance and damages that can result if frost heave should affect foundations, most designers and constructors are of the opinion that it is worth the relatively small extra expense to take precaution at the time of construction to install foundations for exterior elements below frost depth. Figure 6-27 shows maximum frost depth data prepared from information obtained across the United States.

For grade-level structures such as building slabs and highway pavements (no deep foundations), other approaches to the control of the frost-heave problem are taken. Control usually requires achieving at least one of the following:

1. Remove existing frost-susceptible soils to about frost depth.
2. Remove or cut off the water source that could feed ice-lens growth.
3. Protect the susceptible soil from freezing temperatures.

For road or airfield pavements, the effort for control or protection has frequently included removing the frost-susceptible soils to frost

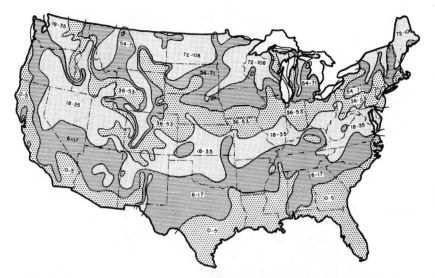

Figure 6.27. Maximum depth of frost penetration in inches (from Frost Action in Roads and Airfields, HRB Publication 211, Ref. 47).

depth and replacing them with a non-trost-susceptible material, or providing drainage facilities to keep the water table low in the vicinity of the pavement. For building slabs, replacing the frost-susceptible soil is frequently a practical solution.

Non-frost-susceptible soils for replacement purposes include coarse-grained soil (sand and sand-gravel mixtures) and clays. Coarse-grained soils are preferred because they are easier to place and compact, are less difficult to work with in inclement weather, and have other desirable properties when in place, such as easy draining. To be non-frost-susceptible, studies and experience indicate that a *well-graded* granular soil should contain not more than 3 percent of particle sizes smaller than 0.02 mm,[5] while *uniform* granular soils can have up to about 10 percent of particle sizes smaller than 0.02 mm. Some ice-lens formation may still result with these soils, but such development will generally be of limited extent and tolerable.

Other methods for control of frost heave have included installation of a barrier at about frost depth to prevent upward capillary movement that could feed ice lenses. The barrier can be a very coarse soil of a thickness through which capillary rise cannot occur, an impermeable material such as a densely compacted clay layer, or a plastic or other sheeting material through which water cannot penetrate. These

---

[5]To retain a perspective on sizes, it is noted that the #200 sieve has openings of 0.074 mm; a #400 sieve has openings of 0.037 mm.

methods appear most suitable for protecting limited areas, such as building slabs, or for highway fill sections. The methods are not considered practical for highway construction in areas of cut section or at-grade construction because of the expense of excavating and replacing subgrade soils that would be associated with installation of a barrier. Further, frost-susceptible subgrade soils may be difficult to replace and compact, especially in wet weather.

Successful protection has also been achieved by wrapping pavement subgrade soils within an impermeable membrane after they have been properly compacted, so as to keep the soil at a desired optimum moisture content and to prohibit ice-lens formation. A disadvantage of this method of control relates to costs of installing the membrane and placing the soil.

If the soil beneath a slab or pavement can be protected so that it does not freeze, there will be no frost-heave problem. In the construction of frozen-storage warehouses, ice rinks, and similar facilities, insulation is placed between the soil subgrade and the floor of the facility, to protect the soil from the cold temperatures prevailing in the structure. Similar methods are adaptable for protecting exposed ground-level slabs for buildings. Various methods that have been used are illustrated in Fig. 6-28. For highways, some work has been done with providing insulating material beneath the pavement.

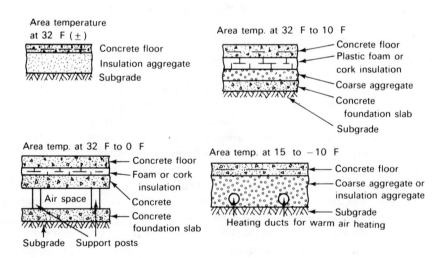

Figure 6-28.   *Some methods that have been used to insulate floor systems for refrigerated areas.*

Soil Percolation Rate

Soil percolation rate refers to the ease or difficulty with which soil accepts fluid, generally water, into it. It is a condition frequently of interest for planning private underground sewage disposal systems, such as the leach fields and leach pits for septic tanks. With a septic tank system, all sewage passes into the septic tank. The solids are retained, but the liquid is passed on for absorption into the earth.

The percolation rate can be related to the coefficient of permeability of a soil; the coarser, permeable soils are capable of accepting larger volumes of water and at a faster rate than fine-grained soils having low permeability. However, "percolation" and "permeability" do *not* refer to the same property of a soil. For example, a coarse soil having a high coefficient of permeability may have a low rate of percolation if the groundwater table is high and the soil is not capable of accepting additional water.

Procedures to check the percolation rate for a soil generally stipulate testing of the area and depth where the septic tank leach field or leach pit is to be located. For the common requirement that leach field drain lines be installed about two feet below the ground surface, percolation tests are performed about 30 inches (75 cm) below the surface. A small test hole, usually between 6 inches and 12 inches (15 to 30 cm) in diameter, is dug at least six inches deep. The test hole area is saturated by filling the hole several times with water and letting it drain completely before beginning the actual test. For the test, the hole is filled with water and the time for the water level to drop a specified distance (e.g., from six inches deep to five inches deep, or from 15 cm to 12.5 cm) is recorded. A fast rate of drop indicates "good" percolation. A slow rate or no drop indicates poor percolation. Areas having poor percolation rates should not be used as leach fields. Details of the applicable methods for performing percolation tests should be obtained from local health department authorities, to ensure that procedures are in accord with their requirements.

PROBLEMS

1. A canal is located in an area where its route parallels that of a nearby river. The elevation of the water surface in the canal is +375 ft. In the river, the water surface is at elevation +345 ft. A stratum of coarse-grained soil ten ft thick is sandwiched between relatively impermeable fine-grained soils, and extends between the

canal and the river. The distance between the river and the canal is 500 ft, and the coefficient of permeability of the buried coarse stratum is 0.01 ft per hour.

(a) Make a sketch of the described conditions.

(b) Calculate the seepage loss from canal to river, in gallons per day per mile of river-canal length.

2. What are flow nets and why are they used?

3. A concrete gravity dam is 100 ft long from upstream to downstream end. It is constructed on a stratum of sand-silt soil 40 ft thick. The coefficient of permeability for this soil is 0.1 ft per day. Rock underlies the soil stratum. The bottom of the dam is six feet below the soil surface.

(a) For the condition where the height of water against the upstream face of the dam is 45 ft above the soil surface, and the water level at the downstream end is just at the soil surface, draw the flow net for seepage beneath the dam and calculate the quantity of seepage in gallons per day per foot of width of dam.

(b) For the condition where the height of water against the upstream face of the dam is 35 ft above the soil surface, and the water level at the downstream end is 5 ft above the soil surface, draw the flow net for seepage beneath the dam and compute the seepage in gallons per day per foot of width of dam.

4. The vertically upward flow of groundwater (an artesian condition) occurs through a sand deposit where the void ratio is 0.60 and the specific gravity of soil particles is 2.65. What hydraulic gradient is necessary for a quicksand condition to develop?

5. The hydraulic gradient for an occurrence where there is a vertically upward flow of water through a sand mass is 0.95. If the specific gravity of the soil particles is 2.75 and the void ratio is 0.65, determine if a quicksand or condition of erosion could develop.

6. Well points are to be used to dewater a large trench excavation that extends 35 ft below the surface of the water table. Assume that the bottom width of the trench excavation is 70 feet.

(a) Develop a simple sketch of the necessary well-point system. Use separate diagrams to show the progress of the necessary stages of installation.

(b) Assume that half the necessary lowering of the water table elevation is achieved with the outer row of well points. If the soil being dewatered is a fine to medium sand, approximately at what distance beyond the well-point location is the original groundwater table not affected by well-point pumping.

7. Indicate the intended functions for foundation drains, and the major requirements for design and installation so as to achieve a properly operating system.

8. What are the intended functions and limitations of interceptor drains as conventionally used for highway and airfield construction?

9. What are the primary reasons for having limitations on the sizes of particles used to construct drainage filters?

10. Briefly describe the conditions necessary and the resulting sequence of occurrences that typically take place in a frost-heaving situation. Specify the soil type considered most susceptible to large frost heaves.

11. What methods are commonly used to prevent the occurrence of, or to offset the effects of, frost heave?

CHAPTER 7
Combined Stresses in Soil Masses: Stress at a Point and Mohr's Circle

When a body or mass is subjected to external loading, various combinations of *internal* normal and shear stresses are developed at the different points within the body or mass. Generally, data concerning internal stress conditions are used to determine deformations (in soils work, deformation is most frequently associated with settlement) and to check for the possibility of a material's (soil's) failing because its strength is exceeded. In performing a stress study, it is convenient to use methods from engineering mechanics for analyzing stress at a point. With these methods, stresses acting on any plane passing through the point can be determined. The combinations of normal and shear stress that develop will vary, depending on the plane being analyzed and the magnitude of the external loading. The "stress at a point" analysis is also applicable for determining the weakest plane or potential plane of failure in a material, not always easily evident, and indicating the magnitude of the stresses that act on this plane. This type of analysis has particular application for soil and foundation studies, since stability failures in soil masses are the result of the shear strength of the soil being exceeded.

Stress at a Point—
Analytical Development

The stresses acting on any plane passed through a point consists of a normal stress (compression or tension) and a shearing stress. Depending on the type of external loading causing the stress condition, it is possible for the shear *or* the normal stress, or both, to be zero on some planes. In soil problems, most external loadings are compression.

141

(The downward weight of a structure supported by a soil mass would be a compressive loading.) For a situation where the loading is compressive, normal stresses that develop on any plane would almost always have a value other than zero; some shear stress would act on all planes with the exception of two planes, where it will be zero (discussed in the following paragraphs).

If the combination of normal and shearing stresses acting on any two mutually perpendicular planes (orthogonal planes) are known, the combination of stresses acting on any other plane through the same point can be determined.

In analyzing stress at a point, it is convenient to assume an incremental element that represents the stress conditions at the point and show the known stresses acting on it, as indicated in Fig. 7-1.

For equilibrium, the sum of *forces* (not stresses) acting in any direction must equate to zero, and the rotational moments about any axis caused by forces similarly must equal zero. To satisfy this latter requirement, shear stresses acting on orthogonal planes must be equal in magnitude.

At the same point, but on a differently oriented element (actually a different plane passing through the identical location) the combination of normal and shearing stresses that act will be different; see Fig. 7-2. However, as will be shown, there is a relation between the normal and shear stresses acting on all planes (or orientation of elements) that pass through the same point.

Consider all the planes that can be passed through a point. On one particular plane, not yet defined, the shear stress will be zero, whereas the normal stress will be the maximum possible value of all the normal stresses acting through that point. On a plane perpendicular to the plane just referred to, the shear stress will also be zero, but the

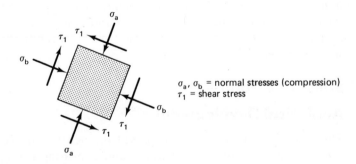

σ_a, σ_b = normal stresses (compression)
τ_1 = shear stress

Figure 7-1. *Incremental element with representative stresses assumed for analysis of "stress at a point."*

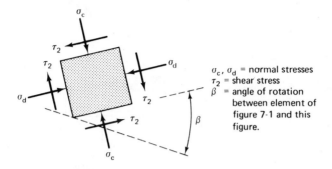

σ_c, σ_d = normal stresses
τ_2 = shear stress
β = angle of rotation
between element of
figure 7-1 and this
figure.

Figure 7-2.   *Method of rotating incremental element to determine stresses on different planes of study.*

normal stress that acts will be the least, or the minimum, of all the normal stresses acting on different planes through the point. These *maximum* and *minimum normal stresses* are called *principal stresses.* The planes on which they act are *principal planes.* The shear stress on a principal plane is *always* zero.

In many practical soil problems, the principal stresses act in the vertical and horizontal directions (or on horizontal and vertical planes), and are easily calculated. For instance, where the incremental element represents a point within a soil mass where the ground surface is horizontal, the vertical stress is due to the weight of the soil overburden at that point. This would be the effective unit weight of soil multiplied by the depth of the point below the ground surface. The horizontal stress is proportionate to the vertical stress. Thus, the magnitudes of the principal stresses are known, as are the orientations of planes on which the stresses act.

If the major and minor principal stresses are σ_1 and σ_3 respectively, the magnitude of a normal stress σ_n and shear stress τ_n on any other plane can be determined. Referring to the stressed element in Fig. 7-3(a) and letting the area on the cut plane be unity [Fig. 7-3(b)], it is apparent that the area of the plane on which σ_1 acts becomes (1) × (cos θ), and the area on which σ_3 acts becomes (1) × (sin θ). The total normal *force* on the cut plane is (σ_n) × (1). The total force on the vertical surface is (σ_3) × (sin θ), while the total force on the horizontal surface is (σ_1) × (cos θ).

To determine σ_n and τ_n in terms of σ_1 and σ_3, the forces acting on the horizontal and vertical planes are resolved into components parallel and perpendicular to the cut plane; see Fig. 7-4. Summing forces parallel to σ_n (normal to the cut plane) gives

$$\sigma_n = \sigma_1 \cos\theta\cos\theta + \sigma_3 \sin\theta\sin\theta$$
$$= \sigma_1 \cos^2\theta + \sigma_3 \sin^2\theta$$

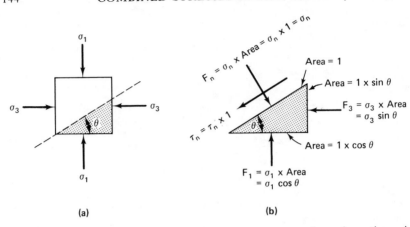

(a)   (b)

Figure 7-3. *Basic step in determining stresses on a random plane through a point in terms of known principal stresses.*

and by trigonometric identity

$$\sigma_n = \sigma_1 \cos^2 \theta + \sigma_3 \sin^2 \theta = \frac{\sigma_1 + \sigma_3}{2} + \frac{\sigma_1 - \sigma_3}{2} \cos 2\theta \qquad (7\text{-}1)$$

Summing forces parallel to the cut plane in the direction of τ_n gives

$$\tau_n = \sigma_1 \cos \theta \sin \theta - \sigma_3 \sin \theta \cos \theta = (\sigma_1 - \sigma_3) \sin \theta \cos \theta$$

and by trigonometric identity,

$$\tau_n = (\sigma_1 - \sigma_3) \sin \theta \cos \theta = \left(\frac{\sigma_1 - \sigma_3}{2}\right) \sin 2\theta. \qquad (7\text{-}2)$$

From examination of Eq. (7-2) it should be recognized that the

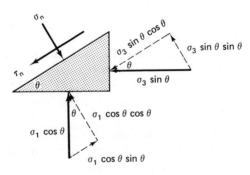

Figure 7-4. *Resolution of principal stresses into components parallel and perpendicular to a random plane. This step is in preparation for determining stresses on the plane in terms of principal stresses.*

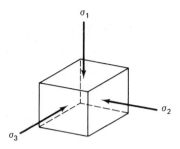

Figure 7-5. *Three-dimensional incremental element showing principal stresses in mutually perpendicular directions.*

maximum shear stress will occur on a plane that is 45 degrees from the major principal plane (θ = 45 deg), and will have a magnitude equal to $\frac{1}{2}(\sigma_1 - \sigma_3)$. On this plane, the normal stress is *always* $\frac{1}{2}(\sigma_1 + \sigma_3)$.

The previous developments apply to a two-dimensional stress analysis (stresses in a plane). In actuality, for soils problems the study of stress at a point involves a three-dimensional study. An incremental element with three principal stresses acting—σ_1, σ_2, and σ_3—is shown in Fig. 7-5. For many practical conditions of loading, the intermediate stress, σ_2, is equal to σ_1 or to σ_3. For some situations however, σ_2 may be a value intermediate between σ_1 and σ_3. As far as practical effects of the intermediate stress are concerned, it appears that there is some influence on the strength and stress-strain properties of the material, but its effect is not clearly understood. However, to keep a proper perspective, it is pointed out that the methods available for making determinations of the stresses within a soil mass that result from external loading are not highly refined. As a result, for most practical problems the degree of accuracy does not appear to be significantly diminished by neglecting the effect of the intermediate stress and working with the simpler two-dimensional condition (σ_1 and σ_3 only).

Mohr's Circle

The varying values of normal stress and shear stress corresponding to differing values of θ can be determined by using Eqs. (7-1) and (7-2). If the combination of normal and shear stress resulting from each value of θ is plotted as a point on a coordinate system where the horizontal axis represents normal stress and the vertical axis represents shear stress [Fig. 7-6(a)], the locus of many plotted points will form a circle [Fig. 7-6(b)].

This fact can be used to advantage. By working with simple properties of a circle, a graphical or pictorial method for solving for normal

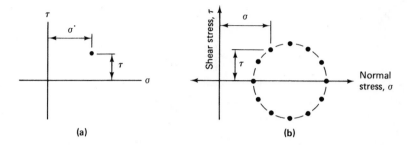

Figure 7-6. *Method of representing the combination of normal and shear stresses acting on any plane through an incremental element on shear stress-normal stress coordinates: (a) single point represents stress combination on one plane; (b) stress combinations occurring on different planes through an incremental element provide points which form a circle.*

and shear stresses on any plane once the stresses on orthogonal planes are known is easily developed. The method simplifies the calculation necessary for determining stresses, and eliminates the need to work with cumbersome equations. The method is referred to as the Mohr's circle for determining stresses, after Otto Mohr (1835–1918), who is credited as the developer of the method.

In using Mohr's circle, a sign convention is required. For soils problems, compressive stresses are conventionally assumed to be positive, and shearing stresses that provide a clockwise couple are also considered positive; see Fig. 7-7.

For a stress combination like that shown in Fig. 7-8(a), σ_1 and σ_3 are the major and minor principal stresses. There is no shear stress acting on the planes. To construct the Mohr's circle for this combina-

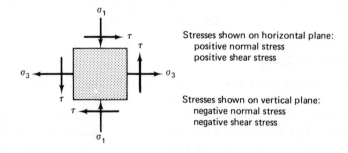

Figure 7-7. *Sign convention assigned to stresses for the Mohr's circle analysis.*

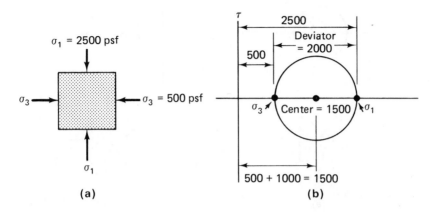

Figure 7-8. Representation of principal stresses acting at a point and the related Mohr's circle plot: (a) stresses acting on incremental element; (b) Mohr's circle plot.

tion [Fig. 7-8 (b)], locate σ_1 (or 2500 psf)[1] and σ_3 (or 500 psf) on the *horizontal axis* of the coordinate system (since the shear is zero). Next, establish the location (or value) for the center of the Mohr's circle, knowing that the diameter of the circle has a value equal to σ_1 minus σ_3. (This numerical difference between σ_1 and σ_3 is called the *deviator stress*). For this illustration, the deviator stress is 2500 psf minus 500 psf, or 2000 psf. The radius of the circle is then 1000 psf, and the center of the circle will plot at 1500 psf, which is the value of σ_3 plus the radius. After the center of the circle and the diameter are established, the circle itself can be constructed; thereafter, the stress combination on *any* plane can be determined.

In working with the Mohr's circle, it is convenient, whenever possible, to reference the plane under study to the major principal plane. Remember that the major principal stress acts on the major principal plane; the angular measurement to the plane in question is made on the Mohr's circle by starting from the point representing the major principal stress. If the angle to be measured is formed by two radii, the angle is measured at the center of the circle. Because of the properties of a circle, the central angle on the circle must be twice the value of the angle θ measured on the original element; see Fig. 7-9 (a). The direction of measurement from the principal plane to the cut plane on the element, clockwise or counterclockwise, is the same direction to be used on the Mohr's circle; see Fig. 7-9 (b). By reference to the Mohr's circle

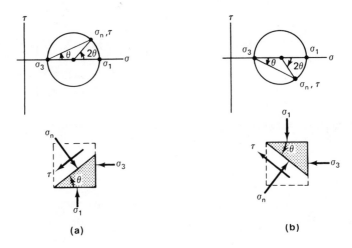

Figure 7-9. Relation of positions of planes on the incremental element to points on the Mohr's circle: (a) reference angle measured counterclockwise; (b) reference angle measured clockwise.

diagram in Fig. 7-10, the shear and normal stress for any point (representing any plane) can be reaffirmed. The normal stress coordinate is the value of the center of the Mohr's circle plus or minus the horizontal projection of the radius. As shown in Fig. 7-10, the normal stress would then be

$$\sigma_n = \frac{\sigma_1 + \sigma_3}{2} + \frac{\sigma_1 - \sigma_3}{2} \cos 2\theta$$

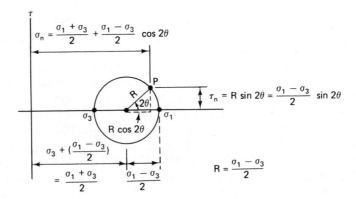

Figure 7-10. Establishment of the general equations for normal stress and shear stress from the Mohr's circle.

which agrees with Eq. (7-1). The shear stress coordinate is the vertical projection of the radius, or

$$\tau_n = \frac{\sigma_1 - \sigma_3}{2} \sin 2\theta$$

which agrees with Eq. (7-2).

As an illustration, to determine the magnitude of the normal and shear stress acting on a plane 45 deg from the major principal plane (Fig. 7-11), the angle measured at the center of the Mohr's circle would have to be twice 45 deg, or 90 deg. Note that, on the element, the angle θ is 45 deg measured counterclockwise, regardless if the top or bottom of the original element is used as the reference plane. On the Mohr's circle the central angle, 2θ or 90 deg, is also measured counterclockwise. The principal stress equations from the previous illustration being used, by simple mathematics if the Mohr's circle is used pictorially or by scaling if the graphical method is used, the value of the normal stress is found to be 1500 psf and the shear stress is 1000 psf.

From examination of the Mohr's circle, it should be evident that the maximum shear stress (in this case 1000 psf) always acts on the plane that is 45 deg from the major principal plane. Further, the maximum shear stress always has a magnitude equal to the radius of the Mohr's circle.

As a second illustration, with the same principal stresses used previously, assume that it is desired to find the stresses acting on a plane that is 60 deg clockwise from the major principal plane (Fig. 7-12). On the Mohr's circle, the central angle to be measured is 120 deg. Mathematically, from a pictorial Mohr's circle, the shear stress τ_n equals $(R) \times (\sin 60°)$, or $(1000 \text{ psf}) \times (0.866)$, which is 866 psf. The normal

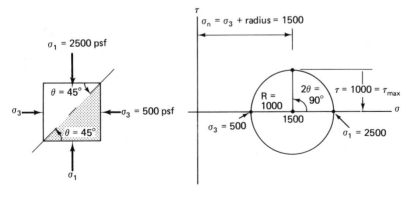

Figure 7-11. Use of Mohr's circle to determine the shear and normal stress acting on a plane 45 degrees from the major principal plane.

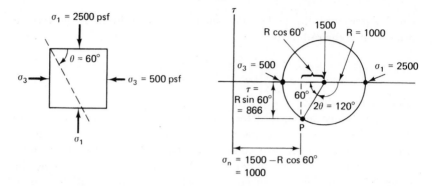

Figure 7-12. Use of Mohr's circle to determine stresses on a plane 60 degrees clockwise from the major principal plane.

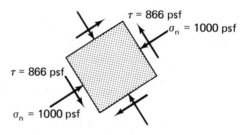

Figure 7-13. Magnitude and direction of stresses acting on the plane studied in the Mohr's circle analysis of Fig. 7-12.

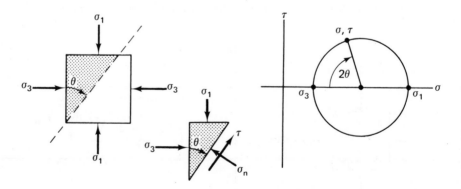

Figure 7-14. Method of locating the stress-combination point on the Mohr's circle when the minor principal plane is used as a reference.

stress σ_n is 1500 psf minus $(R) \times (\cos 60°)$, or 1000 psf, as shown in Fig. 7-12. The 1500 psf value locates the center of the circle. On the desired plane, the stresses are as shown in Fig. 7-13.

If the stress combination is to be determined on a plane measured with reference to the minor principal plane, the related angular measurement on the Mohr's circle is made from the minor principal stress, σ_3, as shown in Fig. 7-14. The direction of the angle will be the same on the Mohr's circle as on the incremental element.

The Mohr's circle pictorial or graphical method can similarly be used for determining principal stresses if the stress conditions for any two orthogonal planes other than principal planes are known. For an element where the normal and shear stresses are as shown in Fig. 7-15 (note that the shear stresses on the mutually perpendicular planes are equal in magnitude), the Mohr's circle construction proceeds as follows (see Fig. 7-16):

1. On the Mohr's circle coordinates, establish the points corresponding to σ_a and σ_b.

2. Connect these points, thereby establishing the diameter of the Mohr's circle [Fig. 7-16(a)], and determine the coordinate for the center of the circle.

3. Calculate the value for the radius of the Mohr's circle [Fig. 7-16(b)].

4. The value of σ_1 (the major principal stress) is the circle radius added to the value established for the circle center. The value of σ_3 (the minor principal stress) is the radius subtracted from the coordinate for the circle center [Fig. 7-16(c)].

5. The angle 2θ on the Mohr's circle is obtained from simple geometry, as shown in Fig. 7-16(d).

6. The orientations of the principal planes with respect to the original element are shown in Fig. 7-16(e).

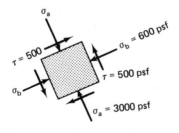

Figure 7-15. An incremental element with stresses that are not principal stresses acting on principal planes.

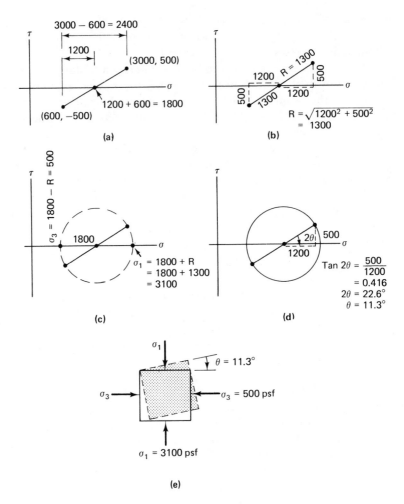

Figure 7-16.   *Procedure for applying the Mohr's circle analysis to solve for principal stresses.*

7.   Since the complete Mohr's circle is established, stresses on any other plane can be determined by using the methods covered previously.

Illustration 7-1:   The major and minor principal stresses acting at a point are 50 psi compression and 10 psi compression, respectively. Draw the Mohr's circle for this stress combination, and determine the magnitude of shear and normal stress on the plane where shear is a maximum.

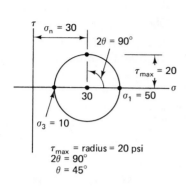 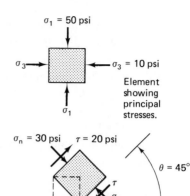

Illustration 7-2: At one point in a soil mass, the minor principal stress is 1000 psf. What is the maximum value possible for the major principal stress at this point if the shear stress cannot exceed 2000 psf?

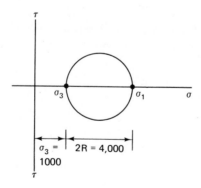

The maximum shear stress at a point is the value of the radius of the Mohr's circle, i.e.,

$$\text{Radius} = 2000 \text{ psf}$$

Since the major principal stress σ_1 is $\sigma_3 + 2R$,

$$\sigma_1 = 1000 + 2(2000) = 5000 \text{ psf}$$

PROBLEMS

1. At a point, the major principal stress is 5000 psf compression and and the minor principal stress is 2000 psf compression.

    (a)   Draw the Mohr's circle for this stress combination.

    (b)   Find the maximum shear stress, and also indicate the value of normal stress acting on this same plane.

    (c)   Determine the values of the normal and shear stress acting on a plane that is 60 deg from the major principal plane.

2. At a point, the major principal stress is 80 psi compression and the minor principal stress is 20 psi compression.

    (a)   Draw the Mohr's circle for this stress combination.

    (b)   What is the maximum shear stress acting at the point, and what value of normal stress acts on this same plane?

    (c)   Determine the value of the normal and shear stress acting on a plane that is 60 deg from the minor principal plane.

3. The major and minor principal stresses acting at a point are 80 psi compression and 20 psi tension.

    (a)   Draw the Mohr's circle for this stress combination.

    (b)   What is the maximum shear stress acting at the point and what value of normal stress acts on the plane?

    (c)   Determine the value of the shear and normal stress acting on a plane that is 30 deg from the major principal plane.

4. At a point, the normal and shear stresses acting on one plane are 8000 psf compression and 2000 psf, respectively. On a perpendicular plane, the normal and shear stress combination is 2000 psf compression and 2000 psf.

    (a)   Draw the Mohr's circle.

    (b)   Determine the value of the principal stresses.

    (c)   Find the angle between the plane on which the 8000 psf compressive stress acts and the major principal plane.

    (d)   What is the maximum shear stress acting at the point?

5. The normal and shear stresses acting on one plane passing through a point in a soil mass are 120 psi compression and 25 psi, respectively. On an orthagonal plane the stresses are 40 psi compression and 25 psi.

    (a)   Draw the Mohr's circle for this stress condition.

    (b)   What are the principal stresses?

    (c)   Determine the angle between the plane on which the 120-psi stress acts and the major principal plane.

    (d)   What is the maximum shear stress acting at the point?

6. The major principal stress and deviator stress at a point are 100 psi and 60 psi, respectively. What is the value of the maximum shear stress, and the normal stress acting on the same plane?

7. For a minor principal stress of 50 psi, what is the maximum value possible for the major principal stress if the maximum shear stress is not to exceed 40 psi?

8. In a soil where the ground surface is horizontal, the vertical stress acting at a point is 2000 psf and the horizontal stress is 1000 psf. If these stresses represent the principal stresses, what is the maximum shear stress acting at the point?

9. Given a soil mass with a level ground surface and a material having a unit weight of 120 pcf. The vertical pressure at any depth represents the major principal stress at that particular point. The lateral pressure acting at any point is one-half the vertical pressure and represents the minor principal stress.

   (a) Determine the principal stresses acting at a depth of 10 ft below the ground surface and draw the Mohr's circle for these conditions.

   (b) What is the value of the maximum shear stress at this depth?

10. A soil deposit exists with a level ground surface. The saturated unit weight of the soil is 125 pcf, and the water table is at the ground surface.

    (a) Determine the principal stresses acting at a depth 15 feet below the ground surface that results from the effective (submerged) soil weight if the lateral pressure is 0.45 of the vertical pressure.

    (b) What is the magnitude of the maximum shear stress at this depth?

CHAPTER 8
Subsurface Stresses

At a point within a soil mass, stresses will be developed as a result of the soil lying above the point and by any structural or other loading imposed onto the soil mass. The magnitude of the subsurface stress at a point is affected by the groundwater table if it extends to an elevation above the point.

In most foundation design problems, the safe bearing capacity of the soil (the ability to support structural load) and settlement (the soil volume change resulting from loading) that will develop under a given intensity of structural loading are major items of concern. For analysis, the significant stresses are considered to be those acting in the vertical direction. In the design of vertical structures such as retaining walls, underground walls, sheeting for braced excavations and waterfront structures, and some types of pile foundations, the soil stresses acting in the horizontal or lateral direction are the most significant.

Stresses Caused By the Soil Mass

VERTICAL STRESSES / In a soil mass having a horizontal surface, the vertical stress caused by the soil at a point below the surface is equal to the weight of the soil lying directly above the point. Vertical stress thus increases as the depth of the soil overburden increases. Where the stress is determined for a unit area, as is conventional, the unit stress will be equal to the weight of a "column" of soil extending about this unit area, [Fig. 8-1 (a)]. For a homogeneous soil having a wet unit volumetric weight of γ_t (normally expressed as pounds per cubic

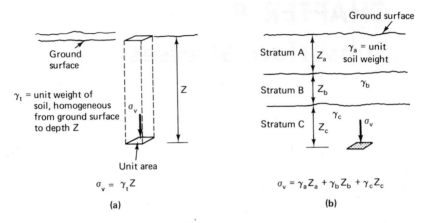

Figure 8-1.   *Vertical subsurface stress resulting from the soil mass.*

foot),[1] the stress σ_v (normally in pounds per square foot) at a depth Z below the ground surface is

$$\sigma_v = \gamma_t Z \qquad (8\text{-}1)$$

If the soil mass is comprised of strata of different soil types and the unit weights of the soil in each stratum are different, the vertical stress at a depth Z will be equal to the total weight of the different segments of the soil "column," as indicated by Fig. 8-1 (b).

EFFECT OF GROUNDWATER TABLE / When a soil exists below the groundwater table, the submerged soil particles are subject to a buoyant force resulting from the hydrostatic water pressure, the same phenomenon that acts on any submerged solid. The submerged weight of the soil, γ_{sub}, is termed the *effective soil weight*, and the subsurface stress that results is termed the *effective stress*. Effective stress represents the actual intergranular pressure that occurs between soil particles. This effective stress is the stress that influences shear strength of the soil and volume changes or settlements.

If a condition exists where the *water table is at the ground surface* and the soil mass is homogeneous, the effective stress $\bar{\sigma}_v$ at a depth Z is

$$\bar{\sigma}_v = \gamma_{sub} Z \qquad (8\text{-}2)$$

If the total weight of soil as it exists above the water table is γ_t (the total weight before the buoyant effects of submergence are considered), the effective stress is

$$\bar{\sigma}_v = \gamma_t Z - \gamma_w Z \qquad (8\text{-}3)$$

---

[1] In SI units, $\gamma = kN/m^3$, $\sigma_v = kN/m^2$.

where γ_w is the unit weight of water. (It is conventional to assume 62.4 pcf or 1.0 gm/cc for γ_w.) The last term of this equation is the total water pressure at the depth Z. Total water pressure at a point is termed the neutral stress u, for it acts equally in all directions. Neutral stress refers to any water pressure that develops at a point as caused by hydro-static conditions. It is important to recognize that the neutral stress acts to *reduce* the intergranular stress that develops between soil particles. This condition frequently has an adverse effect on the strength of a soil. The effective stress for conditions just described can be expressed as

$$\overline{\sigma}_v = \gamma_t Z - u \qquad (8\text{-}4)$$

To compute the effective stress for a condition where the *ground-water table lies below the ground surface*, and for the condition where strata of soils of different types and weights exist, two different approaches are possible. One approach involves determining the total soil pressure (buoyancy effects being disregarded) and then subtracting the hydro-static pressure (the neutral pressure) at the point being analyzed. The neutral pressure u is the unit weight of water γ_w, multiplied by the depth below the water table. A second approach is to determine directly the effective stress of the column of soil above the point by using the effective or submerged weight of all soil in the "column." Above the water table, the effective soil weight is the total soil weight, including the pore water; below the water table, the effective soil weight is the sub-merged or buoyant weight; see Fig. 8-2 (a).

Where the soil surface is below water (such as in oceans and lakes), the effective stress should be computed by using the submerged or effec-tive soil weight multiplied by the depth measured from the soil surface [Fig. 8-2 (b)].

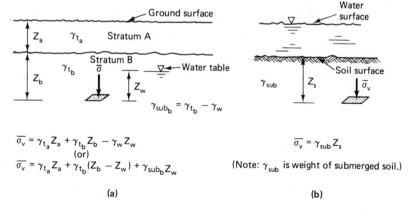

(a)

(b)

Figure 8-2. *Subsurface stress below water.*

A summary of the analytical procedures for computing subsurface stresses resulting from various soil and water table conditions, and numerical illustrations are presented in Fig. 8-3.

HORIZONTAL (LATERAL) STRESSES / The magnitude of vertical stress is relatively simple to determine when the ground surface is level. When this condition does exist, it is also convenient to indicate horizontal (lateral) stresses that exist in a soil mass in terms of the vertical stress. The ratio of lateral stress to vertical stress, K, is termed the *coefficient of lateral earth pressure*. Mathematically,

$$K = \frac{\text{horizontal soil pressure } \sigma_h}{\text{vertical soil pressure } \sigma_v} \qquad (8\text{-}5)$$

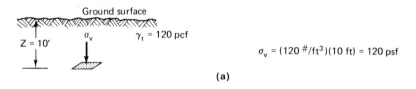

$$\sigma_v = (120 \text{ #/ft}^3)(10 \text{ ft}) = 120 \text{ psf}$$

(a)

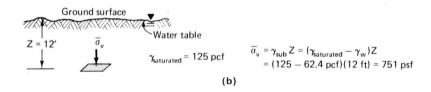

$$\bar{\sigma}_v = \gamma_{sub} Z = (\gamma_{saturated} - \gamma_w)Z$$
$$= (125 - 62.4 \text{ pcf})(12 \text{ ft}) = 751 \text{ psf}$$

(b)

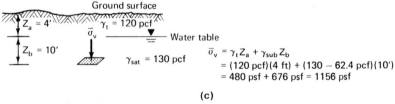

$$\bar{\sigma}_v = \gamma_t Z_a + \gamma_{sub} Z_b$$
$$= (120 \text{ pcf})(4 \text{ ft}) + (130 - 62.4 \text{ pcf})(10')$$
$$= 480 \text{ psf} + 676 \text{ psf} = 1156 \text{ psf}$$

(c)

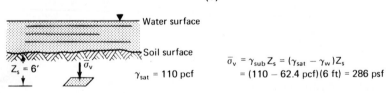

$$\bar{\sigma}_v = \gamma_{sub} Z_s = (\gamma_{sat} - \gamma_w)Z_s$$
$$= (110 - 62.4 \text{ pcf})(6 \text{ ft}) = 286 \text{ psf}$$

(d)

$$1 \text{ psf} = 0.048 \frac{kN}{m^2} \; ; 1 \text{ pcf} = 0.157 \frac{kN}{m^3}$$

Figure 8-3.   *Summary-Method to compute subsurface stress in a soil mass.*

In a horizontal, uniform soil mass of infinite extent, the lateral movement of the soil at any depth is not possible, because the confining pressure is equal in all horizontal directions. Thus a state of static equilibrium exists and the soil is in the *at-rest condition*. The coefficient of lateral pressure for the at-rest condition is indicated by K_0.

The magnitude of K_0 for a given soil mass is affected by the soil deposit's stress history. Soils that have been subjected to heavy loading at some time in their history, such as now dense granular (sand or gravel) deposits and hard overconsolidated[2] clays, would have had to develop resistance to high lateral stress in order to maintain stability. Deposits that have not been exposed to heavy loading, such as loose granular soils and soft normally consolidated or underconsolidated clays, would not have developed high lateral strength. Typically, then, dense granular soils and hard clays end up having lower values of h_0 than do loose granular soils and soft clays. Such typical values are presented in Table 8-1.

TABLE 8-1 / TYPICAL VALUES OF K_0

Soil Type	K_0
Granular, loose	0.5 to 0.6
Granular, dense	0.3 to 0.5
Clay, soft	0.9 to 1.1 (undrained)
Clay, hard	0.8 to 0.9 (undrained)

Illustration 8-1: The unit weight of the soil in a uniform deposit of loose sand is 100 pcf. Determine the horizontal stress that acts within the soil mass at a depth of 10 feet.

$$\sigma_v = \sigma \, \mathcal{Z} = (100 \text{ pcf})(10 \text{ ft}) = 1000 \text{ psf}$$

$$\sigma_h = K_0 \sigma_v = (0.5)(1000 \text{ psf}) = 500 \text{ psf}$$

where K_0 is attained from Table 8-1.

In granular soils below the water table, determination of the *total* lateral pressure requires that the hydrostatic pressure due to the water be added to the at-rest soil pressure computed by using a value from Table 8-1 and the submerged or effective soil weight.

---

[2]Clays which at some past time were subject to loading greater than the weight of all presently existing overlying soil.

Illustration 8-2: A concrete basement wall for a structure extends below the groundwater table. For conditions indicated by the sketch, calculate the total lateral pressure acting against the wall at a point eight feet below the ground surface.

Ground surface
γ = 120 pcf
Medium dense sand, $k_o = 0.4$
$Z_w = 4'$
$\gamma_{sub} = 60$ pcf
σ_h

Soil pressure, σ_h at 8 ft $= K_\sigma \gamma Z$

$$= (0.4)[(4 \text{ ft} \times 120 \text{ pcf}) + (4 \text{ ft} \times 60 \text{ pcf})]$$

$$= 288 \text{ psf}$$

Total pressure $= K_0 \gamma Z + \gamma_w Z_w = 288 \text{ psf} + (4 \text{ ft})(62.4 \text{ pcf})$

$$= 538 \text{ psf}$$

Stress Within the Soil Mass Resulting From Vertical Surface Loading

UNIFORM HOMOGENEOUS SOILS / When a vertical loading from a structure or other body is applied at the surface of a soil mass, new stresses are created within the mass. Because of shearing resistance developed within the soil, loading transferred to the soil mass will be spread laterally with increasing depth from the point or area of application (see Fig. 8-4). With increasing depth the area over which new stresses develop will increase, but the magnitude of the stresses will decrease. For an equilibrium condition, the sum of the new vertical stresses developed in the soil mass on any horizontal plane must equal the weight or force of the surface loading.

The manner in which the stresses become distributed throughout the soil mass is affected by the properties of the soil, including modulus of elasticity and Poissons ratio, and any stratification of different soil types.

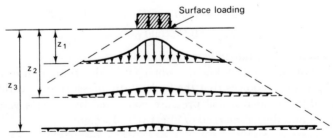

Figure 8-4.   *Variation of vertical stress at different depths z.*

BOUSSINESQ   STRESS   DISTRIBUTION / One of the methods in common use for calculating stresses that result in a soil mass from a surface loading is based upon the work of Boussinesq, a nineteenth-century French mathematician.   Boussinesq assumed a homogeneous, isotropic material (properties the same in all directions) of semi-infinite extent (unlimited depth) and developed equations for the stress distribution resulting from a point load.   Adapted to soil masses, the described conditions are as indicated by Fig. 8-5.   The vertical stress increase $\Delta\sigma_v$ resulting at a depth z and a distance r, measured horizontally from where a point loading Q is applied, becomes

$$\Delta\sigma_v = \frac{3Q}{2\pi}\frac{z^3}{(r^2 + z^2)^{5/2}} = \frac{Q}{z^2}\frac{3}{2\pi\left[1 + \left(\dfrac{r}{z}\right)^2\right]^{5/2}} \qquad (8\text{-}6a)$$

This equation indicates that as the depth increases the stress decreases. Similarly, the stress decreases as the horizontal distance from the line of loading is increased.   For a given depth, the intensity of stress is greatest directly beneath the point of load application.

With the Boussinesq equation for vertical stress, the modulus of elasticity and Poissons ratio are not required, which indicates that the stress is independent of these properties so long as the material is homogeneous and isotropic.

The suitability of using the Boussinesq equation for determining subsurface stresses in a foundation analysis depends on how closely the actual soil conditions and properties resemble the theory's original assumptions.   For practical problems, conditions of a homogeneous and isotropic material are commonly assumed for homogeneous clay deposits, for man-made fills where the soil fill has been placed and compacted in thin layers, and for limited thicknesses of uniform granular soil deposits.

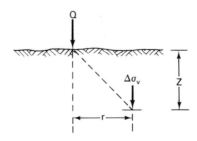

Figure 8-5.   Definition of terms applicable to Boussinesq and Westergaard equations.

WESTERGAARD STRESS DISTRIBUTION / Some sedimentary soil deposits consist of alternating thin layers of sandy soil (coarse, relatively incompressible material) and fine-grained silt-clay soils (compressible material), for example, stratified deposits such as varved clays and laminated clays. For such conditions, the Westergaard equations provide a better means for evaluating the subsurface stresses. In his development, Westergaard assumed that thin layers of a homogeneous and anistropic material were sandwiched between closely spaced, infinitely thin sheets of rigid material that would permit compression but no lateral deformation. For the case where the Poissons ratio is zero, the equation for subsurface stress resulting from a concentrated point loading reduces to

$$\Delta\sigma_v = \frac{Q}{\pi Z^2 \left[1 + 2\left(\frac{r}{Z}\right)^2\right]^{3/2}}$$ (8-7a)

The terms Q, R, and Z are as defined for the Boussinesq equation, and as shown in Fig. 8-5.

COMPUTATIONAL AIDES / With both the Boussinesq and Westergaard equations, the subsurface stress resulting from a given point load will be related to the Z and r distances, or the r/Z ratio. If the Boussinesq equation is written in terms of a stress influence factor I_B, which is related to r/Z as follows,

$$\Delta\sigma_v = \frac{Q}{Z^2}\frac{3}{2\pi\left[1 + \left(\frac{r}{Z}\right)^2\right]^{5/2}} = \frac{Q}{Z^2}I_B$$ (8-6b)

calculations for values of I_B for different r/Z ratios can be computed and presented as shown in Fig. 8-6. The effort to determine a subsurface stress then becomes greatly simplified. Similarly, the Westergaard equation can be written in terms of an influence factor I_w as follows:

$$\Delta\sigma_v = \frac{Q}{Z^2\pi\left[1 + 2\left(\frac{r}{Z}\right)^2\right]^{3/2}} = \frac{Q}{Z^2}I_w$$ (8-7b)

and values of I_w versus r/Z ratios can be developed as presented in Fig. 8-6.

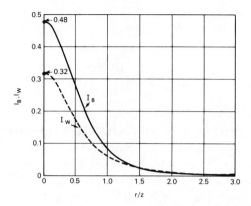

Figure 8-6. *Values of I_B and I_W for calculating vertical stress resulting from surface load Q.*

Illustration 8-3: For Boussinesq conditions, what subsurface stress will result at a point 10 ft below where a 10,000 lb point load is applied?

$$r = 0, \quad Z = 10, \quad r/Z = 0$$

For $r/Z = 0$, obtain $I_B = 0.48$

$$\Delta\sigma_v = \frac{Q}{Z^2} I_B = \frac{10,000 \text{ lb}}{(10 \text{ ft} \times 10 \text{ ft})} (0.48) = 48 \text{ psf}$$

Illustration 8-4: For the Westergaard conditions, what subsurface stress will result 10 ft below, and 10 ft horizontally, from where a 10,000-lb concentrated load is applied?

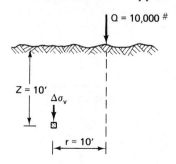

$$r = 10, \quad \mathcal{Z} = 10, \quad r/\mathcal{Z} = 1.0$$

For $r/\mathcal{Z} = 1.0$, obtain $I_w = 0.065$

$$\Delta\sigma_v = \frac{Q}{\mathcal{Z}^2} I_w = \frac{10,000 \text{ lb}}{(10 \text{ ft} \times 10 \text{ ft})} (0.065)$$

$$= 6.5 \text{ psf}$$

APPLICATION FOR FOUNDATION LOADING / In construction practice, the condition of a concentrated point loading is rarely encountered. More commonly, building loads are supported on foundations that cover a finite area (square, rectangular, or round footing), or the structure itself directly imposes loading over a finite area (earth structures such as dams and dikes). The subsurface stresses that result from loads acting over an area can be determined by integration of Eq. (8-6) or (8-7), where the loading on infinitely small increments of the foundation area can be assumed as point loads. The subsurface stress at a point is the summation of the effects resulting from all of the applied point loadings. Such integrations have been accomplished for uniform loads acting on square, rectangular, strip, and circular areas, and also for uniformly varying loads such as those developing from the weight of an earth structure having a sloped cross section (dams and dikes, for instance). Results are available in the form of charts, tables, and graphs so as to provide generalized solutions. In these presentations, the subsurface stresses are expressed as a percentage of the foundation loading intensity. Measurements of the depth and horizontal distance at which a subsurface stress acts are expressed in terms of the dimensions of the loaded foundation area. Subsurface stress conditions indicated by Boussinesq and Westergaard equations for commonly occurring foundation loadings are presented in Fig. 8-7 and 8-8. The following example problems serve to illustrate the use of these figures.

Illustration 8-5: A foundation supported on the surface of a uniform, homogeneous soil is 5 ft square and carries a loading of 125 kips. What subsurface stress increase occurs beneath the center of the foundation at a depth of 5 ft?

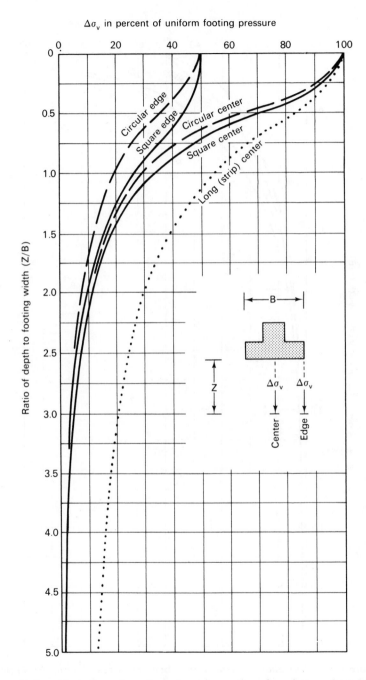

Figure 8-7. *Variation of vertical stress beneath a foundation; Boussinesq analysis.*

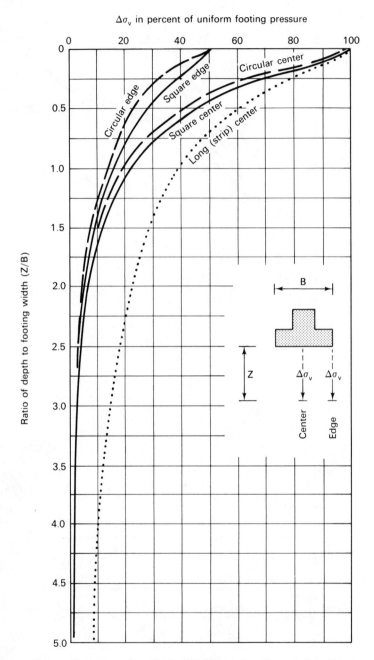

Figure 8-8. *Variation of vertical stress beneath a foundation; Westergaard analysis.*

For uniform homogeneous soil, use Boussinesq conditions.

$$q = \frac{Q}{A} = \frac{125 \text{ kips}}{(5 \text{ ft} \times 5 \text{ ft})} = 5 \text{ ksf} = 5,000 \text{ psf}$$

From Fig. 8-7, $\Delta\sigma_v$ in terms of q is 34 percent for $\dfrac{\text{depth}}{\text{width}} = \dfrac{5 \text{ ft}}{5 \text{ ft}} = 1.0$

$$\therefore \ \Delta\sigma_v = 0.34\,(5000 \text{ psf}) = 1750 \text{ psf}$$

Illustration 8-6: A circular storage tank is supported on soils that satisfy the Westergaard assumptions. What subsurface stress increase develops 10 ft beneath the edge of the tank? The tank is 20 ft in diameter, and the stored fluid material causes a pressure of 1000 psf at the tank base.

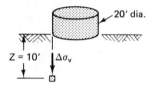

Surface load $q = 1000$ psf

Ratio of depth/width $= \dfrac{10 \text{ ft}}{20 \text{ ft}} = 0.5$

From Fig. 8-8, $\Delta\sigma_v = 23$ percent of q

$$\therefore \ \Delta\sigma_v = 0.23\,(1000 \text{ psf}) = 230 \text{ psf.}$$

The stress curves in Figs. 8-7 and 8-8 show that the stresses beneath the center of a loaded foundation area will be greater than stresses beneath the edge of the foundation until a depth of about twice the foundation width is reached. Below this level, the stresses beneath the center and edge become practically equal. Consequently, in making determinations of the stresses resulting from a foundation loading, it is possible to assume a concentrated point loading, if convenient, where the subsurface depth is greater than twice the foundation width.

SIXTY-DEGREE APPROXIMATION / A method in wide use for making rough estimates of subsurface stresses resulting from a loaded foundation area is the so-called 60-degree approximation. In this method, it is assumed that the subsurface stresses spread out uniformly with depth, the stressed area increasing at a slope of 1 foot horizontally for each 2 feet of depth as measured from the edges of the

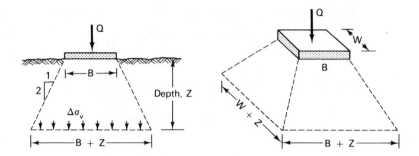

Figure 8-9. *Method for approximating vertical stress increase resulting beneath a loaded foundation (60 degree approximation).*

foundation. These assumed conditions are sketched in Fig. 8-9. At a given depth, the subsurface stress $\Delta\sigma_v$ is assumed to be uniform over the area stressed. The method differs from the Boussinesq and Westergaard theories in this respect. The stress at a depth Z then becomes

$$\Delta\sigma_v = \frac{Q}{(B + Z)(W + Z)} \tag{8-8}$$

This approximation method is incorrect in representing subsurface stresses as being uniform across a plane area. In a homogeneous soil, the computed subsurface stresses directly beneath the foundation will be less than indicated by the Boussinesq analysis, whereas at distances beyond the edges of the foundation the computed stress will be greater. This method's best application may be for estimating stress conditions in deep layers below a foundation and for determining an order of magnitude in a preliminary analysis.

LAYERED SOILS HAVING DIFFERENT PROPERTIES / The condition of soil deposits having uniform properties over a great vertical distance is not always found in practice. There frequently is encountered the situation where, within the zone (depth) that will be affected by structural loads, two or more layers of significantly different soils exist. For the condition of two different strata, the possibilities of relative properties include a firm upper layer overlying a soft layer, and the reverse.

For the condition where a firm layer overlies a soft layer (a dense sand above a compressible clay, or a surface zone of firm dessicated clay overlying the still saturated, softer and compressible lower zone of the clay), the firm layer tends to bridge over the softer layer, spreading out the area over which stresses are transferred into the softer layer. This results in creating lower stresses in the soft material than would result for a homogeneous soil.

With a soft layer overlying a firm layer, the stresses reaching the deeper layer in the area directly beneath the foundation will be greater than that indicated for a uniform homogeneous deposit.

A diagram for determining how the variation in subsurface stress beneath a uniformly loaded circular area is affected by a two-layer soil mass appears in Fig. 8-10. This diagram is for the special case where

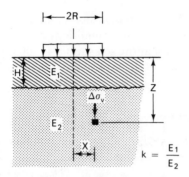

$$k = \frac{E_1}{E_2}$$

E = modulus of elasticity for soil

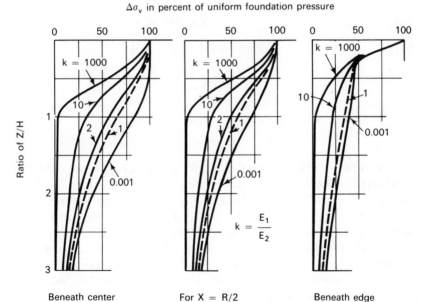

Figure 8-10. *Variation of vertical stress beneath a circular foundation. Two-layer subsurface condition for condition where R = H.*

the thickness of the upper soil layer is equal to the radius of the loaded area.

The curves of Fig. 8-10 can be used for approximating stress conditions beneath a square area as well as the circular area by assuming foundation dimensions giving the same area as the circular foundation. The illustrative example presented below indicates the use of Fig. 8-10.

Illustration 8-7: A circular storage tank 20 ft in diameter is located in an area where soil conditions are as shown by the sketch. If the tank pressure at the soil surface is 500 psf, what stress results beneath the center of the tank at a depth of 15 ft?

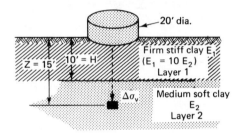

$$H = \text{radius} = 10 \text{ ft}$$

$$\text{Ratio of } \frac{z}{H} = \frac{15 \text{ ft}}{10 \text{ ft}} = 1.5$$

$$k = \frac{E_1}{E_2} = 10$$

From Fig. 8-10, $\Delta\sigma_v$ is 20 percent of surface load q

$$\therefore \Delta\sigma_v = 0.20(500 \text{ psf}) = 100 \text{ psf}$$

EFFECT OF FOUNDATION INSTALLATION BELOW FINISHED GRADE / In practical building design and construction, foundations are seldom placed directly upon the ground surface. More usually, foundations are placed at some depth below the ground surface as protection against frost and seasonal soil volume changes, erosion, or other factors. In other words, some excavation is performed before the foundation can be constructed. The subsurface stresses developed in the soil below the foundation are determined by using the previously described methods, and the depth for computing the stress increase is measured from the base of the foundation.

The total stress at a point in the soil below the foundation is the total of the stress caused by the soil overburden plus the stress due to the foundation loading. In many cases, a foundation excavation is backfilled to the ground or interior floor surface after the foundation is constructed. For practical purposes, then, the stress caused by the soil remains equal to the original soil overburden.

If an excavation is made and not backfilled, the effect is to reduce the subsurface stresses. The reduction in stress can be computed by assuming that the excavation is a "foundation" whose loading is the weight of the excavated soil. The size of the "foundation" is the plan area of the excavation. The subsurface stress is determined as for a real foundation but is handled as a negative stress to be subtracted from the stress caused by the actual foundation that is ultimately installed in the excavation.

Illustration 8-8:   A heavy materials storage building 30 ft by 30 ft is constructed with a basement that extends 6 ft below the ground surface. A foundation for an interior supporting column is constructed at floor level in the middle of the basement. The column footing is 4 ft square and carries a loading of 50 kips. What is the net stress change in the soil 5 ft below the middle of the basement floor? (For this example, neglect the weight of the concrete footing and floor.)

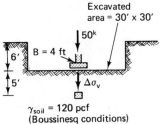

A "negative" foundation loading results from the basement excavation. Negative load is 6 ft deep × 120 pcf = 720 psf. For a foundation 30 ft × 30 ft at a depth 5 ft below the floor,

$$\frac{\text{depth}}{\text{width}} = \frac{5 \text{ ft}}{30 \text{ ft}} = 0.17$$

and the stress decrease $\Delta\sigma_v$ = 95 percent × 720 psf = −685 psf (Fig. 8-7). Stress increase from column footing,

$$\left(\text{where } q = \frac{50 \text{ kips}}{4 \text{ ft} \times 4 \text{ ft}} = 3.14 \text{ ksf} = 3,140 \text{ psf}\right)$$

is $+\Delta\sigma_v$ = (25 percent)(3,140 psf) = +780 psf (Fig. 8-7),

$$\text{for the ratio of } \frac{\text{depth}}{\text{width}} = \frac{5 \text{ ft}}{4 \text{ ft}} = 1.25$$

Therefore, actual stress increase,

$$\Delta\sigma_{v_{net}} \;=\; +780 - 685 = 95 \text{ psf.}$$

EFFECT OF CHANGING THE SURFACE GRADE / On large construction sites, the surface grade may be changed to improve the topography and to make the surface more suitable for the needs of the project. If the surface grade is lowered, this results in a reduction in the subsurface stresses. Where earth fill is placed to raise the surface level, an increase in the subsurface stresses results. When a uniform thickness of material is placed or removed, over a large area, the subsurface stress increase or decrease is the product of the unit weight of the soil added or removed and the thickness involved. This subsurface stress change is constant with depth; it does not diminish with increasing depth, as for foundation loadings.

Where the thickness of fill or soil removal is variable, as is frequently the case, a limited area can be analyzed and the average thickness of soil added or removed can then be used to calculate the subsurface stress change for the area.

Illustration 8-9:   A compacted fill for a shopping center development is placed over an area where soil conditions are as shown by the sketch. What stress increase results in the middle of the clay layer from the weight of the fill?

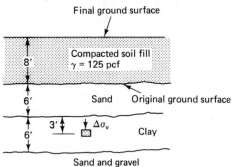

Since this is an areal fill (fill placed over a large area) the stress increase in the clay layer is

$$\Delta\sigma_v \;=\; (\gamma_{\text{fill}})(\text{height of fill})$$
$$= 125 \text{ pcf} \times 8 \text{ ft} = 1{,}000 \text{ psf}$$

Subsurface stresses caused by foundations subsequently installed in fill or cut areas are computed by using the methods already presented. It is important to remember that the final total subsurface stress must reflect the changes caused by filling or lowering the surface of the site as well as the foundation loading.

Illustration 8-10: Assume that a building foundation is constructed on the fill discussed in the previous illustration. If the footing is 5 ft square and carries a loading of 75 kips, what net stress results in the middle of the clay layer beneath the center of the footing (consider effects of fill *and* foundation loading)?

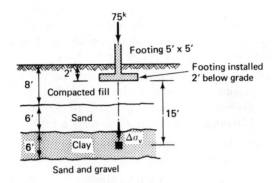

$$q_{\text{footing}} = \frac{75 \text{ kips}}{5 \text{ ft} \times 5 \text{ ft}} = 3 \text{ ksf} = 3,000 \text{ psf}$$

$$\text{Ratio of } \frac{\text{depth}}{\text{width}} = \frac{15 \text{ ft}}{5 \text{ ft}} = 3$$

Estimate stress increase by either Boussinesq or Westergaard, $\Delta\sigma_v \approx 5$ percent $\times q$ (from Fig. 8-7 or Fig. 8-8).

$$\therefore \Delta\sigma_v = 0.05\,(3,000 \text{ psf}) = 150 \text{ psf from foundation.}$$

Stress increase from fill $= 1000$ psf (previous illustration)

Total stress increase $= 150 + 1000$ psf $= 1150$ psf.

PROBLEMS

1. At a planned construction site, subsurface sampling indicates that the wet unit weight of the soil is 123 pcf.

    (a) Determine the effective vertical stress at the 12-foot depth if the water table is deep.

    (b) Determine the effective vertical stress and the neutral stress at the 12-foot depth if the water table rises to within six feet of the ground surface.

2. At an offshore location, the soil surface is 30 ft below the water
    surface. Weighing of soil samples obtained in the investigation
    indicates the saturated unit weight to be 118 pcf. Determine the
    effective vertical pressure at a depth 55 ft below the water surface.

3. Determine the lateral earth pressure at a depth 20 ft below the
    ground surface in a loose sand deposit. The wet unit weight of the
    sand is 115 pcf. What would the total lateral pressure be if the
    water table rose to the ground surface?

4. A deep basement for a building is constructed in dense granular
    soil whose unit weight is 130 pcf. For a distance 10 ft below the
    ground surface, determine the increase in total lateral pressure
    that results by having the water table change from a location be-
    low the basement level to the ground surface.

5. A 20-kip concentrated (point) load acts on the surface of a soil
    mass. Determine the vertical stress 10 ft below the ground surface
    at locations directly beneath the load, 10 ft horizontally from the
    load, and 20 ft horizontally from the load, for the

    (a)   Boussinesq conditions.
    (b)   Westergaard conditions.

6. A four-foot-square foundation located on the surface of a soil mass
    supports a total load of 96 kips. Determine the vertical stress re-
    sulting from the foundation loading at a depth 6 ft below the
    ground surface for locations beneath the center of the footing and
    beneath the edge of the footing, assuming

    (a)   The Boussinesq conditions apply.
    (b)   The Westergaard conditions apply.

7. A four-foot-wide long (strip) footing carries a wall loading of
    20,000 pounds per foot of wall length. What vertical stress in-
    crease results below the center of the footing at depths of 4, 8 and
    12 ft, if

    (a)   The Boussinesq conditions apply?
    (b)   The Westergaard conditions apply?

8. An oil storage tank 50 ft in diameter imposes a maximum loading
    of 2500 psf onto the ground surface where the tank is supported.
    The soil underlying the tank has a unit weight of 125 pcf. For a
    depth 25 ft below the ground surface, compute the effective ver-
    tical stress when the tank is empty and when the tank is full, at
    locations beneath the center and the edge of the tank. Assume
    that the Boussinesq conditions apply.

9. Compare the stress increase resulting eight feet below the center of a 10-foot-square foundation supporting 5000 psf when the 60-degree approximation is assumed and when the Boussinesq conditions are assumed.

10. A circular foundation 12 ft in diameter imposes a pressure of 8000 psf onto the soil. At the 12-ft depth, determine the vertical stress increase beneath the center and the edge of the loaded area, assuming

    (a)   The Westergaard conditions apply.
    (b)   The 60-degree approximation.

11. A 30-foot-diameter storage tank is supported on the ground surface at a site where a 15-foot-thick layer of dense sand overlies a very thick clay layer. The modulus of elasticity for the upper layer is ten times the modulus for the lower layer. The tank imposes a pressure of 3 ksf.

    (a)   Determine the increase in vertical stress below the center of the tank and at a depth of 30 ft.
    (b)   Determine the stress increase at the 30 ft depth if subsurface conditions were homogeneous (assume Boussinesq) and compare the result with (a).

12. For a construction project, the ground surface at a building location is lowered five feet. A four-foot-square footing carrying a loading of 4000 psf is then constructed at the level of the new surface. What net stress increase results in the soil mass four feet below the center of the foundation? Assume that the soil unit weight is 120 pcf.

13. For a construction project, ten feet of compacted earth fill is placed in preparation for erecting a building. At one of the building's interior locations a machine foundation 20 ft square will be installed at floor level to support equipment weighing 2000 kips. What net stress increase results below the center of the machine in the original soil mass, five feet under the natural ground surface? Assume a unit weight of 125 pcf for the compacted fill.

CHAPTER 9
Settlement: Soil Volume Change and Consolidation

The weight of any structure supported on the earth will result in stresses being imposed on the soils below the level of the base or foundation of the structure. The deformations in the soil occurring with these stresses cause volume changes in the soil, with the result that the structure undergoes settlement. The amount of settlement that will actually occur is related to the pressures (stresses) imposed on the soils, and the stress-strain properties of the soil.

For the typical situation, loadings imposed by a structure at foundation level act in a downward vertical direction. The resulting stresses developed within the soil mass beneath the structure act vertically but in other directions as well. However, for the generally occurring situation, where the stresses that are developed are well below the ultimate strength of the soil, most of the settlement of a structure results because of the stresses acting in the vertical direction. Reasonable accuracy in predicting settlements can be obtained by simply considering only the effect of vertical stresses.

Settlement is caused both by soil compression (or vertical squeezing together of soil particles) and lateral yielding of the soils located under the loaded area. The settlement factor that is most significant for cohesive soils is different from the factor that is most significant for cohesionless soils. In cohesive soils, compression is the cause of most of the settlement. For homogeneous cohesionless soils, lateral yielding is the more significant cause.

In practice, determination of the settlement characteristics of a soil is frequently attempted by one of several methods:

1. Working backwards from observations of the behavior of structures in the area near the planned building site.

2. Prior to construction, performing large-scale field-load tests at the actual building site.

3. Performing laboratory compression tests on soil samples obtained from borings or test-pits made at the proposed construction site.

4. Estimating compressibility or volume change characteristics on the basis of index property tests performed on samples from borings or test-pits. (Such tests typically include classification, moisture-density determinations, liquid limit and plastic limit determination, and relative-density determinations.)

There are inherent shortcomings to methods 1 and 2. For 1, it can be difficult accurately to ascertain the actual amount of settlement that an existing structure has experienced, and, perhaps more important, soil conditions may not be identical at the different building sites. However, checking the behavior or performance of existing buildings in an area is good practice, provided that the procedures discussed hereafter are also followed. For 2, field-load tests seldom truly represent the loading conditions that will be developed by a structure, and the resulting settlement or compression data may be seriously misleading. Data from field-load tests may be grossly in error when cohesive soil strata underlie the test area. Further, full-scale load tests are relatively expensive to perform.

Evaluation of a soil's compressibility or volume change characteristics is more accurately determined from laboratory tests. With the requirement for making soil borings or other form of subsurface investigation an accepted necessity for construction projects, it becomes relatively simple and economical also to obtain soil samples for purposes of performing laboratory tests. Compression tests provide the most accurate information (item 3), but index property tests can frequently provide reliable information, and are quicker and less costly to perform.

Compressibility

Compressibility is the term applied to one-dimensional volume changes that occur in cohesive soils that are subjected to compressive loading. This property of a fine-grained soil can be determined directly by performing a laboratory compression test, frequently called a *consolidation test*. In this test, an undistorted sample is fit into a ring or cylinder so that the sample is confined against lateral displacement, and compressive loading is imposed to the soil (Fig. 9-1). For known magnitudes of load, the amount of compression and the time required for compression to occur are recorded. The test is usually performed by

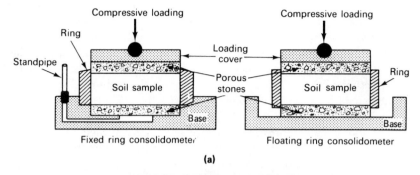

Fixed ring consolidometer   Floating ring consolidometer

(a)

(b)

Figure 9-1.   *(a) Schematic representation of conventional types of laboratory consolidometers; (b) typical laboratory consolidation equipment.*

imposing a series of increasing compressive loadings and determining time-rate-of-compression data for each increment of loading. The entire body of data permits the stress-strain characteristics of the soil to be determined, as discussed more fully in later sections of this chapter.

In soil compression, volume decreases are due principally to decreases in void spaces between soil particles as the particles rearrange so to develop resistance to the new external loading. Practically, there is no decrease in the actual volume of soil particles.

The speed or rate of time that is required for volume changes to occur differs significantly for the coarse (cohesionless) soils and the fine-grained (cohesive) soils. The cohesionless soils experience compression relatively quickly—frequently instantaneously—after a loading is imposed. Conversely, clay soils generally require a significant period before full compression under an applied loading results. Relating compression with the *time period* necessary for the compression to occur (time-rate of compression) is consolidation. The effects of this are discussed in a later section of the chapter.

Laboratory compression tests are seldom performed on cohesionless soils. Part of the reason for this is the practical difficulty associated with obtaining suitably undisturbed samples of granular soil from a construction site. Also, because the rate of settlement for these soils is rapid, they do not cause the problems of postconstruction settlement that cohesive soils do. When information on the volume change or settlement characteristics of a granular soil is needed, it is most frequently obtained indirectly, from an in-place density determination, relative density tests, or other correlations.

PRESENTATION AND ANALYSIS OF LABORATORY COMPRESSION TEST DATA / Compression test data are presented in any of several ways, usually depending on the preferences and experiences of the individual using the data. Because compression is due to changes in the void spaces in the soil, methods in common use frequently indicate compression as a change in the void ratio. Using arithmetic coordinates, a typical test result for change in void ratio versus the increase in loading pressure is as shown in Fig. 9-2. The slope of the curve at any point is a_v, the *coefficient of compressibility*. Mathematically, $a_v = de/d\sigma_v$. Because of the constantly changing slope of the curve, it is somewhat difficult to use a_v in a mathematical analysis, as is

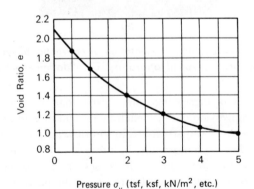

Pressure σ_v (tsf, ksf, kN/m^2, etc.)

Figure 9-2. *Presentation of compression test data on arithmetic coordinates.*

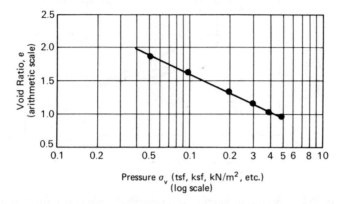

Figure 9-3. *Presentation of compression test data on semilog coordinates.*

desired in order to make settlement calculations. When semilog coordinates are used and the void ratio is plotted versus the logarithm of pressure, the data will plot approximately as a straight line (or, as described later, a series of straight lines); see Fig. 9-3. In this form, the test data are more adaptable to analytical use.

Another method of presentation shows unit change in sample thickness (or strain) versus the logarithm of pressure (Fig. 9-4). This method gives essentially the same results as the method of Fig. 9-3, but it has the advantage of minimizing the amount of work associated with reducing the test data. It may also make for simpler settlement calculations. With this method, no conversion to void ratio is necessary. The compression data for the sample are converted to strain by dividing total compression by the original sample thickness, and strain is then plotted against the logarithm of pressure.

If, during the compression test, the pressures on a sample are increased to a certain magnitude, unloaded to a lesser value, and then re-

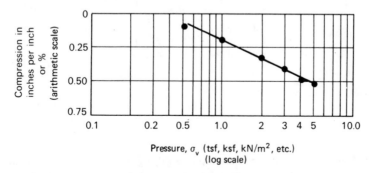

Figure 9-4. *Method of indicating soil compression as strain instead of void ratio.*

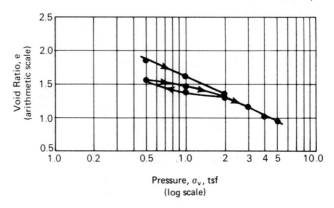

Figure 9-5.   *Results of loading-unloading-reloading cycle applied to a soil.*

loaded and increased to magnitudes greater than previously, results
like those shown in Fig. 9-5 are obtained.   Note that the soil does not
expand to its original volume when pressure is removed.   Some of the
volume change due to external loading is permanent.   Soil, therefore, is
not an elastic material.   Upon reloading, the resulting slope of the compression curve is less steep than the original slope.   These factors of
soil behavior have significant effect on the settlement of structures.

Consider a soil sample obtained from a site where conditions
are as shown in Fig. 9-6(a).   The ground surface overlying the sample
has never been above the existing surface, and there never was extra
external loading acting on the area.   For this condition, the maximum
vertical pressure ever imposed on the sample being considered is the
current weight of overlying soil, σ_{v_0}.   The result of a compression test

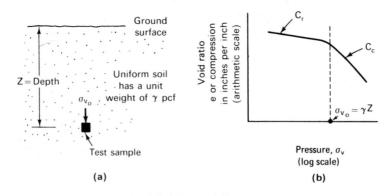

Fig. 9-6.   *Description of conditions applying to compression test sample:
(a) location of soil sample obtained for compression test; (b) result of compression test.*

performed on the sample is as shown in Fig. 9-6(b). For laboratory loading less than σ_{v_0}, the slope of the compression curve is less than it is for loads greater than σ_{v_0} since, to the soil, this represents a reloading. The slope of the curve at loading greater than σ_{v_0} is termed the *virgin compression curve*, and the slope of the curve is the *compression index* C_c. The slope of the curve at values less than σ_{v_0} is the recompression slope C_r. Mathematically, the slope of the curve, for either C_c or C_r, is $\Delta e / \Delta \log \sigma_v$.

From examination of Figs. 9-5 and 9-6, it should be concluded that a change in the slope of the compression curve results when the previous maximum pressure ever imposed onto the soil is exceeded. If the ground surface had at some time in past history been above the existing surface and eroded away, or if the weight of a glacier had been imposed on the area at some time in history, σ_{v_0} (the existing overburden) is not the maximum pressure that has been imposed on the soil sample. The greatest pressure that previously existed on the soil is $\sigma_{v_{max}}$, which would be the total pressure that developed from the existing soil overburden *and* the weight of the eroded soil or glacier. Compression test results for this occurrence would be as shown in Fig. 9-7.

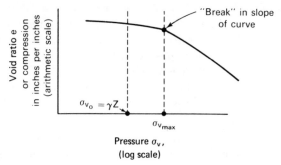

Figure 9-7. Compression test results where past pressure on soil has exceeded existing overburden pressure.

The effect of this soil property on the settlement of structures is shown by a comparison of conditions from Figs. 9-6(b) and 9-7. Assume that the weight of the structure causes an additional stress of $\Delta \sigma_v$ to act on the sample being studied. The total soil compression that will occur within the soil deposit, and therefore the settlement that the structure experiences, is related to the compression occurring in the test sample. Comparative results are shown by Fig. 9-8(a) and (b). It is seen that, for the same magnitude of structural loading, soil conditions of Fig. 9-8(a) cause more compression and, therefore, more building settlement than the conditions of Fig. 9-8(b). This indicates that the stress history

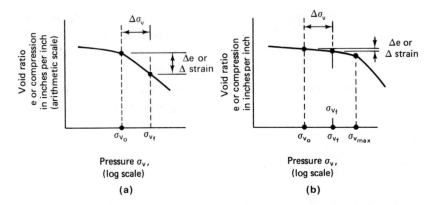

Figure 9-8. *Comparison of compression that occurs for (a) normally consolidated and (b) overconsolidated soils at loading above the overburden pressure.*

of a soil may be more significant than other soil properties insofar as settlement of structures is concerned.

Soil deposits whose condition is represented by Figs. 9-6 (b) and 9-8 (a) are termed *normally loaded* or *normally consolidated,* meaning that the present overburden pressure is the greatest pressure that has been imposed on the soil. Soil whose condition is represented by Figs. 9-7 and 9-8 (b) are termed *precompressed, preconsolidated,* or *overconsolidated,* meaning that at some time in past history there were imposed pressures greater than those that currently exist.

If the virgin curve of a laboratory compression test begins at a loading less than σ_{v_0}, the soil is termed *underconsolidated,* which means that the soil has not fully adjusted or stabilized under the current overburden pressures. (This would represent a "new" soil deposit.)

For reliable results, undisturbed soil samples are required for the compression test. If the soil sample to be tested is disturbed because of the soil boring techniques used to obtain the sample or during transportation and handling, the test data may be in error. Typical test results occurring from disturbed samples are shown in Fig. 9-9.

ESTIMATING COMPRESSIBILITY FROM INDEX PROPERTIES / Laboratory compression tests on fine-grained soils normally require days to complete (about two weeks in the period conventionally used). It is frequently necessary to obtain information about the compressibility of a soil in as short a time as possible. Or in the interest of job economy it may be desirable to limit the number of compression tests but still to evaluate the compressibility characteristics of many soil-boring samples. For these situations, it is possible to use correlations that have been established between compression properties

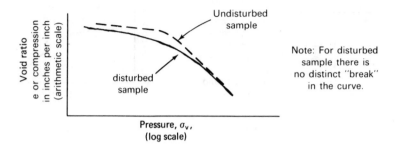

Figure 9-9. *Typical compression test curve shapes for disturbed and undisturbed test samples.*

and some more easily or more quickly determined properties of soils. These correlations, which permit *estimates* of the compressibility of silts and clays to be made, are

$$C_c = 0.54 (e_0 - 0.35) \qquad (9\text{-}1a)$$

where e_0 is the in-place void ratio,

$$C_c = 0.0054 (2.6w - 35) \qquad (9\text{-}1b)$$

where w is the in-place water content,

$$C_c = 0.009 (LL - 10) \qquad (9\text{-}1c)$$

where LL is the liquid limit.

Equation (9-1c) should be applied only for normally consolidated clays. In a normally consolidated clay, the natural water content is about at the liquid limit. If a clay has its natural water content significantly less than the liquid limit, it is preconsolidated.

The shear strength of a clay soil can also be used to indicate if the soil is normally consolidated or overconsolidated. The magnitude of the maximum pressure imposed on a soil is from three to four times its shear strength. By determining the shear strength and then $\sigma_{v_{max}}$, the magnitude of $\sigma_{v_{max}}$ can be compared with the overburden pressure that acted on the sample. Close agreement would indicate a normally consolidated soil.

SETTLEMENT DUE TO SOIL COMPRESSION / Assume that a building is to be constructed at a location where soil conditions are as indicated by the sketch of Fig. 9-10(a). The results of a laboratory compression test performed on a sample obtained from the center of the clay stratum is shown in Fig. 9-10(b). The value of σ_{v_0}, the overburden pressure, is equal to $y\gamma_{sand} + (H/2)\gamma_{clay}$ where γ_{sand}

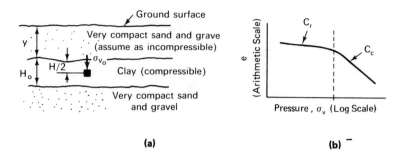

(a)   **(b)**

Figure 9-10. *Description of conditions applying to compression test sample: (a) subsurface profile indicating source of compression test sample, and (b) compression test results.*

and γ_{clay} are the unit weights (in pcf, kg/m³ or kN/m³) of the sand and clay soils, respectively.

The foundation load imposed by the new structure results in an additional pressure, $\Delta\sigma_v$, acting on the clay. Figure 9-11 uses the block diagram method to illustrate changes that the clay sample experiences under the increased loading. The change in volume is equal to the change in void ratio. The relation of the change to the original volume is

$$\frac{\Delta V}{V_T} \quad \text{or} \quad \frac{\Delta e}{1 + e_0}$$

If the clay sample represents the *average* volume change that occurs throughout the clay stratum of Fig. 9-10(a), the following proportion will apply:

$$\frac{\Delta H}{H_0} = \frac{\Delta e}{1 + e_0}$$

H_0 is the original thickness of the clay layer and ΔH is the compression

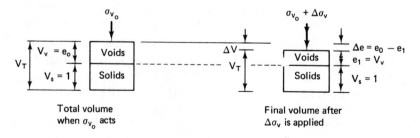

Figure 9-11. Block diagram illustrating change in soil volume that occurs with increase in loading.

that this layer will experience. The settlement of foundations supported above the clay layer will be equal to the compression that occurs, or

$$\text{Settlement} = \Delta H = H_o \left(\frac{\Delta e}{1 + e_0} \right) \qquad (9\text{-}2)$$

As previously noted, the slope of the curve from the compression test for loadings greater than the overburden pressure is C_c, where

$$C_c = \frac{\Delta e}{\Delta \log \sigma_v}$$

Rearranging gives

$$\Delta e = C_c [\Delta \log \sigma_v]$$

Substituting in the equation for settlement gives

$$\Delta H = \frac{H_o}{1 + e_0} C_c (\Delta \log \sigma_v) \qquad (9\text{-}3a)$$

or

$$\Delta H = \frac{H_o}{1 + e_0} C_c (\log \sigma_{v_f} - \log \sigma_{v_0}) \qquad (9\text{-}3b)$$

In this equation, σ_{v_0} is the overburden soil pressure, and σ_{v_f} is the sum of the overburden soil pressure σ_{v_0} and the pressure caused by the weight of the structure $\Delta \sigma_v$.

If the compression test data are presented in the form shown by Fig. 9-4, the settlement equation is

$$\Delta H = H_o \times \text{(change in percentage of compression for the load change } \Delta \sigma_v)$$

or

$$\Delta H = H_o C_c' (\log \sigma_{v_f} - \log \sigma_{v_0}) \qquad (9\text{-}4)$$

where C_c' is the slope of the compression vs. $\log \sigma_v$ plot (ref. Fig. 9-4).

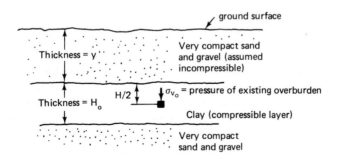

Figure 9-12. *Soil conditions for problem of settlement due to compression in a buried clay layer.*

Illustration 9-1:   Referring to the conditions shown by Fig. 9-12, assume that y is twelve ft, H_o is eight ft, and γ_{sand} and γ_{clay} are 135 pcf and 110 pcf, respectively. For the clay, e_0 is 1.20 and the compression index C_c is 0.20 (both dimensionless values).   The weight of the structure causes a stress of 600 psf at mid-height of the clay layer. (Therefore, $\Delta\sigma_v = 600$ psf.) [*Note:* Values of stress increase, $\Delta\sigma_v$, are computed by using methods described in Chapter 8.]

The settlement would be

$$\Delta H = \frac{H_o}{1 + e_0} \, C_c (\log \sigma_{v_f} - \log \sigma_{v_0})$$

where $\sigma_{v_0} = (12 \text{ ft} \times 135 \text{ pcf}) + (4 \text{ ft} \times 110 \text{ pcf})$

$$= 2060 \text{ psf}$$

and $\sigma_{v_f} = \sigma_{v_0} + \Delta\sigma_v = 2060 + 600$

$$= 2660 \text{ psf}$$

$$\Delta H = \frac{8 \text{ ft}}{1 + 1.20} \, (0.20)(\log 2660 - \log 2060)$$

$$= \frac{(8)(0.20)}{2.20} \, [3.426 - 3.315] = 0.0806 \text{ ft} = 1 \text{ in.} (\pm)$$

If the groundwater table is at the soil surface, the soil overburden pressure σ_{v_0}, is due to the submerged, or effective, weight of the soil. If the unit weights of 135 pcf and 110 pcf represent saturated unit weights, the submerged unit weights will be

For the sand:

$$\gamma_{sub} = \gamma_{sat} - \gamma_w = 135 \text{ pcf} - 62.4 \text{ pcf} = 72.6 \text{ pcf}$$

For the clay:

$$\gamma_{sub} = \gamma_{sat} - \gamma_w = 110 \text{ pcf} - 62.4 \text{ pcf} = 47.6 \text{ pcf}$$

If the soils are not fully saturated, it is generally sufficiently accurate to assume that a submerged effective soil weight is about half its weight when not submerged. Using the values calculated, the overburden pressure σ_{v_0} is then

$$\sigma_{v_0} = (72.6 \text{ pcf} \times 12 \text{ ft}) + (47.6 \text{ pcf} \times 4 \text{ ft}) = 1062 \text{ psf}$$

and $\sigma_{v_f} = \sigma_{v_0} + \Delta\sigma_v = 1062 + 600 = 1662 \text{ psf}$

(Note that $\Delta\sigma_v$ is not affected by the water table or submergence.)

$$\Delta H = \frac{H_o}{1 + e_0} C_c[\log\sigma_{v_f} - \log\sigma_{v_0}]$$

$$= \frac{(8\text{ ft})(0.20)}{2.20}[3.222 - 3.028] = 0.14\text{ ft} = 1.7\text{ in.}(\pm)$$

Thus, the effect of a high water table is to cause more settlement.

If the soil is an overconsolidated soil, the equation for settlement calculations is modified to:

$$\Delta H = \frac{H_o}{1 + e_0} C_r(\log\sigma_{v_f} - \log\sigma_{v_0}) \qquad (9\text{-}5)$$

provided that $\sigma_{v_{\max}} > \sigma_{v_f}$. (Note that $\sigma_{v_{\max}}$ is the pressure where the slope of the compression test plot changes from C_r to C_c.[1] For an over-consolidated soil where σ_{v_f} is greater than $\sigma_{v_{\max}}$, the settlement equation should be used as:

$$\Delta H = \frac{H_o}{1 + e_0} C_r(\log\sigma_{v_{\max}} - \log\sigma_{v_0})$$

$$+ \frac{H_o}{1 + e_0} C_c(\log\sigma_{v_f} - \log\sigma_{v_{\max}})$$

or

$$\Delta H = \frac{H_o}{1 + e_0}\left[C_r(\log\sigma_{v_{\max}} - \log\sigma_{v_0}) + C_c(\log\sigma_{v_f} - \log\sigma_{v_{\max}})\right] \quad (9\text{-}6)$$

Illustration 9-2:   Assume that a buried stratum of clay six feet thick will be subjected to a stress increase of 700 psf at the center of the layer. The magnitude of the preconstruction soil overburden pressure is 1000 psf at the center of the layer. A laboratory compression test indicates that the clay is overconsolidated, with $\sigma_{v_{\max}}$ equal to 1500 psf. The value of C_c is 0.30, and the value of C_r is 0.05. What change in thickness results in the clay layer due to the stated conditions?

---

[1]To obtain a value of C_r from the laboratory compression test for use in this equation, the recommended practice is as follows: Impose loading on the sample in increments to above the precompression stress, unload or rebound to at least the existing overburden; then reload the sample, and use the last recompression slope for determining C_r.

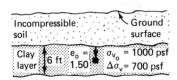

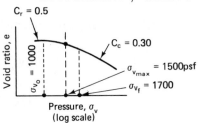

The change in thickness of the clay layer, from Eq. 9-6 is

$$\Delta H = \frac{H_o}{1 + e_0} [C_r(\log \sigma_{v_{max}} - \log \sigma_{v_0}) + C_c(\log \sigma_{v_f} - \log \sigma_{v_{max}})]$$

$$= \frac{72 \text{ inches}}{2.50} [(0.05)(\log 1500 - \log 1000)$$

$$+ (0.30)(\log 1700 - \log 1500)]$$

$$= 0.73 \text{ in.}$$

If the thickness of the compressible soil layer is great, it should be broken into thinner layers for purposes of making the settlement calculations. The increment of settlement for each layer should be handled by using the methods discussed. The value of σ_{v_0} is the overburden pressure at mid-height of each layer, and $\Delta \sigma_v$ is the increase in pressure from the weight of the structure, also acting at the midpoint of each layer. The total settlement will be the sum of increments occurring in all the layers. The depth of soil that is considered to be affected by foundation loading is limited to where $\Delta \sigma_v$ is approximately one-tenth of the soil overburden pressure.

Illustration 9-3:   A settlement calculation is to be made for a structure that is planned for a site that is underlain by a thick deposit of normally consolidated clay soil. Soil properties are as indicated on the sketch. What building settlement is expected?

For this condition, the settlement calculation should be performed by dividing the thick clay deposit into sublayers, and calculating the compression that occurs in each sublayer. The overburden pressure is calculated for the center of each layer, and the stress increase due to the structure is also calculated for the center of each layer. Results of these computations are shown on the sketch. The related settlement calculations follow.

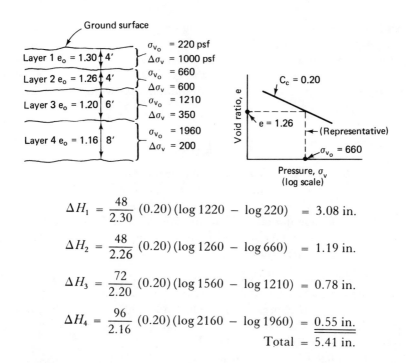

$$\Delta H_1 = \frac{48}{2.30}\,(0.20)\,(\log 1220 - \log 220) = 3.08\ \text{in.}$$

$$\Delta H_2 = \frac{48}{2.26}\,(0.20)\,(\log 1260 - \log 660) = 1.19\ \text{in.}$$

$$\Delta H_3 = \frac{72}{2.20}\,(0.20)\,(\log 1560 - \log 1210) = 0.78\ \text{in.}$$

$$\Delta H_4 = \frac{96}{2.16}\,(0.20)\,(\log 2160 - \log 1960) = \underline{\underline{0.55\ \text{in.}}}$$

$$\text{Total} = 5.41\ \text{in.}$$

Volume Changes in Sand

In the illustrated problem of Fig. 9-12, it was assumed for simplicity that the cohesionless soil layer overlying the clay was compact enough to be considered incompressible (i.e., compression of the sand layer would be negligible). This is not always a true condition. As stated earlier, the settlement of structures supported on cohesionless soil generally occurs immediately upon application of load. This settlement is primarily the result of volume changes caused by lateral yielding or shear strain that occurs in the soil. The exception is where the foundation or loaded area is large, in which case soil compression becomes more significant and should not be ignored.

In many practical situations, building loads developed gradually during the construction period will produce settlement whose effects can be absorbed by the structure without causing distress. However, settlement caused by postconstruction conditions such as live loads could result in distressing occurrences such as wall cracks, uneven floors, and ruptured buried utility lines.

In uniform cohesionless soil deposits, the soil rigidity (or modulus of elasticity) increases with depth because the overburden pressure and confining pressure, which affect the rigidity, increase with depth. For the condition where the increase in rigidity is uniform with depth, the following expression can be used to estimate settlements, provided that the *depth* of the foundation below the ground surface is less than the width of the foundation, and that the foundation width is not greater than 20 feet:

$$\Delta H = \frac{4 q B^2}{K_v (B + 1)^2} \tag{9-7}$$

where B = width of the foundation in feet (or meters).

         q = pressure imposed by the foundation in kips/ft² (or kN/m²).

         K_v = modulus of vertical subgrade reaction for one-foot-square (0.3-meter-square) plate bearing on the ground surface, kips/ft³ (or kN/m³).

The modulus of vertical subgrade reaction K_v can be determined directly from a plate bearing test performed in the field at the planned location for the structure. In this test, loads are applied to the plate in increments, and the plate settlement is recorded for each load. The results are plotted as shown in Fig. 9-13.

The tangent to the initial straight line portion of the plot is K_v, the modulus of vertical subgrade reaction. Mathematically,

$$K_v = \frac{\text{pressure}}{\text{settlement}} = \frac{\text{kips/ft}^2}{\text{ft}} = \frac{\text{kips}}{\text{ft}^3}$$

Field plate-bearing tests are time-consuming and costly. For purposes of estimating settlements, representative values of K_v that have been correlated with other soil properties can be used; see Table 9-1.

If the subsurface conditions consist of a cohesionless soil stratum

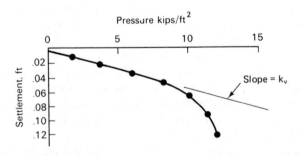

Figure 9-13. *Plot of load-settlement data from a plate-bearing test, to determine modulus of vertical subgrade reaction.*

TABLE 9-1 / TYPICAL VALUES* OF K_v
FOR DIFFERING SAND DENSITIES

Condition	Relative Density, %	Representative Values of Dry Unit Weight		Values of K_v	
		PCF	kN/m^3	$kips/ft^3$	$\dfrac{kN/m^3}{\times\ 10^3}$
Loose	less than 35	below 90	below 14	100	15
Medium dense	35 to 65	90–110	14–17	150–300	25–50
Dense	65 to 85	110–130	17–20	350–550	55–85
Very dense	above 85	above 130	above 20	600–700	95–110

*For the condition where the groundwater table is at a depth greater than 1.5B. If the water table is at the base of the foundation, use $\frac{1}{2}\ K_v$. Interpolate for intermediate locations of the water table.

underlain by a cohesive layer, it may be necessary to calculate the settlement of the granular layer and the clay layer separately, and then sum up the separate values to obtain the total expected settlement.

Illustration 9-4: A footing is to support a total column loading of 280,000 lb. Half of this will be live load. The building site is underlain by a thick layer of sand. Laboratory tests on samples from soil borings indicate that the dry density of the sand varies between 115 and 120 pcf. The design calls for the footing to be located 2 ft below the ground floor level of the building, and for an allowable soil-bearing pressure of 8000 psf to be used. Estimate the foundation settlement.

The footing area required is 280,000 lb/(8000 lb/ft²) = 35 ft². If a square footing is used, the width $B = \sqrt{35} = 5.9$ ft $\longrightarrow$ use 6 ft. For soil densities of 115–120 pcf, estimate a K_v value as 400 kips/ft³ from Table 9-1.

Estimated settlement:

$$\Delta H = \frac{4\,q\,B^2}{K_v(B+1)^2} = \frac{(4)\,(8\ \text{kips/ft}^2)\,(6\ \text{ft} \times 6\ \text{ft})}{(400\ \text{kips/ft}^3)\,(6+1)\,(6+1)\,\text{ft}^2}$$
$$\simeq 0.06\ \text{ft} = 0.75\ \text{in.}\,(\pm)$$

The postconstruction settlement (settlement due to live loading only) would be in the ratio of the live load to the total load. Since half of the total load is live load, the postconstruction settlement will be about $\frac{1}{2}(0.8\ \text{in.}) = 0.4$ in., or less than $\frac{1}{2}$ in.

Settlement Resulting from Earth Fill

It has been indicated that volume changes in subsurface soils should be expected as a result of any new loading imposed on the soil. Filling of construction sites with compacted earth so as to raise the grade of a low area or make an uneven site level is now a relatively common undertaking in the construction industry. Areal settlements due to fills should always be considered. And where structures are to be supported on the compacted earth fill, it should be recognized that the total settlement for the structure could well be the sum of the volume changes occurring in the natural soil from the weight of the fill plus the volume changes in the fill and natural soil caused by the weight of the structure itself. Only if the completed fill is left in place a sufficient time so that full compression of the natural soils occurs before the structure is built can the effect of settlement due to fill be neglected. Settlement of granular soils will still, of course, occur almost instantaneously. However, compression of fine-grained soil does usually require a time period.

Areal settlement due to the placement of fill can be computed by using methods already presented. However, the increase in stress $\Delta\sigma_v$ caused by the weight of the fill is assumed to remain constant throughout the full depth of the natural or original soil. This is in contrast to the condition that occurs beneath a foundation area, where the value of $\Delta\sigma_v$ decreases with increasing depth below the foundation.

The settlement of fill can be assumed to result from *compression* that occurs in the original soils underlying the fill. For cohesive soils, the necessary value of C_c or C_r is determined or estimated from laboratory tests. For granular soils, the expression for soil compression is also used. Appropriate values of C_c can be estimated from index properties, or from the following:

For loose uniform sands (D_R from 25 to 40 percent)

$$C_c = 0.05 \text{ to } 0.06 \qquad (9\text{-}8a)$$

For dense or compact sands (D_R from 60 to 80 percent),

$$C_c = 0.02 \text{ to } 0.03 \qquad (9\text{-}8b)$$

Consolidation

Compression of clay soil occurs gradually when new loading is applied. The causes for this are related to the low permeability of these fine-grained soils and the condition of their usual occurrence in nature. Most clays were deposited under water. All void spaces were occupied with water. Even with the passage of time, these sedimentary materials remained fully or almost fully saturated. In soil, volume decreases are due to decreases in the void spaces between the soil particles as the particles rearrange in order to support the additional pressure caused by external loadings. For compression to occur in a saturated soil, water in the voids must first be expelled in order for the decrease in void spaces to occur. This behavior may be pictured as similar to that of a saturated sponge having the water squeezed out of it by a press.

When external load is applied to a saturated clay, the water in the voids, or pores, becomes subject to an immediate and equal increase of pressure. This is the same principle that is in effect for any hydraulic loading system, where an external force results in pressure developing in the fluid throughout the system. The trapped water starts to flow towards areas of lesser pressure, away from the area where loading is occurring. But because of the low permeability of fine-grained soil, the movement or escape of this pore water takes place slowly. Only as water escapes can soil particle rearrangement and decrease in void spaces take place. The result is that, for saturated clay soils, compression is a gradual occurrence. As water escapes, the pressure caused by the external loading is transferred from this water to the soil. The process of load transfer to the soil as pore water escapes is the *consolidation* process.

The rate of consolidation for a stratum of clay soil is affected by several factors, including:

1. The permeability of the soil.
2. The extent or thickness of the compressible soil, and the distance that pore water in the soil must travel to escape from the zone where pressures due to foundation loading exist.
3. The in-place void ratio of the compressible soil.
4. The ratio of new loading to the original loading.
5. The compression properties of the soil.

The consolidation properties of a soil (not the soil stratum) are dependent on items 1, 3, 4, and 5. The effect of these factors can be grouped together to obtain a property termed the *coefficient of consolidation* c_v, which indicates how rapidly or slowly the process of consolidation takes place. The coefficient of consolidation is

$$c_v = \frac{k(1 + e)}{a_v \gamma_w} \qquad \left(\textit{Note: } \text{Units for } c_v \text{ are } \frac{\text{length}^2}{\text{time}} \right)$$

where
e = void ratio (dimensionless value).
k = coefficient of permeability (cm/sec, ft/min, ft/day, etc.).
a_v = coefficient of compressibility (dimensionless).
γ_w = unit weight of water (gm/cm^3 or lb/ft^3).

   Values for k, e, and a_v can be determined separately and substituted into the equation to determine c_v. However, it is common practice to determine c_v directly from a laboratory consolidation test.[2] The consolidation test is a compression test, but with the added provision that time readings (amount of compression versus time, for each applied loading) also are obtained.

   As previously indicated, the consolidation process requires two interrelated occurrences. One, the soil particles will be adjusting to a denser condition so that the soil mass can become strong enough to support the imposed loading. Simultaneously, pore water in the void spaces is escaping so that the required decrease in void ratio just referred to can occur. Referring to Fig. 9-14, note that initially the stress increase

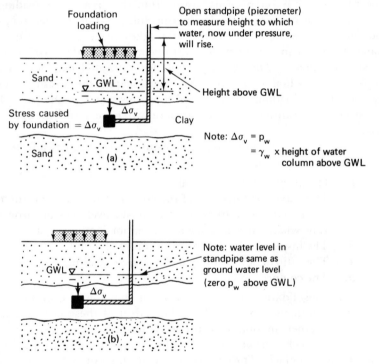

Figure 9-14. *Change in pore water pressure occurring as clay soil consolidates: (a) hydrostatic pressure in clay layer when foundation loading is first applied; (b) hydrostatic pressure in clay layer when 100 percent consolidation has occurred.*

---

[2] The procedure for performing the laboratory consolidation test, including the determination of c_v values, is described in **Appendix 1**.

due to structural loading $\Delta\sigma_v$ is imposed fully on the trapped pore water. This means that, at first, none of the new loading is imposed to the soil. Only after water is squeezed from the voids does the loading start getting transferred to the soil.

The end of the consolidation process has been reached when the water pressure in the voids drops to normal and no more water is being squeezed out. Full consolidation (100 percent), or a condition of equilibrium, is assumed to have occurred when the full loading of the structure is carried by the soil particles (no excess pressure in the pore water) and the total expected compression has taken place. Values of consolidation less than 100 percent indicate that, for the applied loading, compression (and settlement) is still occurring. Strictly speaking, however, compression resulting after trapped pore water under pressure has escaped is primary consolidation. Percentage of primary consolidation is expressed by the term U percent. Some additional long-term compression from what is termed secondary consolidation also takes place as soil particles continue to adjust under the applied loading. However, for most inorganic soil deposits the effects of secondary consolidation are small and can be neglected.

For given field conditions, where the distance that trapped pore water must travel to escape from the compressible layer (H_{dr} in Fig. 9-15) and c_v are known, the time period t required for a given percentage of consolidation to occur is

$$t = \frac{T_v H_{dr}^2}{c_v} \tag{9-10}$$

where T_v = time factor for consolidation due to vertical drainage.
  H_{dr} = drainage distance for escaping pore water.
  c_v = coefficient of consolidation.

The value T_v is a time factor that is a constant for a given percentage of consolidation and for a given type of subsurface condition. T_v is dimensionless. For the conditions indicated in Fig. 9-15, referred to as "double drainage" because pore water can move downward and

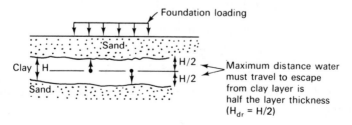

Figure 9-15.   "Double drainage" conditions in consolidation theory.

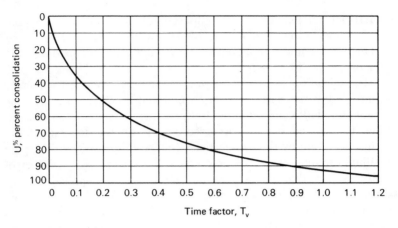

Figure 9-16. *Variation of time factor T$_v$ with percentage of consolidation U.*

upward to escape, the values of T_v for a given percentage of consolidation (U percent) can be determined from the curve of Fig. 9-16.

> **Illustration 9-5:**   Referring to Illustration 9-1 and Fig. 9-12, assume that the laboratory consolidation test data indicate that, for the range of loading applied to the soil, c_v is 0.2 ft^2 per month.
>
> (a) How long will it take for half of the estimated settlement to occur?
>
> Half of the estimated settlement is a U of 50 percent.   Referring to Fig. 9-16, for U equals 50 percent, obtain the value $T_v = 0.20$.
>
> $$t = \frac{T_v H_{dr}^2}{c_v} = \frac{(0.20)(8/2 \text{ ft})^2}{0.2 \text{ ft}^2/\text{month}} = 16 \text{ months}$$
>
> [Note that for this type of problem, H_{dr} is one-half of the clay layer.]
>
> (b) How much settlement will occur in one year?
>
> $$t = 1 \text{ year} = 12 \text{ months}$$
>
> $$T_v = \frac{tc_v}{H_{dr}^2} = \frac{(12 \text{ months})(0.2 \text{ ft}^2/\text{month})}{(8/2 \text{ ft})^2} = 0.15$$
>
> From Fig. 9-16, for $T_v = 0.15$ we get $U = 43$ percent, and therefore
>
> $$\text{Settlement } \Delta H = 43 \text{ percent of 1 inch} = 0.43 \text{ in.}$$

Surcharging

The situation may exist where, due to compressible or weak sub-soils, a proposed structure would undergo settlements that would exceed tolerable limits. Alternatively, stresses resulting in the weak soil from the weight of the structure could cause the weak soil to fail. If the site is to be used, the problem soil somehow must be circumvented (remove if possible, or use pile foundations penetrating to below the poor soil) or improved so that the strength is adequate and the compressibility reduced. For construction projects extending over large areas, such as highways, improvement of poor subsoils is always given consideration because of possible cost savings.

The improvement of poor soils may possibly be accomplished by utilizing a surcharge program. Surcharging of a soil, in its simplest intent, is merely imposing an external loading for a long enough duration to cause desirable changes in the soil before a structure is erected and supported on the soil. •

Under the weight of a surcharge load, a poor soil will compress and increase in strength. The surcharge is made to the desired weight or loading and continued for a sufficient period to achieve results satisfactory for the need of the planned structure. Generally, this might include having the surcharge cause the magnitude of soil compression (settlement) that the planned structure would have caused, so that after the surcharge is removed and the structure is built, its settlement will be negligible, or, having the surcharge cause a sufficient increase in the strength of the poor soil so that it can safely support the weight of the planned structure.

Any material that will impose pressure onto the subsoils that require improvement can be used for the surcharge loading. Most conventionally, the material used is soil (borrow or fill). Soil is utilized because of its general availability and low cost, its ease of handling, and the absence of problems about deterioration.

Surcharging for improvement is most commonly applied for cohesive and organic soils—soil types that require a time period for compression (consolidation) and strength gain to occur. Surcharging programs *are* applicable for improving loose, granular soil deposits, but other methods, frequently quicker and more economical, can be utilized for these soils (methods that do not have application with cohesive soils).

Surcharges for improving the strength of weak soils have to be applied slowly or in increments so that the weak soils are not overstressed (failed) *before* they have time to improve. With cohesive soils, the compression-consolidation process involves a readjustment of soil particles as water is squeezed from the voids of the soil. A gain in strength occurs because of the consolidation, but the strength gain re- sults only as rapidly as water is squeezed from the soil (an implication from the strength curve shown in Fig. 9-17). Consequently, a properly performed surcharge program requires field monitoring to establish a safe rate for increasing the surcharge and to determine how long it should remain in place. Monitoring normally includes instruments such as piezometers to measure excess pore water pressures developed within the weak soil, to learn of the rate at which consolidation is occurring, and to indicate when additional increments of the surcharge can be safely added. Settlement plates located in or on the surface of the original soil are used to monitor the rate and amount of settlement caused by a surcharge. For accurate records, any reference bench mark to measure settlement must be located outside of the area that is settling.

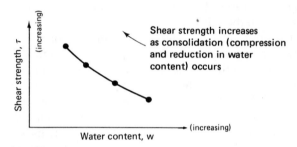

Figure 9-17. *Change in soil strength and water content as consolidation under surcharge occurs.*

In some cases, the structure itself can be used to apply a surcharge load in increments. An example of this is with ground level storage tanks, such as those used to hold petroleum or other fluids. For many situations, the tolerable settlement of storage tanks is relatively large, provided that large *differential* settlements that might cause the tank to rupture do not occur. A fluid, possibly the material to be stored but more frequently water, is used to partially fill the tank. The tank is left partially filled until the subsoil achieves the necessary increase in strength. Then additional liquid is pumped in, and the next waiting period is allowed to elapse. This procedure continues until the subsoils supporting the tank reach the desired degree of stability. The tank can then be put into permanent use. Permanent piping connections should not be made until all, or most of, the settlement has occurred.

With this method, a laboratory test program should be performed before field operations are begun to ensure that the weak subsurface soils will be able to reach the strength required to support the tank and to obtain an estimate of the time required to complete the consolidation process. A laboratory program involves consolidating samples of the weak subsoil under the same vertical pressures that the tank will impose onto the soil, and then performing shear tests to determine if adequate shear strength (Chapter 10) is being obtained.

ACCELERATION OF THE SOIL IMPROVEMENT PROCESS / Utilization of a surcharge program requires that the surcharge be applied and left in place for a sufficient time to obtain the desired compression or strength gain. Depending on the properties and thickness of the poor soils, a surcharge may remain in place for months or longer before it accomplishes its purpose and can be removed. Construction cannot begin until the surcharge is removed, of course. If time is an important factor, means of accelerating the consolidation period can be undertaken.

One method to reduce the time for a desired amount of compression (or settlement) to result is to use an excessive surcharge, provided that the subsurface soils have sufficient strength. A given amount of settlement will occur more quickly under a heavy surcharge than under a light surcharge.

Illustration 9-6: Assume that an area is underlain by a stratum of compressible soil 16 ft thick. The weight of the proposed structure, say an earth embankment for a new highway, will result in an increase of 600 psf pressure to the compressible soil. The settlement under this loading would be 12 in. The compressible layer is doubly drained, and the soil has a c_v value of 3 ft² per month. Calculations indicate that it would take approximately 22 months for the settlement to occur under the weight of the proposed embankment. How long would be required for a surcharge imposing a pressure of 1200 psf to cause the 12 in of settlement?

The first step of the solution requires that the total settlement to be expected from the 1200-psf surcharge be computed. Using Eq. (9-3), we find that the settlement would be 20 in. To obtain but 12 in. of settlement, 60 percent consolidation (12 in./20 in. × 100 percent) has to occur. The T_v value for 60 percent consolidation is 0.3, and the time required is

$$t = \frac{T_v H_{dr}^2}{c_v} = \frac{(0.3)(8 \text{ ft} \times 8 \text{ ft})}{3.0 \text{ ft}^2/\text{month}} \cong 6\tfrac{1}{2} \text{ months}$$

SAND DRAINS OR DRAIN WELLS / When a surcharge is used, another method for accelerating the consolidation process is to shorten the length of the drainage path for the pore water escaping from the consolidating soil. This can be accomplished by providing vertical drains or wells at spacings closer than the drainage distance for vertical flow. Frequently, too, the horizontal coefficient of permeability for cohesive soil deposits is many times greater than the coefficient of permeability for the vertical direction, which further increases the speed of consolidation. The vertical drain wells can be holes that are drilled into the compressible soil and filled with sand (frequently called *sand drains*). Since the sand is very permeable material, pore water under pressure because of the surcharge will flow to the drain wells and be forced upward (or downward, too, if the well extends to a deeper permeable stratum underlying the compressible soil). After flowing upward to the surface, the water is carried away from the surcharge area; see Fig. 9-18(a). If the drain wells are located as indicated in Fig. 9-18(b), the zone of drainage can be assumed to be circular.

The expression for the time period necessary to achieve consolidation when this condition exists is

$$t = \frac{T_h(\gamma_w a_v)}{k_h(1 + e_0)} = \frac{T_h d_d^2}{c_h} \qquad (9\text{-}11)$$

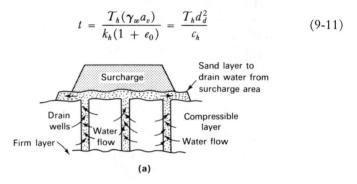

(a)

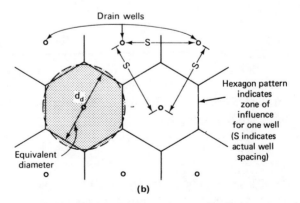

(b)

Figure 9-18. *Cross section and plan of typical sand-drain installation: (a) cross section of installed drain wells and surcharge; (b) drain-well pattern. (Ref. 12).*

where t = elapsed time.

T_h = time factor for horizontal drainage (dimensionless).

c_h = horizontal coefficient of consolidation (*Note:* units are length^2/time).

d_d = diameter of the zone draining to a drain well, ft.

The value of T_h varies according to the ratio of the draining area to the escape (drain-well) area. If d_d is the diameter of the draining area, and d_w is the drain-well diameter, n can be the ratio of d_d to d_w. Curves of T_h, similar to those for T_v used for vertical consolidation, are available for different ratios of n; see Fig. 9-19.

The drain-well spacing affects the time for consolidation more than the drain-well size. Quicker results will be obtained by using a small drain-well spacing. However, there is the practical problem of constructing the wells close together without collapsing or otherwise damaging them. There is also the potential problem that the sand drains may shear along their length if the compressible soils experience large settlement or lateral movement during compression. The result is that water flowing from the consolidating soils to the drain well cannot escape to the surface, thus invalidating the drain-well system. Because of the specialized techniques and equipment required for a sand-drain installation, it is generally not economical to use the method for improving small projects.

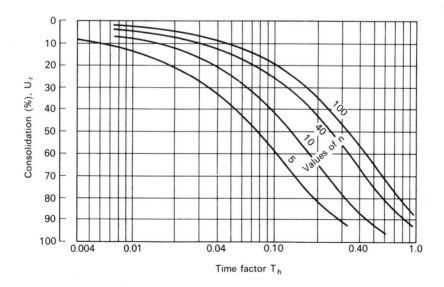

Figure 9-19. *Variation of time factor* T_h *with percentage of consolidation* U *for sand-drain installations. (Ref. 12)*

Illustration 9-7: The settlement problem of Illustration 9-6 examined the effects of using a surcharge to improve the subsurface soils for a planned highway embankment. The surcharge program involves the placement and removal of an amount of soil greater than that required for the embankment, with a related expense. It is desired also to study the effect of utilizing a sand-drain installation, on the possibility that it may be less costly than surcharging (because less surcharge is required) and faster results may be provided.

Laboratory testing of samples indicates that the coefficient of permeability for horizontal flow ranges from five to fifteen times the coefficient for vertical flow; the coefficient of consolidation for horizontal drainage would be similarly effected. For a preliminary analysis, use a value of c_h that is seven times the value of c_v. Since c_v was 3 ft²/month, c_h will become about 20 ft²/month.

It is planned to install sand drains that are 2 ft in diameter and on a center-to-center spacing of 10 ft, as indicated in Fig. 9-18(b).

Settlement calculations indicate that an 800-psf surcharge would cause a settlement of $13\frac{1}{2}$ in. This is 12/13.5, or 90 percent of the settlement expected from the necessary embankment fill.

The ratio of d_d to d_w is $\frac{10}{2}$, or 5. With reference to Fig. 9-19, the value of T_h for 90 percent consolidation and n equal to 5 gives a time factor of 0.28. Substitution of appropriate values into Eq. (9-11) gives

$$t = \frac{T_h(d_d)^2}{c_h} = \frac{(0.28)(10 \text{ ft} \times 10 \text{ ft})}{20 \text{ ft}^2/\text{month}} \cong 1\frac{1}{2} \text{ months}$$

Whether or not sand drains are more economical than a surcharge program depends on the relative cost of the sand-drain installation and price of embankment fill. Factors to make a sand-drain installation more attractive would include the possibility for the surcharge to be adequately compacted during placement so that it can be left in place as the structural embankment when the desired settlement has been reached.

PROBLEMS

1. Provide reasons why field plate-loading tests performed at a planned construction site may give erroneous settlement information if the area is underlain by cohesive soils. (*Hint:* Field loading

tests are frequently performed with bearing plates smaller in area than the foundation footings will be. Also, consider the effects of consolidation.)

2. Indicate advantages related to performing laboratory compression tests on samples obtained from strata underlying a building site, for determining soil compression data.

3. Briefly review the manner in which laboratory compression tests are conventionally performed so as to obtain load-deformation data.

4. For soils undergoing compression, what is one-dimensional volume change?

5. In conventional laboratory compression testing, what is the cause of soil volume decrease?

6. In a laboratory compression test, the void ratio of the test sample changes from 1.55 to 1.36 as the loading increases from 2000 to 4000 psf.

   (a) What is the value of the coefficient of compressibility for these conditions?

   (b) What is the compression index for this loading range?

7. Referring to occurrences observed in a typical field or laboratory compression test, explain why a soil mass does not expand to its original volume when loading is released.

8. Briefly indicate the difference between a normally consolidated and an overconsolidated soil deposit.

9. A cohesive soil sample obtained from a known normally consolidated clay deposit is found to have a liquid limit of 80 percent. Approximately, what would the compression index for this soil be?

10. A seven-foot layer of clay is buried beneath a ten-foot stratum of very compact granular soil. Compact sand underlies the clay. The layer of granular soil is composed of material having a unit weight of 130 pcf. The clay weighs 105 pcf. A laboratory compression test on a sample of the clay indicates a compression index of 0.40 and a natural void ratio of 1.30. A planned building loading will cause a 550-psf stress increase at the middle of the clay layer.

    (a) What amount of compression occurs in the clay layer from the indicated conditions?

    (b) How much compression of the clay layer would result if the groundwater table were at the ground surface (all other conditions remain the same)?

    (c) How much clay layer compression would occur if the clay

were an overconsolidated material, the past maximum pressure were 2000 psf, and the C_r value were 0.10? Assume a deep water table.

11.   A planned construction site is underlain by a thick deposit of normally consolidated clay soil. A building foundation six feet square will be located on the ground surface and carry a total loading of 180,000 lb. Determine the foundation settlement by analyzing the volume changes in layers that are 2, 4, and 6 ft thick, respectively, from the foundation level downward. For simplification, assume a soil unit weight of 115 pcf constant with depth, and an in-place void ratio of 1.05 and a compression index of 0.35 for each layer analyzed.

12.   A plate-bearing test is performed on soil near the surface of a granular stratum in order to determine the modulus of vertical subgrade reaction. The load versus settlement information appears as a straight line on a graphical plot. At a point where the load is 5 tons/ft^2, the indicated settlement is $\frac{1}{2}$ in. What is the value for the modulus of vertical subgrade reaction?

13.   A 5-foot-square footing supported on a sandy stratum is to carry a column loading of 250,000 lb. The unit weight of the sands underlying the footing location vary between 120 and 125 pcf. Estimate what foundation settlement would be expected. (Neglect the weight of the footing.)

14.   Describe the occurrences that take place when a clay soil undergoes consolidation.

15.   What is the physical meaning of the coefficient of consolidation c_v?

16.   A buried clay layer ten feet thick is sandwiched between strata of free-draining granular soil so that double drainage during consolidation can occur. Calculations indicate that, due to a planned building loading, an ultimate settlement of 2 in. is expected as a result of compression in the clay layer. The coefficient of consolidation for the clay is 0.01 ft^2/day.

   (a)   How long will it take for 90 percent of the estimated settlement to take place?
   (b)   How much settlement will occur in one year?
   (c)   How long will it take for one inch of settlement to take place?

17.   A sand-drain installation is proposed for an airport project to accelerate the consolidation of a thick deposit of fine-grained soil underlying the construction area. It is planned to install sand

drains 1.5 ft in diameter at a staggered spacing so that each sand drain will handle a plan area 15 ft in diameter. Laboratory tests indicate the coefficient of consolidation for horizontal drainage of the soil is 0.5 ft²/day. Computations indicate that the surcharge load will cause an eventual settlement of 12 in. How long a period will be required for 90 percent of this settlement to take place?

CHAPTER 10
Shear Strength Theory

The ability of a soil mass to support an imposed loading or for a soil mass to support itself is governed by the shear strength of the soil. As a result, the shearing strength of the soil becomes of primary importance in foundation design, highway and airfield design, slope stability problems, and lateral earth problems that deal with forces exerted on underground walls, retaining walls, bulkheads, and excavation bracing.

In the study of the shear strength of soils, it is common to consider the two major categories of soil types—cohesionless and cohesive—separately. Overall, the factors that can affect the shearing strength of both soil types are the same. Practically, however, the factors that have the *most* influence on the shear strength that is or will be developed by either soil type are different.

The shearing strength and related deformations (or stress-strain relationship) of a foundation or construction soil is conventionally studied in the laboratory by testing soil samples obtained from the construction site and using established testing procedures. Additionally, field test methods have also been developed for determining the shear strength of soil in its natural location, for reasons of expediency and economy, and sometimes necessity if samples for testing cannot be obtained. Typically, field determinations of shear strength are quick procedures so that many soil samples (from different borings or test pits and different depths) can be checked easily and economically. With field tests, strength values only (no deformation data) are generally obtained.

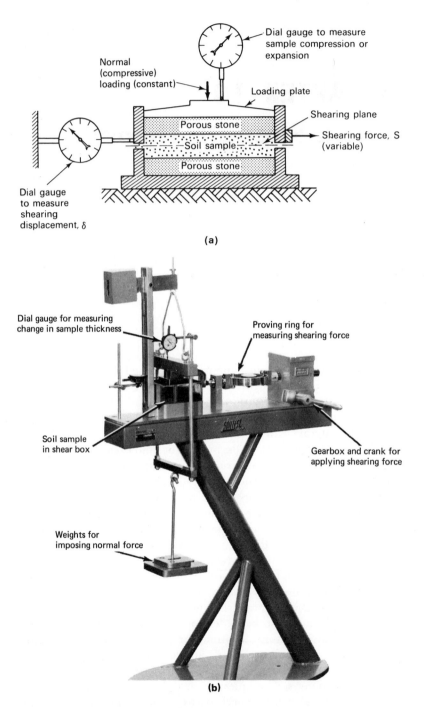

Figure 10-1. (a) Schematic diagram of direct shear apparatus; (b) laboratory direct shear equipment (manually operated). (Courtesy of Soiltest, Inc.)

212

Laboratory Tests

The most widely used laboratory tests for studying the shear strength and related deformations of soils include the direct shear test (single and double shear), the triaxial compression test, and the unconfined compression test. For cohesive soils, vane shear tests can be used in the laboratory and the field to determine the cohesive strength, or cohesion, of soils. (Cohesion is related to shear strength.)

Currently, the most preferred type of strength test is the triaxial test. The unconfined compression test is a type of triaxial test that can be used only for cohesive soils. The use of the direct shear test has decreased since the development of the triaxial test, but it still represents the basic approach for studying the stress-strain characteristics of a soil during shearing and possesses much merit as a learning tool.

DIRECT SHEAR / The direct shear apparatus for performing single shear is essentially a rectangular box having separated lower and upper halves; see Fig. 10-1 (a). After the sample to be tested is placed in the apparatus, a normal loading is applied to compress the soil. The upper half of the apparatus is then moved laterally by a recorded shearing force, forcing the sample to shear across the plane between the two havles of the apparatus. The normal (compressive) force is kept constant during the test. The shearing force starts at zero and increases until the sample fails (is sheared). Usually, a record of the magnitude of the shearing force and the resulting lateral movement is kept so that shearing strength stress versus shearing strain can be computed and plotted graphically. Changes in sample thickness that occur during the shearing process are also recorded so that volume change versus shearing stress or shearing strain can be studied. Typically, results of shearing stress versus shearing strain are as shown in Fig. 10-2. The initial portion of the diagram is curvilinear (constantly changing slope), which

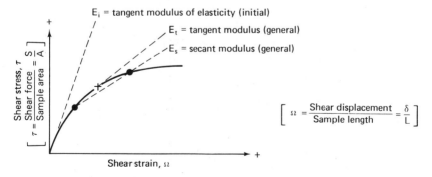

Figure 10-2. *Typical plotted representation of shearing stress versus shearing strain data.*

continues until the maximum shear is reached, after which continuing deformation occurs with no increase in loading. Failure is considered to have occurred at the maximum shearing value.

The slope of the stress-strain curve is the modulus of elasticity E. The modulus of elasticity is an indication of the stiffness or resistance to deformation of the material; the higher the value of E, the stiffer or stronger the material. A *tangent modulus* is the slope of a line drawn tangent to a point on the curve. A *secant modulus* is the slope of a line connecting any two points on the stress-strain curve. By reference to Fig. 10-2, it should be clear that a tangent modulus or a secant modulus will not be constant for all parts of the curve. Both the tangent modulus and secant modulus are used in problem analysis.

In direct shear testing, shearing can be accomplished by either controlling the rate of strain or the rate of stress. For the *strain-controlled* test, the shearing deformation (lateral movement) occurs at a controlled rate, usually continuously and at a constant rate. With this type of test the shearing force necessary to overcome the resistance within the soil is automatically developed.

In the *stress-controlled* test, the magnitude of the shearing force is the controlled variable. The stress is increased at either a uniform rate or in established increments. For each increment of shearing force, it is applied and held constant until the shearing deformation ceases.

The strain-controlled shear test appears to be the most widely used of the two methods, probably because a mechanically operated strain-controlled apparatus is the simplest to devise. *Double shear* testing is similar to the single shear test, except that two parallel surfaces are sheared. This type of test is usually performed on soil samples obtained from test borings that have utilized special tube equipment designed to

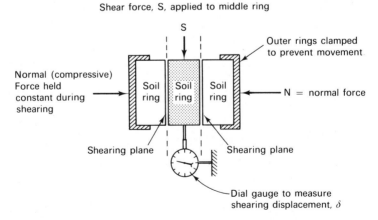

Shear force, S, applied to middle ring

Figure 10-3.   *Schematic diagram of double-ring shear.*

fit directly into a shearing apparatus, so that a minimum of sample handling is required. As an illustration of one method, the soil-sampling apparatus used to extract samples from test borings is such that the soil is obtained in a tube whose interior lining consists of a series of rings. In the laboratory, three rings of soil are taken for the shear test. The shearing takes place as indicated in Fig. 10-3. The stress-strain data from the double shear test are similar to the data from the single shear test.

A variation of the direct shear test is a torsional test, where one section of the tested sample is twisted relative to the other. The torsional resistance developed on the failure plane is related to the shear strength of the soil. This test is not in common use. Similarly, some past studies have been made of using an extrusion-type test for cohesive soils, but such tests are not in common use.

TRIAXIAL COMPRESSION TEST / The triaxial test is currently the most popular test for determining the shearing strength of soils. Though not as simple a test as the direct shear test, it has several advantages that are of practical importance. Among these are (a) the loading conditions in the triaxial test can be made to simulate more accurately the loading conditions that the soil was (or will be) subjected to in its natural state; (b) failure is not forced to occur across a pre-determined plane, as occurs for the direct shear test.

In the triaxial test, a cylindrical soil sample (as conventionally obtained in soil borings) is wrapped in a rubber membrane for protection and placed in a chamber where an all-around, or confining, pressure can be applied. The confining pressure is usually applied through water or air introduced into the sealed chamber. The sample sits on a fixed pedestal, and a cap attached to a vertical piston rests on the top of the sample. In testing, the confining pressure is applied all around and to the top of the sample and, generally, held constant. An axial (vertical) load is subsequently applied to the sample through the piston, which passes through the top of the chamber. The axial load is steadily increased until failure of the sample occurs. Figure 10-4 (a) shows schematically the triaxial testing arrangement.

Analysis of a three-dimensional element within the tested triaxial specimen shows that a lateral normal stress equal to the confining pressure p_c acts on orthogonal vertical planes, and a vertical stress acts on the horizontal plane. The vertical stress is equal to the sum of the confining pressure p_c plus that pressure increase resulting from the axial loading, Δp; see Fig. 10-5.

In practical problems, the confining pressure is frequently selected to be equal to the confining or lateral pressure that acted on the soil in its natural location within the soil mass, and the axial pressure that eventually causes the sample to fail is assumed to be representative of

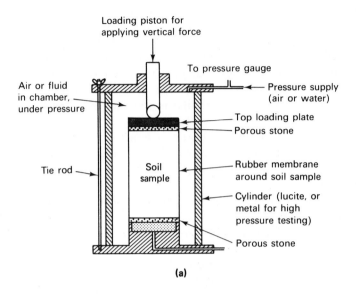

(a)

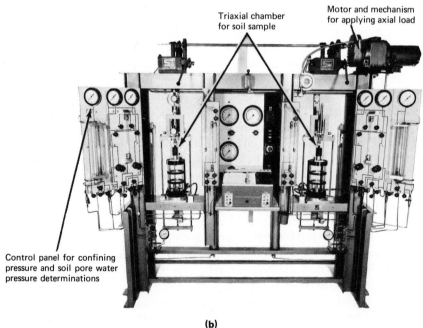

(b)

Figure 10-4. *(a) Schematic diagram of triaxial compression test apparatus; (b) laboratory triaxial equipment (dual-chamber, motorized).* (Courtesy of Soiltest, Inc.)

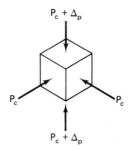

Figure 10-5. *Stress combination acting on incremental element of soil subjected to triaxial testing.*

the maximum vertical pressure that can be imposed on the soil at the depth at which the confining pressure acts. By relating test pressures to subsurface stresses, the confining pressure represents σ_3, the minor principal stress, while the *total* axial stress represents σ_1, the major principal stress.

In the triaxial test setup, porous discs or stones are placed at the top surface and bottom surface of the sample so that pore water in the sample can be drained, if and when desired. The pedestal and loading cap are provided with drainage tubes for the water to escape, and from which the pore water pressure can be determined. If the problem under study requires, water can also be introduced into the sample to determine the effect of increased water content on the strength.

The numerical difference between the total axial pressure σ_1 and the confining pressure σ_3 in the triaxial test is termed the *deviator stress*. When stress-strain data are graphed, it is the deviator stress that is conventionally plotted against strain. Typical qualitative results are shown in Fig. 10-6.

UNCONFINED COMPRESSION TEST / The unconfined compression test (Fig. 10-7) is a triaxial compression test, but where the

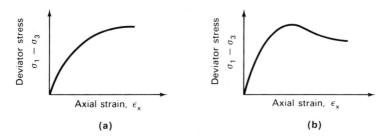

Figure 10-6. *Typical triaxial test results: (a) for loose sand and normally consolidated clays, (b) for dense sands and overconsolidated clays.*

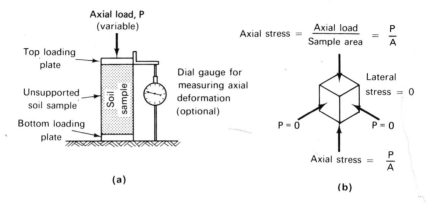

Figure 10-7.   *Representation of unconfined compression test: (a) test arrangement, (b) stresses acting on incremental element.*

all-around confining pressure is zero. The axial force represents the only source of external pressure imposed onto the soil. Because it is necessary that the soil sample under test be capable of standing in the testing apparatus under its own internal strength, the test is limited to use for soils possessing cohesion. Test results are generally similar to conventional triaxial test results.

VANE SHEAR TESTS / Vane shear tests are coming into wider use for determining the in-situ strength of cohesive soils. The vane shear apparatus consists of thin-bladed vanes that can be pushed into the soil with a minimum of disturbance. A torque applied to rotate the vanes is related to the shear strength of the soil. A vane shear apparatus as pictured in Fig. 10-8 can be attached to a long vertical rod and inserted into borings, so that the in-place strength of the soil can be determined without removing the soil from its natural state in the

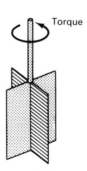

Figure 10-8.   *Vane of vane-shear apparatus, for determining in-place shear or cohesive strength. (Also refer to Fig. 11-43.)*

ground. Similar equipment is also used in the laboratory. It is specifically useful for determining the cohesive strength of soft or sensitive clays that could possibly be effected by the handling that the conventional laboratory shear tests require.

An adaptation of the vane shear apparatus is the torvane (Fig. 10-9), used for determining the cohesive strength of samples in the field or laboratory.

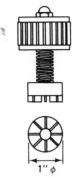

Figure 10-9. Torvane for determining shear or cohesion of soil in field or laboratory.

SHEAR TEST RESULTS PLOTTED ON MOHR CIRCLE COORDINATES / Results of the direct shear and triaxial tests can be plotted graphically (or semi-graphically) on a shear stress versus normal stress coordinate system. They are the same coordinates used for Mohr circle plots (Chapter 7). The stress combination that represents the condition at the maximum shearing strength or the ultimate shearing strength (at failure) is generally used for making the plot.

For the direct shear test [Fig. 10-10(a)] the results of *each* test would plot as a point, as indicated in Fig. 10-10(b).

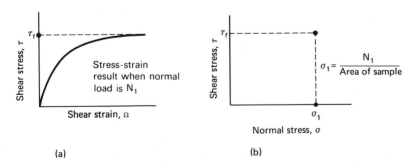

Figure 10-10. Direct shear results for one test: (a) stress-strain plot; (b) result of test on normal stress-shear stress coordinates.

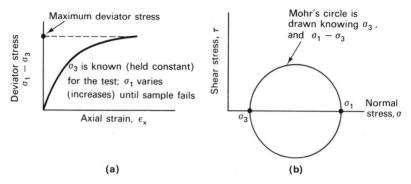

Figure 10-11. *Triaxial results for one test: (a) stress-strain plot; (b) result of test on normal stress-shear stress coordinates.*

With triaxial test results, the values of σ_3 and σ_1, the principal stresses, are known, and plotting these points permits the Mohr's circle for these stress conditions to be drawn; see Fig. 10-11 (b).

If identical homogeneous soil samples were tested under identical conditions in a direct shear test *and* a triaxial test, the same strength information, for practical purposes, would be obtained. Where the results from a direct shear test are superimposed on the results from a triaxial test, the plotted point of τ versus σ appears as shown in Fig. 10-12 (a) or (b).

FAILURE ENVELOPE / If a series of direct shear tests were performed on different samples of homogeneous soil and the normal loading were different for each test, the results plotted on τ_f vs. σ coordinates could be as shown in Fig. 10-13. A curve passing through the plotted points establishes what is referred to as the *failure envelope curve*. This failure envelope indicates the limiting shear strength that a material will develop for a given normal stress. Since the failure en-

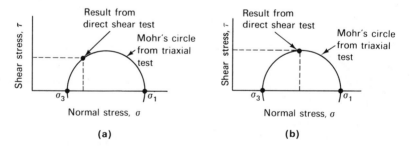

Figure 10-12. *Result of direct shear and triaxial test plotted on same co-ordinates: (a) general location of τ versus σ point from direct shear test on cohesionless soil; (b) general location of τ versus σ point from direct shear test on cohesive soil.*

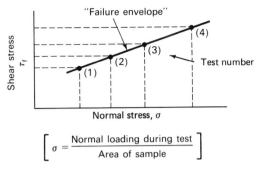

Figure 10-13. *Results of direct shear tests on samples of a homogeneous soil.*

velope is a boundary established from test results, it should be evident that it is physically impossible to have a stress combination whose coordinates plot above the boundary.

The same information can be obtained from a series of triaxial tests performed on a homogeneous soil. In each test the confining pressure σ_3 is different. This consequently effects the maximum axial stress σ_1 that develops at failure. The Mohr's circle test results could be as shown in Fig. 10-14.

A curve that is tangent to all the Mohr's circles establishes the failure envelope curve. The *point of tangency* to each circle establishes the combination of shear and normal stresses that act on the failure plane in the tested soil sample. These should be the same combinations of shear and normal stresses obtained from direct shear tests.

From reference to the Mohr's circle plot, it should be understood that the plane of failure does not necessarily occur on the plane where the shear stress is a maximum. Rather, it is the *combination* of shear and normal stresses developing within the material that is critical. The combination that is critical relates to the slope or angle of the failure envelope curve. If the failure envelope is horizontal (parallel to the normal stress axis), the *maximum* shear stress indicated by the Mohr's circle is the shear stress on the failure plane. Where the failure envelope curve makes an angle with the horizontal axis, the shear stress on the

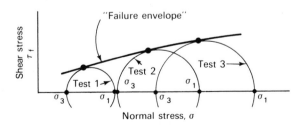

Figure 10-14. *Results of triaxial tests on samples of a homogeneous soil.*

failure plane is *less* than the maximum shear stress represented by a Mohr's circle. In soils, the condition of both the horizontal and the sloped failure envelope exists. This is discussed further under the heading of "Shearing Strength."

Quite frequently in soil analysis, the Mohr's circle plot requires use of only the upper half of the circle, the lower half being considered redundant because it is a mirror image of the upper half. However, the entire circle is best shown when failure envelopes are discussed so as to establish clearly the concept of boundaries for the nonfail and fail combination of stresses; see Fig. 10-15.

For strength conditions indicated by the Mohr's circle of Fig. 10–15, it should be seen that an incipient failure situation develops on two different planes simultaneously. But when it is desired only to learn the combination of principal stresses that will cause failure or the magnitude of shear and normal stresses acting on the failure planes, as is usual, using only half of the Mohr's circle is sufficient.

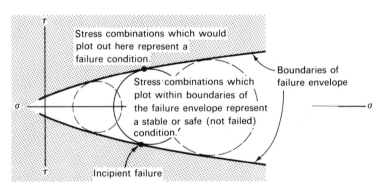

Figure 10-15. *Concept of failure envelope showing combination of stresses representing failure and stable conditions.*

Shearing Strength

Basic understanding of the shearing strength of cohesive and cohesionless soils can be obtained by reference to the results from triaxial or direct shear tests. A discussion relating to the more simple direct shear test is easiest to follow. The behavior described, and explanation of the factors affecting the behavior, apply to triaxial conditions as well as to direct shear conditions.

SHEARING STRENGTH OF COHESIONLESS SOIL / The results of a direct shear test performed on a dry cohesionless soil, where the normal load is held constant during the shearing process, are presented on stress-strain coordinates. Typical results are indicated in Fig. 10-2.

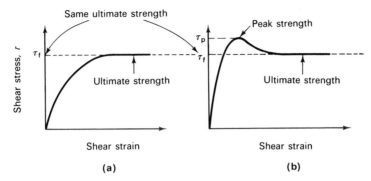

Figure 10-16. *Comparison of shear test results as affected by soil density: (a) cohesionless soil, sample initially loose; (b) cohesionless soil, sample initially dense.*

The actual configuration of the stress-strain curve will be affected by the size and shape of the soil particles and by the density of the sample at the beginning of the test, as well as by the magnitude of the normal loading. A sample that is initially loose will develop a stress-strain curve as indicated in Fig. 10-16(a), whereas an initially dense sample will present a curve as shown in Fig. 10-16(b). The ultimate strength for samples whose only difference is the initial density will be the same. The difference in the behavior is explained by the factors that contribute to the shearing strength of dry cohesionless soil. Resistance to movement across the failure plane, and hence the shearing *strength,* is developed from friction that occurs between particles under the applied normal loading and to interlocking between particles. The extra shearing resistance for the initially dense sample (difference between peak strength and ultimate strength) is attributed to a greater degree of interlocking. This greater interlocking is overcome as shearing displacements increase, indicating that the initially dense sample loosens during shearing. This is verified by recording the thickness (or volume) of the sample during testing and observing that a thickness increase occurs. In other words, the void ratio of the initially dense sample increases during shearing. Conversely, the thickness of the initially loose sample decreases during testing, indicating a decrease in the void ratio. At the ultimate strength, after shearing movement, the final void ratio will be similar, regardless of the initial void ratio. This final void ratio is referred to as the *critical void ratio.* Soils whose natural void ratios are above the critical will attempt to decrease in volume during shearing, whereas the reverse is true for soils whose natural void ratio is below the critical ratio; see Fig. 10-17.

If a series of shearing tests is performed on identical samples, but the normal loading that acts during shearing is different, the ultimate

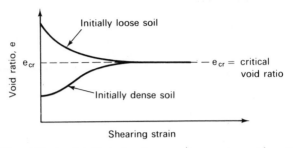

Figure 10-17. *Effect of initial density on change in void ratio during shearing.*

shearing resistances (and peak resistances if the soil is initially dense) will also be different. Greater shearing resistances, or shearing strengths, are obtained at higher normal loadings; see Fig. 10-18.

If the results of the series of direct shear tests are plotted on shear stress versus normal stress coordinates, as for a Mohr's circle analysis, a failure envelope curve for the soil is obtained; see Fig. 10-19.

Through the range of normal stresses usually encountered in foundation problems, the plotted points of τ versus σ establish what is basically a straight line. The angle that the failure envelope creates with a horizontal line is indicated as ϕ, the angle of internal friction for the soil. For dry cohesionless soils, the failure envelope obtained from ultimate shear strength values is assumed to pass through the origin of coordinates. Thus, the relation of shearing strength to normal stress can be calculated from

$$\tan \phi = \frac{\tau}{\sigma}$$

or
$$\tau = \sigma \tan \phi \qquad (10\text{-}1)$$

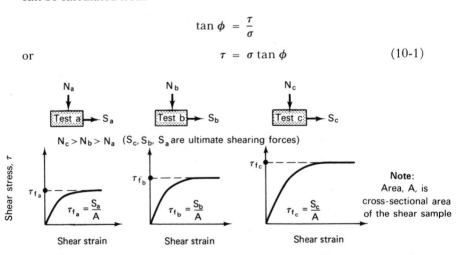

Figure 10-18. *Effect of normal loading on shear strength of cohesionless soil.*

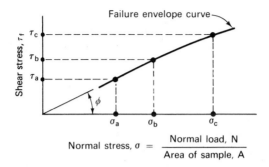

Figure 10-19. Summary representing maximum shear strength versus applied normal stress for cohesionless soil.

If the peak values of shear strength are plotted, values of ϕ somewhat greater than obtained from the ultimate shear strength values will be obtained. Representative values for ϕ are presented in Table 10-1.

TABLE 10-1 / REPRESENTATIVE VALUES OF ϕ FOR COHESIONLESS SOILS

Soil Type	Angle ϕ, Degrees	
	Ultimate	Peak
Sand and gravel mixture	33–36	40–50
Well-graded sand	32–35	40–45
Fine to medium sand	29–32	32–35
Silty sand	27–32	30–33
Silt (nonplastic)	26–30	30–35

The value of ϕ (ultimate or peak) selected for use in practical soil or foundation problems should be related to the soil strains that are expected. If soil deformation will be limited, using the peak value for ϕ would be justified. Where deformations might be relatively great, ultimate values of ϕ should be used.

Illustration 10-1: A sample of dry sand is tested in direct shear. A normal load equivalent to 2 ksf is imposed for the test. The shearing force applied to fail the sample is increased until shearing does occur. The shear stress at failure is 1350 psf. What is the angle of internal friction ϕ for the sand?

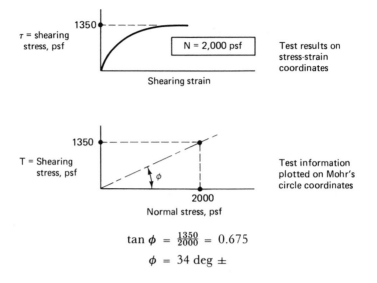

$$\tan \phi = \tfrac{1350}{2000} = 0.675$$

$$\phi = 34 \text{ deg} \pm$$

Illustration 10-2:   A dry cohesionless soil is tested in a triaxial test to determine the angle of internal friction ϕ. A confining pressure equal to 1000 psf is used. The sample fails when the axial load causes a stress of 3200 psf. What is the value of ϕ?

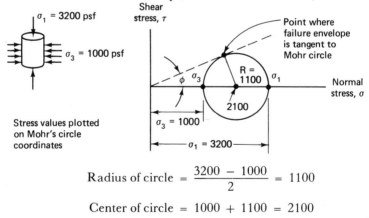

$$\text{Radius of circle} = \frac{3200 - 1000}{2} = 1100$$

$$\text{Center of circle} = 1000 + 1100 = 2100$$

From the Mohr's circle plot,

$$\sin \phi = \tfrac{1100}{2100} = 0.525$$

$$\phi = 31.5 \text{ deg} \pm$$

If moisture is present in the soil samples during shearing, the subsequent plotting of ultimate shearing strength versus the applied normal stress does not present a failure envelope curve that passes through

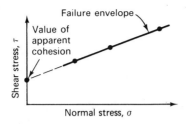

Figure 10-20. *Results of shear tests on moist cohesionless soils indicate an "apparent cohesion" value.*

the origin. Rather, at zero normal stress, the failure envelope intersects the shear stress coordinate, as shown in Fig. 10-20. This indicated shearing strength is referred to as *apparent cohesion*, a shear strength value attributed to factors other than friction developed from the normal stress. With damp cohesionless soils, the extra strength is due to compressive forces exerted on soil particles as a result of surface tensions where water menesci have formed between soil particles (as in capillary water). The extra shearing strength exists as long as the soil retains some moisture. The extra strength would be lost if the soil were to dry out or to become saturated or submerged. For the saturated and submerged case, all voids are filled with water, and all water menisci are lost. For these reasons the extra shear strength attributed to apparent cohesion generally is neglected in foundation studies.

Values of ϕ are not significantly affected when the soil is below a groundwater table or otherwise submerged. However, in applying Eq. (10-1) to determine the shear strength available, it must be remembered that the submerged or effective soil weight must be used in calculating soil overburden and normal pressures.

Illustration 10-3: (a) Samples taken from a uniform deposit of granular soil are found to have a unit weight of 125 pcf and an angle of internal friction of 35 deg. What is the shearing strength of the soil on a horizontal plane at a point 10 ft below the ground surface?

At a depth of 10 ft, the soil overburden pressure, or normal stress, is

$$(125 \text{ pcf}) (10 \text{ ft}) = 1250 \text{ psf}$$

The shearing resistance that can be developed is

$$\tau = \sigma \tan \phi = (1250 \text{ psf}) (\tan 35 \text{ deg})$$
$$= 875 \text{ psf}$$

(b) A proposed structure will cause the vertical stress to increase to 1200 psf at the 10-foot level. Assume that the weight of the structure also causes the shearing stress to increase to 1000 psf on a horizontal plane at this depth. Does this shearing stress exceed the shearing strength of the soil?

The total vertical pressure due to the structure and soil overburden is

$$1200 \text{ psf} + 1250 \text{ psf overburden pressure} = 2450 \text{ psf}$$

The shearing strength that can be developed by the soil at this depth is

$$\tau = (2450 \text{ psf}) (\tan 35 \text{ deg}) = 1720 \text{ psf}$$

This would indicate that the shear strength of the soil is greater than the developed shear stress; therefore, a shear failure does not occur (1720 psf > 1000 psf).

If the water table rose to the ground surface, the effective soil overburden pressure would be reduced to about

$$(\tfrac{1}{2} \times 125 \text{ lb/ft}^3) (10 \text{ ft}) = 625 \text{ psf}$$

The total vertical stress would than be

$$625 \text{ psf} + 1200 \text{ psf} = 1825 \text{ psf}$$

The shear strength available is

$$\tau = (1825 \text{ psf}) (\tan 35 \text{ deg}) = 1275 \text{ psf}$$

This is still greater than the shear stress resulting from the loading conditions (1275 psf > 1000 psf).

The preceding discussion has indicated that the shearing strength that can develop is directly proportional to the effective normal stress that acts on the plane under analysis. For most practical problems, the effective stress in cohesionless soil is calculated by using the effective overburden weight plus any stress increase created by structural loading. However, if the soil is saturated or nearly so, there is the possibility that the stress resulting from loads newly applied onto the soil mass will not result in a related increase to the effective stress acting within the soil. This is the situation of an excess hydrostatic pressure condition developing (pore water in the soil voids becomes subject to a pressure greater than the normal hydrostatic, as discussed in the explanation of the consolidation process, Chapter 9). With this situation, there is the potential danger that the shearing stress resulting from the external loading may increase faster than the shearing strength (which is controlled by the effective stress), and a shear failure may occur. Fortunately, cohesionless soils have relatively high rates of permeability that under most conditions permit rapid drainage of pore water when new loadings and stresses are imposed onto the soil mass. The danger of excess hydrostatic, or excess pore pressure, conditions in cohesionless soils and possible shear failures are generally limited to the situation where loose saturated material is exposed to vibratory, instantaneous, or shock loading such as from explosives, earthquakes, and traveling trains. The occurrence of loss of strength under these conditions is called *liquefaction,* for the soil momentarily liquifies and tends to behave as a dense fluid. The soils found most susceptible to liquefaction are the saturated and loose fine to medium sands having a uniform particle-size range.

APPROXIMATING VALUES OF ϕ FROM BORING DATA / Soil borings for subsurface investigations are frequently sized so that soil samples can be obtained with a "standard" two-inch-diameter split spoon soil sampler. Soil-boring and -sampling procedures are discussed in Chapter 11. With the split spoon sampler, the blow-count required to drive the sampler into undisturbed soil is recorded for all samples obtained. The samples recovered with this equipment are considered to be disturbed and are unsuitable for performing strength tests. Obtaining undisturbed samples of cohesionless soil is generally difficult under many existing subsurface conditions, and may require the use of large-diameter borings and special soil samplers. Because of the wide usage of the two-inch sampler (partially because of the relative economy) correlations between the blow-count N from the standard penetration test and the angle of internal friction have been developed. These correlations are presented in Table 10-2. Because of the generalized nature of the correlation, any application of such data to final foundation designs should be made with caution.

TABLE 10-2 / APPROXIMATE RELATION BETWEEN N^*
AND ϕ FOR COHESIONLESS SOIL

Value of N^*	Relative Condition of Soil	Approximate Value of ϕ, Degrees
10	Loose	$30°^{\pm}$
20	Medium dense	$32°^{\pm}$
30	Medium dense to dense	$35°^{\pm}$
40	Dense	$38°^{\pm}$
50	Dense to very dense	$40°^{\pm}$
60	Very dense	$42°^{\pm}$

*In the so-called standard penetration test, N is the number of blows required to drive a standard 2 in. outside diameter split barrel soil sampler 12 in. into undisturbed soil with a 140-lb. weight falling 30 in. For further information on the standard penetration test, refer to Chapter 11.

SHEARING STRENGTH OF CLAY SOILS / The shearing strength that a clay deposit possesses is related to the type of clay mineral and the water content, but, very importantly, also to the effective stress or consolidation pressure to which the soil has been subjected in its past (the soil's stress history). The change that is possible in a clay's shear strength (such as if loading conditions change due to the weight of a new structure) is affected by these just-mentioned factors, but also by the pore water drainage that can occur as shearing deformations tend to occur. Consideration of drainage is of practical importance, because most clays in their natural condition are close to full saturation, and the low permeability of these soils tends to inhibit changes in pore water content that try to occur during shearing.[1]

The significance of effective stress and drainage during shearing can be explained by reference to a series of shearing tests made on fully saturated clays. Practically, the discussion also applies to nearly saturated clays.

Assume that the soil for all the test samples is obtained from one location in a homogeneous, isotropic, normally consolidated clay deposit. All properties of all samples to be tested are identical; see Fig. 10-21. Samples are to be tested in direct shear.

In each series of direct shear tests, each test sample will have a different normal load applied during shearing to determine the effect. The normal loads are greater than the effective overburden (or consolidation) pressure.

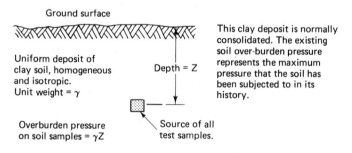

Figure 10-21. *Source of samples for analysis of shear strength of clay soils.*

---

[1]Strictly speaking, stress history and drainage factors can affect the shearing properties of a cohesionless soil, although shape, size, and distribution of particles also have considerable influence. A granular soil that had been subjected to high effective stresses or high overburden pressures in its past would be dense and capable of developing a peak strength, as shown in Fig. 10-16(b). The prevention of pore water drainage during shearing of a saturated soil would lessen the shearing strength, but under most practical conditions there are no restrictions on drainage because of the relatively high permeability.

For the first series of tests, no drainage of pore water is permitted, either when the normal load is applied or during shearing. This "no drainage" control can be achieved by having the sample wrapped in a thin tight impermeable membrane when it is placed in the shearing apparatus. The normal load is applied just prior to beginning the shearing process. Shearing is then completed relatively quickly. These test conditions are referred to as *unconsolidated-undrained* shear tests (U-U), because the samples are not permitted the time or the drainage necessary to consolidate to the pressure exerted by the normal loading, and no drainage or volume change is permitted during the shearing process. For this condition, excess pore water pressures develop in the samples during shearing. Results for this series of tests are shown in Fig. 10-22. For the U-U conditions, the shearing strengths for all samples are the same, since the effective stresses within the soils have not changed from the conditions that existed when the soil was still in the ground. Even though different normal loads are applied during shearing, consolidation to the normal load pressure can only occur if pore water is permitted to drain from the sample. Since drainage and consolidation cannot occur, the internal effective stresses are the same for each of the samples, and, therefore, the shearing strengths will remain the same.

For the second series of tests, different normal loads are applied to each test sample, and full consolidation to the respective new normal load is permitted. After full consolidation has taken place so that the effective stress acting is equal to the applied normal loading, the samples

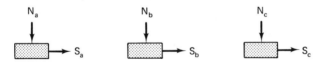

$N_c > N_b > N_a$ (S_c, S_b, S_a represent ultimate shearing forces)

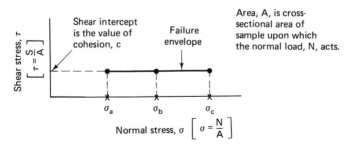

Figure 10-22.   *Shear strength results for unconsolidated-undrained (U-U) tests.*

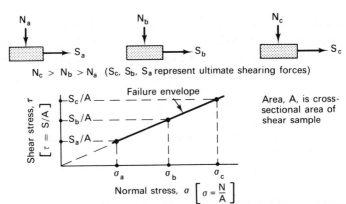

Figure 10-23.   Shear strength results for consolidated-undrained (C-U) tests.

are wrapped in the impermeable membrane, (so that no further drainage or volume change can occur) and are placed in the shearing apparatus. Shearing is performed quickly. These test conditions are referred to as *consolidated-undrained* (C-U), indicating that the samples are consolidated to their respective normal load prior to shearing, but during shearing no drainage or volume change is permitted. For this condition, excess pore water pressures develop in the sample during shearing, but they are not as great as those that occur in the U-U test. Results for such a series of tests are shown in Fig. 10-23. The increase in shearing strength (compared to the U-U conditions) is the result of the increased effective pressure to which the test samples have been consolidated.

For the third series of tests, the samples are fully consolidated to

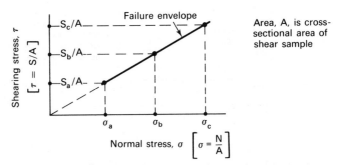

Figure 10-24.   Shear strength results for consolidated-drained (C-D) tests.

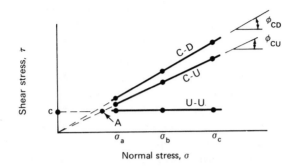

Figure 10-25. *Qualitative comparison of shear strength results for U-U, C-U, and C-D tests.*

the normal loads that will be applied during the shearing test. However, in this series, shearing will take place very slowly, and drainage and volume changes *are permitted* during the shearing process. These test conditions are referred to as *consolidated-drained* (C-D) conditions. No excess pore water pressures develop in the soil during shearing. Results for such a series of tests are shown in Fig. 10-24.

The results from the three series of tests are presented together in Fig. 10-25. A difference in the C-D and C-U curves results because of a soil's tendency to undergo volume decrease during shearing. If volume change is prevented as in the undrained test, excess pore pressures are developed with a subsequent reduction in the effective stress and shear strength.

As for tests on cohesionless soils, the slope of the τ versus σ curve is designated by the angle ϕ.

The stress represented by point A in Fig. 10-25 is approximately the value of the effective stress σ_{v_c} that acted on the soil in its natural subsurface location. If the soil has been normally consolidated, σ_{v_c} is equal to the soil overburden pressure ($\sigma_{v_c} = \gamma Z$; see Fig. 10-21).

If C-U or C-D tests were performed and the applied normal loading was less than σ_{v_c}, the plotted shear strength values would fall close to the failure envelope curve obtained for the U-U test series.

Illustration 10-4: To determine the strength properties of a clay soil, a series of consolidated-undrained direct shear tests is performed. For the first test, the normal pressure is 750 psf and the sample fails when the shear stress is 450 psf. The second sample is tested under a normal load of 1500 psf, and failure occurs when the shear stress is 500 psf. A third sample is tested under a normal loading of 2500 psf, and failure occurs when the shear stress is 800 psf. From these data, estimate the cohesion of the soil in the in-situ condition and the value of ϕ_{cu}.

A problem of this type frequently is best solved graphically.

1. Locate test data on a scaled plot.

2. Draw in a sloped failure envelope line to estimate ϕ_{cu}.

3. For a test having a low normal pressure, assume that the pressure is less than σ_{v_c}, and draw a horizontal failure envelope line through it to obtain the value of in-situ cohesion.

4. The intersection of sloped and horizontal failure envelope lines gives a normal pressure value that is approximately the value of σ_{v_c} (the maximum past overburden stress).

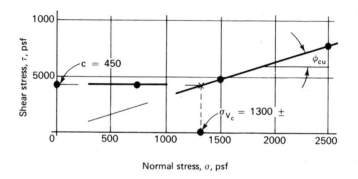

From the scaled plot, $c = 450$ psf, $\phi_{cd} = 17$ deg $\pm$

In almost all practical soils problems involving clays, applied structural loads result in excess pore pressures' being developed. Some drainage does occur, but it is restricted because of the low permeability of the soil. Consequently, the actual shearing conditions that are developed fall somewhere between the U-U and the C-D case. In many soil and foundation design problems, the U-U strength is used where the initial period of loading is critical, since it is realized that the shear strength will become greater (increase) and safer with time as excess pore water drainage occurs.

These test results also indicate that it is possible to improve the shear strength of clay soils by consolidation, provided that time is available for permitting the necessary pore water drainage to take place. In effect, consolidation results in decreasing the water content of the clay, with the subsequent increase in shear strength shown in Fig. 10-26.

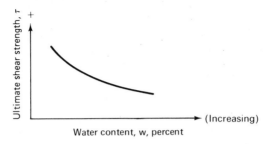

Figure 10-26. *Change in strength for a cohesive soil as water content changes.*

Where the U-U strength is required to be known for analysis of a soil or foundation study, it is convenient to determine the shearing strength from unconfined compression tests. As explained earlier in this chapter, the unconfined compression test is a triaxial test where the confining pressure acting on the sample is zero. This test is quick and easy to perform. The axial stress required to fail the tested soil is designated q_u. The result of an unconfined compression test is plotted on τ versus σ coordinates, as shown in Fig. 10-27. The cohesion, or shear strength, is simply one-half the unconfined axial stress.

$$c = \tfrac{1}{2}q_u \qquad\qquad (10\text{-}2)$$

Some clay deposits have developed cracks or fissures (sometimes called *slickensides*). This condition may be the result of desiccation, seismic effects, or other factors. If tested in unconfined compression, shear failure along the slickensides could give a low and misleading indication of the strength of soil confined in the natural deposit. For such material, a triaxial test is preferred.

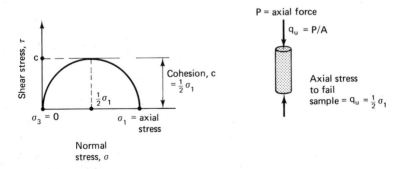

Figure 10-27. *Unconfined compression test data used to determine cohesion c.*

APPROXIMATING VALUES OF COHESION FROM BORING DATA / Because of the wide use of the standard penetration test for making soil borings, approximate relationships between values of cohesion and N have been developed. Values are presented in Table 10-3. Due to the approximate nature of the values, the data should be applied with caution.

TABLE 10-3 / APPROXIMATE RELATION BETWEEN $N*$
AND COHESION OF CLAYS

Value of $N*$	Relative Condition of Soil	Approximate Value of Cohesion, c	
		psf	kN/m^2
2 to 4	soft	250–500	12–24
4 to 8	medium	500–1000	24–48
8 to 15	stiff	1000–2000	48–96
15 to 30	very stiff	2000–4000	96–190
above 30	hard	above 4000	above 190

*N, the standard penetration test, is the number of blows required to drive a standard two-inch outside diameter soil sampler 12 in. into undisturbed soil with a 140-pound weight falling 30 in. For additional information on the standard penetration test, refer to Chapter 11.

SHEAR STRENGTH OF MIXED SOILS / Mixtures of clay and granular soils in nature are not unusual. Material that is predominantly clay, in which all granular particles are surrounded by clay materials, will behave essentially as a clay. Mixtures that are predominantly granular soil with limited clay will present a sloped failure envelope curve on a τ versus σ plot, but the intercept is on the τ axis, as shown in Fig. 10-28. The relationship of shear strength to normal stress can be expressed as

$$\tau = c + \sigma \tan \phi \qquad (10\text{-}3)$$

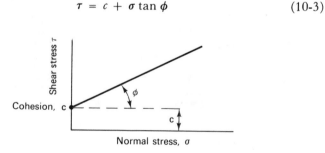

Figure 10-28. *Typical plot of τ versus σ for mixed soils.*

For such soils, the limitations on permeability and drainage should not be overlooked when considering new loadings that are applied to the soil mass. Excess pore pressures may develop and only part of the newly applied stress may represent effective stress, with a subsequent lag in the development of shear strength.

POSITION OF FAILURE PLANE RELATED TO ANGLE ϕ / Homogeneous soils stressed to failure during a triaxial test or unconfined compression test typically develop a distinct plane of failure, as indicated in Fig. 10-29(a). Applying the analysis for stress at a point, Fig. 10-29(b) and (c), a practical relationship between the angle of internal friction, ϕ, and the position of the failure plane is obtained:

$$\phi + 90° + (180 - 2\theta) = 180 \text{ deg, the sum of the interior}$$
$$\text{angles in a triangle.}$$

$$\phi + 90 = 2\theta$$

$$\theta = \frac{90 + \phi}{2}$$

$$\theta = 45 + \frac{\phi}{2} \text{ deg.} \tag{10.4}$$

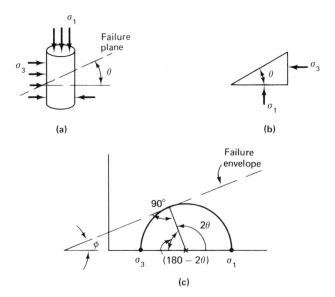

Figure 10-29. *Mohr's circle analysis to relate ϕ and position of failure plane: (a) failure plane on tested sample; (b) incremental element at failure plane; (c) Mohr's circle.*

PROBLEMS

1. A sample of dry sand is tested in direct shear. Under an applied normal stress of 5000 psf, the sample fails when the shear stress reaches 3000 psf. What is the angle of internal friction for this soil?

2. A sand sample is subjected to direct shear testing at its normal (in-situ) water content. Two tests are performed. For one of the tests, the sample shears at a stress of 3000 psf when the normal stress is 4000 psf. In the second test, the sample shears at a stress of 4000 psf when the normal stress is 6000 psf. From these data, determine the value of the "apparent cohesion" and the corresponding angle of internal friction.

3. A sand sample in a triaxial test failed when the confining stress (minor principal stress) was 1000 psf and the axial stress (major principal stress) was 4000 psf. What is the angle of internal friction ϕ for this soil?

4. A sand is known to have an angle of internal friction ϕ equal to 35 deg. What is the maximum major principal stress that the soil can withstand when the minor principal stress is 20 psi?

5. The soil in a granular deposit has an angle of internal friction equal to 33 deg. At a point in the soil mass where the lateral pressure (minor principal stress) is 1000 psf, what maximum vertical pressure (major principal stress) can be applied?

6. Computations indicate that a planned building loading will cause the principal stresses in the soil at a point beneath the building to increase to total values of 4500 psf and 1500 psf (major and minor principal stresses, respectively). If the soil is a sand with an angle of internal friction equal to 35 deg, will the indicated stresses cause a shear failure at the point?

7. Samples of a dry sand are to be tested in a direct shear and a triaxial test. In the triaxial test, the sample fails when the major and minor principal stresses are 140 psi and 40 psi, respectively. What shear strength would be expected in the direct shear test when the normal loading is equal to a stress of 5000 psf?

8. A clay soil is subjected to a triaxial test under unconsolidated-undrained conditions. At failure, the major and minor principal stresses are 3000 psf and 1000 psf respectively.
    (a) What is the cohesion for this soil?
    (b) If this soil were subjected to an unconfined compression test, what axial load would result at failure?

9. A clay soil has a cohesion of 800 psf. How could the strength of this soil be increased?

10. A mixed soil is found to possess a unit cohesion of 500 psf and an angle of internal friction of 30 deg. What shear strength is expected at a point where the normal stress is 1500 psf?

CHAPTER 11
Site Investigations:
Purpose and Methods

Site investigation refers to the procedure of determining surface and subsurface conditions in an area of proposed construction. Surface and subsurface features may influence *what* can be built, and will directly affect the design and construction procedures relating to *how* a structure is built.

Information on surface conditions is necessary for planning construction techniques. Surface topography may affect access to the site with necessary construction equipment, including the ability of equipment to work on and travel across the area. The thickness of vegetation, including the density and height of trees, affect the ease or difficulty associated with preparing a site for construction. Disposal of removed surface material may be a problem, particularly in urban areas. If a condition of surface water develops at times, its presence may hinder construction operations or affect the use of the site after construction. Other factors that could affect construction procedures or postconstruction use of an area include availability of water, availability of electrical power, the proximity to major transportation routes, and environmental protection regulations of various government agencies.

The land drainage pattern, a surface consideration but also partially a subsurface feature, is important, for it might affect construction. A finished grading pattern must be planned so that it does not harm the original areal pattern or cause other environmental changes.

Information on subsurface conditions existing at a site is a critical requirement. It is this information that is utilized to plan and design a structure's foundations and other below-ground work. Construction techniques are planned with the help of data on subsurface conditions. The possible need for dewatering will be revealed by the subsurface investigation. Information necessary to plan and design shoring or bracing of excavations for foundations and pipe trenches is obtained from such explorations. If the construction site is underlain by varying soil conditions, the explorations will be used to indicate "better" areas. For projects where there is flexibility in locating structures, considerable savings in foundation costs may be realized by constructing in the "better soil" area.

In many locations, information about the area is frequently available in the form of maps showing surface topography, and maps and literature that provide general information on subsurface soil or rock conditions. Aerial photographs can also offer useful data on subsurface conditions.

Though maps and aerial photographs of an area may provide much useful information about soil and rock conditions existing at a site, virtually all major construction projects will have on-site subsurface explorations performed in order to obtain the *detailed* information on soil types and properties necessary for designing foundations and for planning construction activities. Most typically, such information is obtained through the use of borings or test pits, or by utilizing geophysical investigative methods. Properly used, these procedures are capable of giving a reliable definition of the type and extent of soil strata underlying an area. Where physical properties of the subsurface soil must be known, soil samples from appropriate depths can be obtained for laboratory testing. Alternatively, in-place testing of soil in its natural location is frequently possible.

These methods of subsurface investigation are also used in exploring for sources of soil fill and for checking material in borrow pit areas.

Frequently, the overall reliability of information obtained from a subsurface investigation is related to the extent of the work performed; the more borings and test pits, or the more geophysical lines run, the greater the detail. The availability of information reduces the need for interpolating between locations of known conditions.

Accurate preplanning of a subsurface exploration program can be difficult to do properly. The *depth* of borings or test pits considered necessary or adequate for design and planning should relate to the types and properties of earth materials revealed by the investigation as well as to the size and type of structure that is planned. The *number* of borings, test pits, or geophysical lines that will be necessary relates to the variation in conditions. A site underlain by uniform conditions normally requires a less extensive investigation than one underlain by highly variable conditions. It is not unusual to have to plan an exploration program as it progresses, information as it becomes available determining the need for additional work.

Maps and Aerial Photographs
as a Source of Information

Information on surface and subsurface conditions in an area is useful to the construction and engineering profession and is frequently present in the form of available maps. In the United States, such sources include U. S. Geological Survey topographic maps, U. S. Department of Agriculture soil conservation maps, and state geologic (rock or soil) maps. Many other areas of the world have similar maps available.

Aerial photographs are also capable of providing considerable information about an area, particularly data on surface conditions and land forms that are not clarified on maps or detectable from ground observations. Aerial photographs frequently serve as the basis for topographic and soil maps. Available sources include the governmental agencies responsible for developing maps, other municipal governments or agencies who have aerial photographs from construction or tax-mapping projects, and private firms who take the aerial photographs and provide consulting services in photogrammetry.

U. S. GEOLOGICAL SURVEY TOPOGRAPHIC QUAD-RANGLE MAPS / These maps provide information on surface features and topography. Included on these maps are elevation contours, bodies of water and water courses, indication of areas covered by forests or other vegetation, and man-made features such as buildings, roads, dams and reservoirs, railroads, power lines, airports, harbor development, military bases, and park preserves. Study of contour elevations and indicated land forms on these maps can relate the geology of subsoil formation and character. These maps are very useful for evaluating the effects that topography and drainage will have on an area. An illustration of a USGS map is shown in Fig. 11-1.

U. S. DEPARTMENT OF AGRICULTURE SOIL CON-SERVATION MAPS / These maps provide information on surficial soils existing in an area. The information usually represents conditions to a depth of five or six feet (1.5–2 m). A coded typing is used to indicate different soil catagories on the map. In literature accompanying the maps various properties and possible uses of the soils are described. Included is the application for construction purposes; AASHO and Unified Soil Classifications are given. General suitability of the various soils for embankment and highway base course materials, compaction characteristics, and permeabilities are also provided. An illustration of a soil conservation map is shown in Fig. 11-2.

GEOLOGIC MAPS / These maps indicate the rock types underlying an area. Frequently, geologic maps are accompanied by descriptive papers that complement the actual map. Information on the

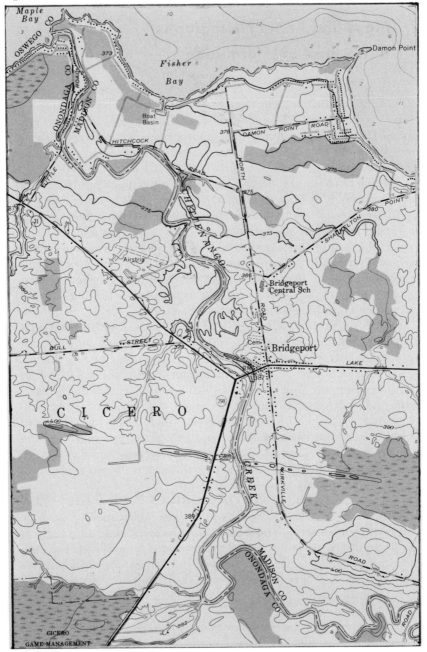

Figure 11-1. *U.S. Geological Survey topographic map (partial section of Cleveland—New York quadrangle). (Note: These maps are printed in color.)*

Figure 11-2. *U.S. Department of Agriculture soil conservation map showing coded soil categories (partial section from southern Herkimer County, N.Y.).*

extent of soil cover overlying the rock and general physical properties of the soil cover and the rock, and data on natural resources such as oil, gas, and commercial minerals are usually included. Developers of geologic maps frequently include state-level governmental agencies. Fig. 11-3 provides an illustration of a geologic map.

AERIAL PHOTOGRAPHS / Air-photo interpretation is the method of providing information on subsurface conditions by relating landform development and plant growth to geology. Applied to the building design and construction profession, air photos have particular value in their ability to provide detailed geologic information over a relatively large area, and as a result can detect conditions that are difficult to observe or evaluate properly from a surface investigation. Illustrations of such special information include the observation of sinkhole cavities in areas underlain by limestone formations, occurrence of an

Figure 11-3. *U.S. Geological Survey geologic quadrangle map (section from Bloomsbury, New Jersey, quadrangle). (Note: These maps are printed in color.)*

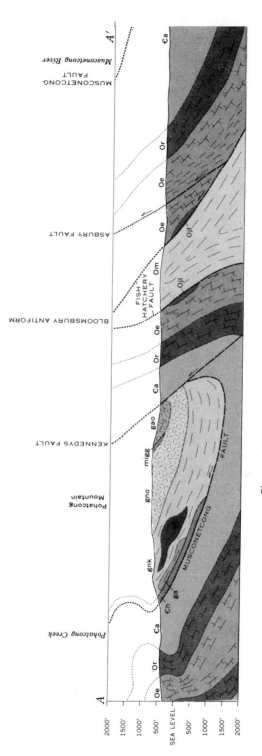

Figure 11-3. (Continued.)

247

area affected by landslides, past meanderings of existing rivers and probable new locations, presence and extent of glacially deposited landforms (such as drumlins, deltas, kettles), changes in beach areas and offshore deposits (such as sandbars), and land drainage patterns.

Not to be overlooked are the benefits of an aerial reconnaisance. A trained observer, or for that matter an experienced designer or constructor, should be capable of detecting items of land formation that can have an effect on design or construction from the view offered by a low and slow flight over a planned construction site. The disadvantage of a flight compared to aerial photographs is that no pictorial record remains for referral in future discussions. Fig. 11-4 provides an example of information shown on aerial photographs.

Borings and Test Pits

Borings and test pits serve to provide a visual identification of soil strata underlying an area by literally penetrating into the earth. Subsurface materials are identified in-place (test pits) or from samples that are taken at a known depth and brought to the surface for examination (borings). In both borings and test pits, soil samples can be obtained for laboratory testing and analysis.

BORING METHODS / With borings, several different techniques are in use for obtaining information on soil conditions. The method utilized depends on the type and extent of information desired from the exploration, the general type of soil conditions in the area being investigated, and the amount of money and time available.

Where only very basic data are necessary, as, for example, if depth to rock (thickness of soil cover) or the thickness of a soft surficial layer (e.g., a marsh deposit) is to be determined, probing by pushing or driving a metal rod until firm resistance or refusal is encountered is a frequently used and economical method. Penetration rods can be handled manually or with mechanical equipment. Usually, manual techniques are practical only where limited depths of probing are necessary.

Hand-operated auger methods (Fig. 11-5) are useful where it is desirable to obtain an indication of various soil types penetrated by the equipment, or where it is desired to create an excavation that can give information on the depth to the groundwater table. Information on soil type at a particular depth is usually determined by noting the soil held on the auger. Continuous-flight augers are frequently used in order to ensure that soil cuttings are carried out of the boring, particularly for deeper borings. When only a limited section of auger is attached to the tip of the rods, the soil does not carry to the surface. Such an occurrence tends to bind the rods and auger in the hole.

Figure 11-4. (a) Aerial photograph showing urban development and surrounding land forms; (b) aerial photo showing landslide in Anchorage, Alaska, area resulting during 1964 earthquake.

Figure 11-5. *Hand auger with recovered soil sample.* (Courtesy of Acker Drill Co.)

Boring contractors and manufacturers have made various refinements to manual penetration and auger equipment that permit small samples of soil to be obtained from a desired depth and brought to the surface for examination, or, the measurement of resistance to penetration offered by the soil at a given depth; see Figs. 11-6 and 11-7.

Where the soil investigation is to extend to some depth, mechanical means of drilling borings are utilized. Two techniques are in wide use: the *auger method* and the *wash boring* method. The auger technique (Fig. 11-8) is basically similar to the hand auger method. Continuous-flight augers are used to drill into the earth, and soil cuttings are carried out of the hole by traveling up the flights of the rotating auger. The main difference between manual and mechanical augering is that drilling machinery is utilized to rotate, and simultaneously mechanically to push downward, the auger sections. Because of the drilling machinery, larger diameter flights than those that are possible with manual methods are used, and the limits to depth are controlled only by the capacity of the equipment.

Subsurface soil types can be detected by noting the soil materials carried to the surface by the augers. In deep drilling, there is the disadvantage of not being able to correlate accurately soil type with depth. In order better to classify soil type with depth, auger samples are used. The procedure consists of augering to a desired depth, withdrawing the augers, inserting an auger sampler to obtain a soil specimen from the

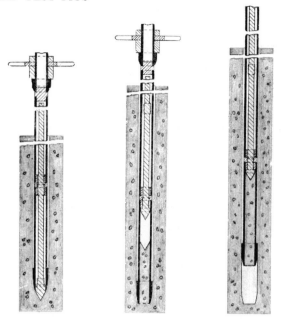

Figure 11-6. *Retractable plug sampler, sequence of operation to recover samples: (1) casing driven with piston extended; (2) piston retracted and casing driven to recover sample; (3) sample recovered and casing withdrawn by hand jack (not shown); (4) brass sample liners (not shown) used in bottom length of casing to facilitate recovery and preservation of sample. (Courtesy of Acker Drill Co.)*

(a)   **(b)**

Figure 11-7. *One type of manual penetrometer: (a) Acker geostick; (b) field use of geostick. (Courtesy of Acker Drill Co.)*

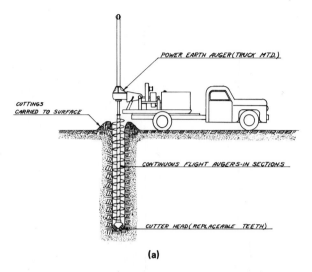

POWER EARTH AUGER (TRUCK MTD.)

CUTTINGS
CARRIED TO SURFACE

CONTINUOUS FLIGHT AUGERS-IN SECTIONS

CUTTER HEAD (REPLACEABLE TEETH)

(a)

(b)

Figure 11-8. *(a) Soil boring using auger method.* (Courtesy of Acker Drill Co.) *(b) Truck-mounted rotary drill rig showing soil augers in use.* (Courtesy of Central Mine Equipment Co.)

Figure 11-9.   *Obtaining soil samples from auger cutting head.* (Courtesy of Acker Drill Co.)

bottom of the hole, withdrawing the sampler after cutting the sample (Fig. 11-9) and then reinserting the augers to drill to the next sampling depth. Soil samples obtained with auger methods are disturbed samples suitable for classification and some laboratory tests (e.g., they are acceptable for grain-size and liquid and plastic limit determinations but not for strength or compressibility tests).

"*Wash boring*" refers to the method of making a boring by applying an up-and-down chopping and twisting motion to a drill bit (or chopping bit) attached at the end of drill rods while simultaneously a stream of water under pressure is directed through the bit to the soil (see Figs. 11-10 and 11-11). This combination of chopping plus wash water serves to loosen soil at the bottom of the boring and to carry the soil cuttings away to the top of the hole. The method has long been popular for making borings because of the limited equipment necessary. Even small-sized engines for lifting drill rods coupled with small water pumps can be used to make deep borings. An advantage to this is that inexpensive but quite maneuverable and portable drilling equipment is possible. (Portability is a factor of importance in difficult-to-reach drilling sites.) Drilling rods are usually inserted into and lifted out of the boring by using a tripod or derrick mast to support a rope and pulley lift.

Borings advanced by wash boring methods through cohesionless soils or below water are subject to cave-in. To prevent this, it is typical to drive a protective casing around the boring. The casing consists of

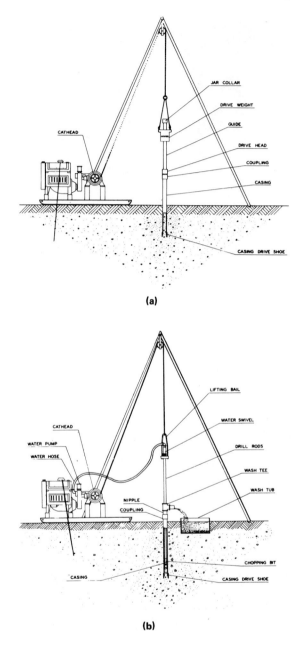

(a)

(b)

Figure 11-10.   *Typical setup for wash boring: (a) driving casing; (b) chopping and jetting.* (Courtesy of Acker Drill Co.)

Figure 11-11. *Wash boring rig in operation in field.* (Courtesy of Acker Drill Co.)

heavy wall pipe that is driven into the ground in sections with a drop hammer (drive weight) as the boring is advanced. The casing is withdrawn for reuse at the completion of the boring. A more recent alternate technique for preventing cave-ins consists of using a heavy liquid slurry ("drillers mud") in the boring as the hole is drilled. The slurry, typically a "bentonite" (a montmorillonite clay) and water mix, is dense enough to exert a lateral pressure adequate to keep the walls of the boring from collapsing inward. Soil cuttings are flushed up through the slurry. At the surface, the entire liquid mixture can be caught, filtered, and recirculated back into the boring. Some of the larger and more sophisticated drilling rigs can make and force compressed air into the boring to prevent cave-ins, instead of using casing or a slurry.

Some information on soil type and changes in strata can be obtained by examining the soil cuttings washed to the surface. However, such soil is significantly disturbed, and there is the danger that mixing of different soil types can occur before the cuttings reach the surface. Because of the significant soil disturbance and the difficulty in establishing boundaries between strata, the determination of subsurface conditions by using wash borings is no longer an acceptable method.

SOIL SAMPLING / In order to develop information on subsurface conditions that is considered accurate, it is now established practice to obtain soil samples that are sufficiently undisturbed to permit accurate classification. Such soil samples are recovered from their natural location in the ground by the utilization of special techniques and sampling equipment. Typically, the boring and sampling procedure involves drilling a hole to a desired sampling depth by using wash boring or auger methods, and then inserting a soil sampler into undisturbed earth to obtain a soil specimen.

A refinement of the auger method to advance a boring involves the use of hollow-stem augers. The soil sampler fits within the hollow stem of the auger, and advances at the bottom of the boring without requiring removal of the augers. Use of this method can considerably speed up the boring and sampling operation.

To obtain a soil sample, the sampler is advanced by driving with a drop hammer (Fig. 11-12), or by pushing with a hydraulic piston or jack (Fig. 11-13). In its basic form, the soil sampler is a section of metal pipe or tube with a cutting edge at one end and attachments to hold the soil in the sampler as it is brought to the surface. Sampler lengths typically range between 18 inches and 3 feet (approximately 0.5 and 1

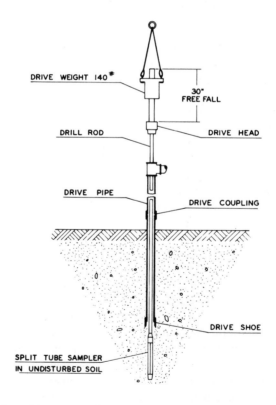

Figure 11-12. *Obtaining soil sample by driving with a drop weight.* (Courtesy of Acker Drill Co.)

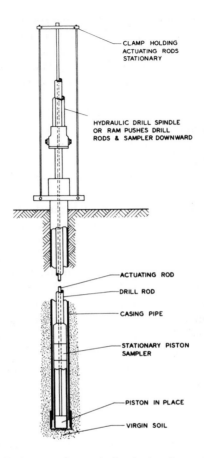

CLAMP HOLDING
ACTUATING RODS
STATIONARY

HYDRAULIC DRILL SPINDLE
OR RAM PUSHES DRILL
RODS & SAMPLER DOWNWARD

ACTUATING ROD

DRILL ROD

CASING PIPE

STATIONARY PISTON
SAMPLER

PISTON IN PLACE

VIRGIN SOIL

Figure 11-13. *Obtaining soil sample by hydraulic pushing.* (Courtesy of Acker Drill Co.)

m). Samplers come in different diameters and with different wall thicknesses. The choice of sampler type selected relates to the type of soil being encountered and what needs to be done with the recovered specimen. Where "in-place" properties need to be determined (such as shear strength and compression-consolidation characteristics), "undisturbed" (or only slightly disturbed) samples are required. If samples are for classification or for performing remolded-type laboratory tests (such as grain size and atterberg limits), "disturbed" samples are permitted. Generally, "undisturbed" soils are obtained by using samplers having a thin wall and a large ratio of inside diameter to wall thickness. Thick-wall samplers and those having a low diameter-to-wall-thickness ratio are considered to provide "disturbed" samples.

One of the most widely known and used soil samplers is the "standard split spoon." This sampler consists of a longitudinally split tube, or barrel, two inches (5.1 cm) in outside diameter and having a $1\frac{3}{8}$-inch (3.5 cm) inside diameter (see Fig. 11-14). The splitting aspect of the sampler permits it to be opened so that the soil specimen can easily be examined and then placed in a container for shipping. The wide use and acceptance of this sampler is due to its size (diameter about equal to typical drill rod diameter), availability and use at the stage of development when investigators and designers started requiring samples for accurate classification, and the comparatively low cost of obtaining this type of sample. The standard split spoon sampler provides samples considered to be "disturbed."

Split barrel samplers are also available in diameters larger than the two-inch "standard split spoon." The larger-diameter samplers are considered to provide better samples for laboratory testing and for sampling soils that include gravel-size particles. Split barrel samplers $3\frac{1}{2}$ inches in diameter for use in 4-inch-diameter borings are included in the typical stock equipment list of many soil-boring contractors. Borings of approximately six-inch diameter, which permit even larger samples to be obtained are sometimes used for projects requiring special soil testing or analysis. However, this size is not typical because of the greater expense involved with drilling borings of larger diameter.

Split barrel samplers can be provided with a "liner," which is a thin metal or plastic tube fitted within the split barrel. The purpose of

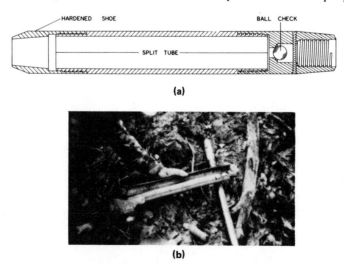

(a)

(b)

Figure 11-14. *Standard split spoon sampler: (a) diagram; (b) sampler opened to expose soil.* (Courtesy of Acker Drill Co.)

the liner is to help hold together and protect the sample during handling, shipping, and storage. The liner may consist of a solid length of tube or of a series of rings.

The split barrel sampler is a heavy-wall sampler and generally is considered to provide disturbed samples (not suitable for shear and consolidation tests). However, in the large-diameter split barrel samplers, disturbance may be limited to just the peripheral zone of the sample.

Samplers for obtaining less disturbed ("undisturbed") specimens are commonly available. Most use a thin-wall tube as the sampler, or as a sampler tip or extension (such as the Dames and Moore Sampler, Fig. 11-15). Methods to carefully advance the sampler into the natural

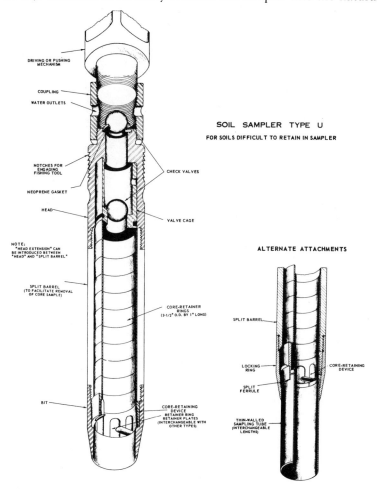

Figure 11-15. Dames and Moore large diameter split barrel sampler. (Courtesy of Dames and Moore)

soil are followed to avoid disturbance. When used by themselves, the thin-wall samplers are typically used only in fine-grained soils (silts and clays). Such samplers are generally not used in coarse-grained soils because of the danger of buckling or crimping the thin wall.

Three of the more widely used "undisturbed" samplers are the Shelby Tube, the Piston Sampler, and the Denison Sampler.

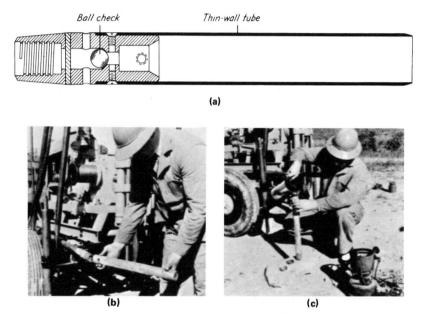

(a)

(b)   **(c)**

Figure 11-16.   *Thin-wall sampler: (a) schematic diagram; (b) tube being removed from head; (c) sealing thin wall tube with paraffin for shipping.* (Courtesy of Acker Drill Co.)

The thin-wall Shelby Tube sampler is a seamless metal tube having a limited wall thickness. To obtain the sample, the Shelby Tube is "pushed" (hydraulically pressed) into the earth to obtain the soil specimen. After recovery, the soil is left in the sampling tube for shipping or is hydraulically pushed out of the sampler and placed in an appropriate shipping container. (See Fig. 11-16.)

The piston sampler (Fig. 11-17) is an adaptation of the Shelby

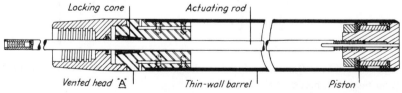

Figure 11-17.   *Stationary piston sampler.* (Courtesy of Acker Drill Co.)

Tube sampling method. This sampler includes a piston device, which serves to push the thin-wall tube into undisturbed soil from the bottom of the boring.

The Denison Sampler, or Denison Core Barrel (Fig. 11-18), is a double-walled sampler that rotates or cores its way into undisturbed soil. The outer barrel rotates to cut into the soil. The sample is obtained with the inner barrel. Because of the coring feature, it has the advantage of being able to obtain samples in hard or cemented soils, where the pressed thin wall would be in danger of buckling.

Figure 11-18. *Denison core barrel.* (Courtesy of Acker Drill Co.)

SAMPLE SPACING IN SOIL BORINGS / The accuracy of information obtained from a particular soil boring relates to the spacing between samples. Samples can be taken continuously; that is, samples will be obtained for virtually each foot of boring depth. However, it is more usual practice to take samples at selected intervals of depth. In the soil-boring industry it is common to sample at five-foot (1.5 m) intervals (one sample for each 5-foot section of boring). Thus, if the sampler is driven 18 in. to recover a sample, the actual distance between samples is $3\frac{1}{2}$ feet. Soil conditions between samples are determined by noting soil cuttings brought to the surface as the boring is made deeper, and feeling resistance to the drilling advance. Where it is detected that a change in soil conditions has been encountered, further drilling should cease and an extra soil sample should be taken.

The drilling of a soil boring, therefore, actually consists of a repetitive series of steps. The drilling equipment is used to make a hole to the depth where a sample is to be obtained. After sampling, the boring is then drilled to the next point where a sample is required. This procedure continues until a boring of the desired depth has been completed.

BORING SPACING AND DEPTH / The number of borings for a project, or the spacing between borings, is related to the type, size, and weight of structure planned, to the extent of the variation in soil conditions that permits "safe" interpolation between borings, to the funds available for the boring program, and possibly to the demands of a local building code. Generally, these same considerations apply to the depth for borings. A primary purpose of borings is to establish the presence, location, and extent of "good" soil suitable for foundation support and to determine the possible existence and extent of "poor" soil that could have an adverse effect on foundation performance. Where poor soils are present, the type of foundation selected would be based on circumventing the location of poor material or would be designed to minimize the effect of its presence.

If an extensive thickness of poor material overlies "good" soil or rock, deep borings will be required. Where a sufficient thickness of good soil exists close to the ground surface, borings may not have to be deep. Even for this condition, however, it is wise to plan some deep borings to reveal conditions at the greater depths, for a deep material that could have an adverse effect on the structure might exist. Where deep borings are necessary to establish the depth at which good foundation soil can be found, soil sampling in the poor material can be minimized or omitted to help reduce drilling costs.

Because most building projects are unique to some degree, certainly with regard to location, no set of rules can be made for planning a boring program that ensures answering all of a designer's or construction contractor's questions. Where there is much uncertainty about subsurface conditions, a preliminary investigation can be made to obtain general information on the area. From this initial information, a more definite or final exploration program can be planned.

For projects extending over a long horizontal distance, such as a highway or a dam, borings for a preliminary investigation may be on 500- to 1000-foot (150- to 300-m) spacings. The final program may eventually require borings on a 100- or 200-foot (30- to 70-m) spacing. With building projects, borings for a preliminary investigation may be located near the planned boundaries of the building area, with a few borings being drilled at intermediate locations. A final program could include borings at building corners plus at important interior locations, with spacings not to exceed 50 to 100 feet (15 to 30 m) depending on conditions encountered and building code requirements with regard to maximum spacing.

PENETRATION RESISTANCE AND THE STANDARD PENETRATION TEST / In situations where soil samples from borings were obtained by using drive methods (the sampler is driven by use of a drop hammer), it became apparent that the number of blows required to advance the sampler (overcoming resistance) was capable of providing a qualitative indication of the in-place properties of the soil. In firm or dense soils more blows would be required to advance the sampler than in soft or loose soil. Consequently, when samples are taken, it is now common practice to include the number of blows necessary to advance the sampler a distance of one foot (30 cm). For the blow-count information to be meaningful, the weight of the drive hammer used to advance the sampler must be indicated, since a heavy weight will drive a sampler with fewer blows than a light weight.

What has become known as the standard penetration test is a soil sampling procedure that is in wide use and is generally accepted as providing some correlation with in-place properties of a soil. The standard penetration test requires that a two-inch split spoon sampler be used in conjunction with a 140-pound (63.6-kg) drive weight. The standard penetration test reports the number of blows N to drive the 2-inch sampler one foot into undisturbed soil by using a 140-pound weight falling 30 inches (0.76 m). In practice, it has become typical to obtain a sample by driving the sampler a distance of 18 inches. The blow count for each six inches of penetration is recorded separately, and the standard penetration test result is the number of blows required for the last 12 inches of driving. Detailed requirements for the taking and care of soil samples using the standard penetration test procedure is presented in ASTM Test Designation D-1586. A correlation between blow count and soil condition is shown in Table 11-1. Further correlations between the standard penetration test and soil properties are presented in Chapters 10 and 14.

TABLE 11-1 / CORRELATION BETWEEN SOIL CONDITIONS AND STANDARD PENETRATION TEST

Sampler .... 2″ O.D. x 1⅜″ I.D.
Hammer................ 140 lb., 30″ fall

Soil	Designation	Blows/Ft
Sand and silt	Loose	0–10
	Medium	11–30
	Dense	31–50
	Very dense	Over 50
Clay	Very soft	Less 2
	Soft	3–5
	Medium	6–15
	Stiff	16–25
	Hard	Over 25

Source: Acker Drill Co.

PENETRATION RESISTANCE AND CONE PENETROMETERS / The cone penetrometer consists of a slender metal rod equipped with cone-shaped tip. The penetrometer is either pushed or driven into the earth. When pushed, usually by hydraulic jack, the penetrometer is classified as a static cone penetrometer. If driven, usually by blows of a drop hammer, the equipment is referred to as a dynamic cone penetrometer. Electric penetrometers also are available. With this type, the cone tip is advanced by an electrical cell constructed within the penetrometer. In use, the penetrometer's resistance to advancing q_c is recorded. The resistance is subsequently related to properties of the penetrated soil.

Cone penetrometers offer the advantage that a continuous resistance record is easily obtained for the full depth investigated. This is in contrast to the usual practice of taking standard penetration test samples or other samples at intervals. The method is applicable to cohesionless and cohesive soils. A main disadvantage with many penetrometers is that no soil samples are obtained.

In the past, cone penetrometers have received limited attention in the English-speaking countries. However, a variety of cone penetrometers have been widely used throughout the European countries, and considerable practical experience has been obtained. Consequently, interest has been developing rapidly in English-speaking areas, particularly for investigating areas underlain by cohesionless soils. With conventional boring and sampling methods, obtaining undisturbed samples of cohesionless soil below the water table is difficult and unreliable.

Of the various cone penetrometers in use, the Dutch (Delft) cone is one of the most widely known. For illustrative purposes, it is representative of many types of static cone penetrometers. The Dutch cone penetrometer consists of a conical point (Fig. 11–19) having a 1.4-inch (3.6-cm) diameter and a base area of 10 cm^2 attached to a rod that fits within a larger-diameter rod. For testing the soil at a particular depth, the outer rod (along with the cone and inner rod) is advanced to the desired elevation. Then, the outer rod is locked in position and the cone is advanced a set distance, usually 5 cm, by applying a force to the inner rod. The force necessary to advance the cone is determined with a load cell or proving ring and recorded. Soil properties are correlated with the penetrating force.

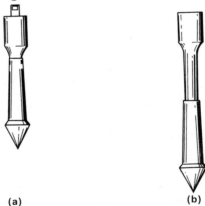

(a)   (b)

Figure 11-19. *Point of Dutch cone penetrometer: (a) position of cone for moving to new depth; (b) cone only advances to determine soil resistance or bearing capacity.*

(a)

(b)

Figure 11-20. *(a) Dutch cone penetrometer conversion unit fitted to conventional truck boring rig.* (Courtesy of Soil Mechanics Equipment Co., Maryland) *(b) Ten-ton trailer-mounted Dutch cone penetrometer. Sketch of helicoidal anchor indicates method used to anchor rig against lifting.* (Courtesy of Soil Mechanics Equipment Co., Maryland)

265

The reaction of the force used to advance the cone penetrometer into the earth has to be resisted by the penetrometer rig. If the penetrometer equipment is truck-mounted [Fig. 11-20(a)], the weight of the vehicle is frequently sufficient to provide the necessary reaction. If lightweight field equipment is used, it is usually necessary to somehow anchor the rig to prevent it from being lifted when the cone is advanced [Fig. 11-20(b)].

A modification of the Dutch cone penetrometer is the friction cone (Fig. 11-21). A friction sleeve is provided above the cone point. Testing at a particular depth consists of determining a cone resistance value but also a separate value for cone plus friction on the sleeve. The sleeve friction is used to help identify the soil type being penetrated.

For proper classification of soil, some calibration is generally required. This typically consists of comparison to a conventional boring where samples were recovered. Once calibration is established, the penetrometer boring normally progresses at a considerably faster rate than the conventional boring.

TEST PITS / In their simplest form, test pits are excavations into the earth that permit visual inspection of the conditions exposed in the walls of the pit. The classification of soil types at various depths is

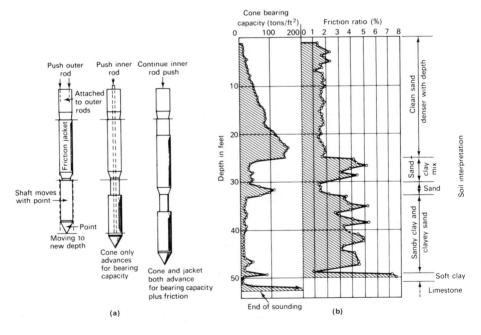

Figure 11-21. *(a) Operation of the friction cone; (b) results from a friction cone sounding. (Ref. 129.)*

possible, along with the opportunity to learn properties of the soil and determine stratum thicknesses. Where desired, soil specimens for testing can be cut from the walls or bottom of the test pit. Test pits can be made manually or with power equipment; backhoes are frequently utilized (see Fig. 11-22).

Test pits provide opportunity for studying subsurface features not possible with borings. Generally, more accurate information as to groundwater elevation is possible (if the pit is deep enough to penetrate to groundwater), and better information on soil variations or unusual features such as presence of underground springs and fissures or cracks in the soil. When the soil deposit includes gravel or boulders, boring samples may not be large enough to recover and identify these items, but such material will be detected in a test pit. If the area under investigation represents a proposed borrow pit, bulk samples are easily obtained for laboratory compaction tests or other analysis.

Natural or man-made cuts in soil deposits offer the same advantage for information as a test pit.

Test pits can be a relatively economical way of obtaining information on subsurface conditions. Frequently, many test pit locations can be dug in one work day with a backhoe. The primary shortcoming of test pit exploration is the limited depth possible with commonly available equipment. Unless a large backhoe or similar equipment is obtainable, the test pits will not penetrate beyond depths of twelve to 15 feet (4 to 5 m). However, where shallow foundations can be used, the uppermost fifteen or so feet of soil is very critical for foundation analysis, and test pits can provide a very useful complement to boring information. When explorations are made for foundation studies, test pits

Figure 11-22. *Test pit excavation with crawler mounted backhoe.* (Courtesy Caterpillar Tractor Co.)

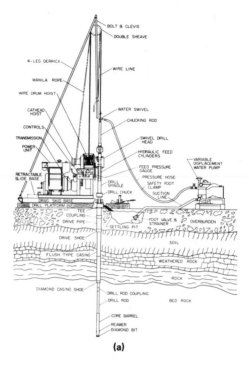

Labels in figure (a):

BOLT & CLEVIS
DOUBLE SHEAVE
4-LEG DERRICK
WIRE LINE
MANILA ROPE
WIRE DRUM HOIST
CATHEAD HOIST
WATER SWIVEL
CHUCKING ROD
CONTROLS
SWIVEL DRILL HEAD
TRANSMISSION
HYDRAULIC FEED CYLINDERS
POWER UNIT
VARIABLE DISPLACEMENT WATER PUMP
FEED PRESSURE GAUGE
RETRACTABLE SLIDE BASE
PRESSURE HOSE
DRILL SPINDLE
SAFETY FOOT CLAMP
DRILL CHUCK
SUCTION LINE
DRAG SKID BASE
DRILL PLATFORM
TEE COUPLING
DRIVE PIPE
FOOT VALVE & STRAINER
OVERBURDEN
SETTLING PIT
DRIVE SHOE
SOIL
FLUSH TYPE CASING
WEATHERED ROCK
ROCK
DIAMOND CASING SHOE
DRILL ROD COUPLING
DRILL ROD
BED ROCK
CORE BARREL
REAMER
DIAMOND BIT

(a)

(b)

Figure 11-23. (a) Diamond core drill setup. (Courtesy of Acker Drill Co.) (b) Skid-mounted rotary drill rig. This type is used in locations having difficult access. Machine shown has soil auger and rock core capabilities. (Courtesy of Central Mine Equipment Co.)

should not be dug at a foundation location, so that the soil eventually required to support the foundation remains undisturbed.

ROCK CORE DRILLING / Rock core drilling relates to the procedure in which rock underlying an area is investigated by coring so as to obtain samples for classification and for determining those properties of rock important to the construction industry.

In a structural design, the close presence of an underground rock surface may be desirable, provided that it does not interfere with construction, for the support capabilities of rock are almost always significantly greater than those of soil. The requirement for rock coring develops with the need to establish that an underground rock material encountered in a boring or test pit is actually bedrock and not a large boulder. Rock coring is commonly performed to determine the quality of rock and to check for possible detrimental properties such as cracks, fissures, and weathering or other deterioration that could affect the strength of the formation. Typically, rock cores are necessary to evaluate a rock's ability to support rock anchors, to determine porosity that could affect the flow of underground water, or to obtain information for a subsurface rock profile and properties should it be necessary to excavate to depths below the top of the rock.

The method used for most rock drilling is similar to the method of rotary drilling in soil. Machinery used for drilling soil borings is frequently the same equipment as that used for drilling rock (Fig. 11-23). For rock work, special bits and core samplers are necessary, however. Cutting bits are hardened steel or steel alloys, or they are diamond chip bits (see Fig. 11-24).

DIAMOND BIT AND ROCK CORE SIZES

Bit Designation	Rock Core Diameter (inches)	(mm)
EX	$\frac{13}{16}$	21
AX	$1\frac{3}{16}$	30
BX	$1\frac{5}{8}$	41
NX	$2\frac{1}{8}$	54
$2\frac{3}{4} \times 3\frac{7}{8}$	$2\frac{11}{16}$	68

Figure 11-24. Standard diamond core bit and summary of common core bit sizes.

Figure 11-25.   *Double-tube rock core barrel.* (Courtesy of Acker Drill Co.)

Figure 11-26.   *Rock core samples being placed in core box for shipping.*
(Courtesy of Acker Drill Co.)

To obtain rock core samples, hollow core barrels are used. Single- and double-tube core barrels (Fig. 11-25) are in wide use. Under rotary action, the core bit advances into the rock. Usually a circulating supply of water is provided to the cutting edge to help flush rock cuttings and dissipate heat. "Core runs" are made to drill the hole in segments, usually up to five feet (1.5 m) in length. The length of core run is limited by the length of core barrel. At the completion of a core run, the barrel and rock sample are brought to the surface; the rock specimen is removed (Fig. 11-26), and the barrel is reinserted for additional drilling.

Presentation of Boring Information

BORING LOGS / Information on subsurface conditions obtained from the boring operation is typically presented in the form of a boring log (boring record). A continuous record of the various soil or rock strata found at the boring is developed. Description or classification of the various soil and rock types encountered, and changes in strata and water level data are considered the minimum information that should constitute a log. Any additional information that helps to indicate or define the features of the subsurface material should also appear on the log. Items such as soil consistency and strength or compressibility can be included. "Field" logs typically consist of the minimum information—classification, stratum changes, and water level readings; see Fig. 11-27(a). A more developed "office" or "lab" log might include laboratory test data presented alongside the boring sample actually tested, so all information pertaining to the various soil or rock types and properties found at a boring location is summarized; see Fig. 11-27(b).

Where subsurface conditions are investigated through use of borings or test pits, the exploration program normally consists of making a number of borings or test pits at different locations. To assist in picturing and understanding underground conditions existing at a site, it has become common to use the boring or test pit series to develop a subsurface profile (Fig. 11-28). This profile helps to define more clearly the subsurface conditions, showing uniformity or variation, and can assist in delineating between good and poor areas. If a subsurface profile is developed as data become available (as the borings are completed), the information can be used for indicating the need and general location for additional borings, or, conversely, indicating that some originally planned borings can be omitted. When subsurface profiles are used, it is important to remember that conditions between borings have been estimated by interpolation, and that actual conditions are known only at boring and test pit locations.

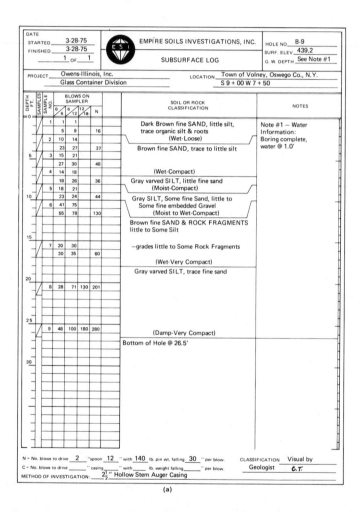

Figure 11-27. (a) Empire Soil Investigations boring log. (Courtesy of Empire Soil Investigations)

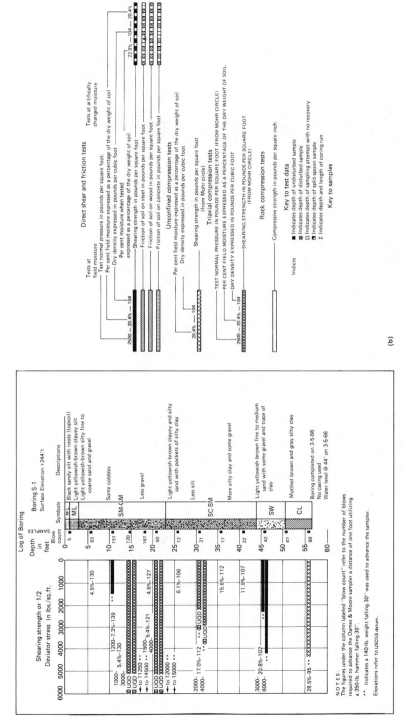

Figure 11-27. (b) Dames and Moore boring log. (Courtesy of Dames and Moore)

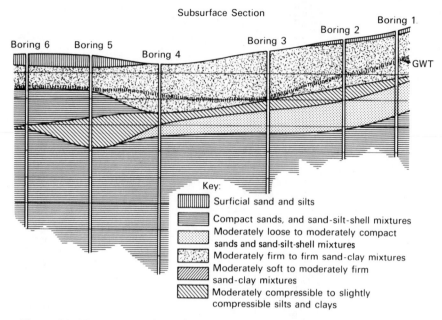

Figure 11-28. *One method of presenting a subsurface profile to summarize boring results.*

Recognizing Limitations of Boring Data

It is accepted that borings are providing information on subsurface conditions only at the actual drilling location, and that the standard practice of interpolating between borings to determine conditions does involve some degree of uncertainty. However, it is also important to recognize that there are limitations inherent to the information shown on the typical driller's log. Members of field crews employed by boring contractors are primarily drilling tradesmen. Such individuals typically have limited experience in detailed soil classification and have virtually no familiarity with the importance of subsurface conditions on the features of building design and construction. Important items of information can be innocently passed over by a driller whose major interest is in the rate of drilling progress, for borings typically are contracted for on a footage basis. Consequently, where an exploration is for a final design, (and often for preliminary work as well) it is becoming increasingly common to have technically trained personnel assigned to the drilling crew, in order to examine and classify recovered soil samples, to direct the depth at which samples should be taken, to select the sequence of drilling borings, and to document other factors relating to surface and subsurface conditions that could have an influence on design or construction. Such an individual is typically from a designer's or foundation consultant's office. After the field work is completed, the individual remains an accessible resource person capable of answering questions if issues develop during the design period.

Geophysical Methods

The determination of subsurface materials through the use of borings and test pits can be time-consuming and expensive. Considerable interpolation between checked locations is normally required to arrive at an area-wide indication of conditions. Geophysical methods involve the technique of determining underground materials by measuring some physical property of the material and, through correlations, using the obtained values for identification. Most geophysical methods determine conditions over a sizable distance. Frequently, this is an advantage over the "point" checking accomplished by borings and test pits. Most geophysical measurements can be rapidly obtained. Thus, the methods lend themselves well to the checking of large areas.

In the engineering-construction profession, two types of geophysical investigation have been found useful. They are the seismic refraction method and the electrical resistivity method. Though these methods

Figure 11-29. *Seismic refraction study being performed in field.* (Courtesy of Soiltest, Inc.)

have proven to be reliable, there are also certain limitations as to the data that can be obtained. Thus, at the present time, subsurface investigations can rely heavily on geophysical methods, but conditions should at least be spot checked with borings or test pits. Typically, when a thorough investigation is made, a number of borings will be required in order to obtain test samples to make accurate determinations of soil properties such as strength and compressibility. It is these borings that can provide the detail required to check and complement the geophysical data.

SEISMIC REFRACTION / When a shock or impact is made at a point on or in the earth, the resulting seismic (shock or sound) waves travel through the surrounding soil and rock at speeds relating to the density and bonding characteristics of the material. In refraction seismology the *velocity* of seismic waves passing through subsurface soil or rock materials is determined, and the magnitude of the velocity is then utilized to identify the material. A seismograph, the instrument used to make a seismic refraction study, consists of a shock- or impact-inducing mechanism, such as an impact hammer or small explosive, plus a receiver to indicate when the seismic wave reaches a point at a particular known distance from impact, and also a timing instrument for measuring the time for the wave to travel the distance from the point of impact to the point of measurement. In shallow refraction seismology, as used for determining subsurface conditions for construction purposes, the shock impact is created with a sledge hammer hitting a striking plate placed on the ground (Fig. 11-29). The seismic wave is then picked up by a sensitive geophone. (The geophone is actually a transducer, an electromechanical device that detects vibrations and converts them into electric signals that can be measured.)

The field survey involves obtaining a series of geophone readings at different distances along a straight line directed from the impact point. For geophone spacings close to the strike plate, the vibrations picked up by the geophone will be from those direct waves traveling through the upper layer of earth material; see Fig. 11-30. For direct waves traveling through the upper layer, the time to reach the geophone is proportional to the distance from the point of impact.

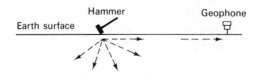

Figure 11-30. Method of imparting sound waves into a soil.

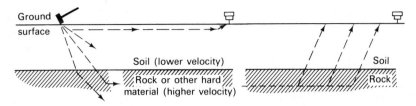

Figure 11-31. *Travel of sound waves through different subsurface materials.*

When the surficial layer is underlain by a harder layer, the seismic waves from the strike plate also progress downward and enter the harder layer. The seismic velocity will be greater in the harder material. Waves traveling through the upper portion of the harder layer transfer energy back into the upper layer through the surface of contact. This energy becomes a refracted wave. For large strikeplate-to-geophone distances, the refracted wave will reach the geophone more quickly than the direct wave. Even though the path of travel is longer, this occurs because part of the path is through the harder, high-velocity material; see Fig. 11-31.

Seismic velocities for the earth materials in the upper and lower strata would be obtained from plotting the measured values of geophone distance and time on a travel-time graph (both coordinate values use an arithmetic scale); see Fig. 11-32. The slope of the first segment of the plot represents the seismic velocity for the material in the upper stratum. The slope of the second segment is the seismic velocity for the deeper layer. To obtain data that properly define the plotted segments, it is recommended that the maximum geophone distance be about five times the depth of investigation.

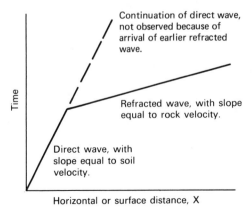

Figure 11-32. *Typical travel-time graph for soil overlying rock.*

Figure 11-33(a) illustrates a soil layer-rock layer subsurface and indicates the type of seismic wave (direct or refracted) picked up at the various geophone positions. The resulting travel-time graph appears in Fig. 11-33(b).

Seismic velocity values representative of different earth materials and conditions are shown in Table 11-2.

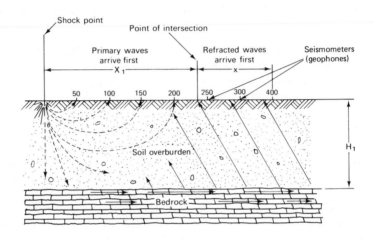

(a)

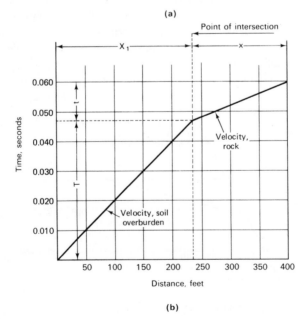

(b)

Figure 11-33. *Seismic refraction conditions and data: (a) subsurface conditions showing direct and refracted sound waves; (b) travel-time plot for conditions in (a).* (Courtesy of Acker Drill Co.)

TABLE 11-2 / REPRESENTATIVE SEISMIC VELOCITY VALUES

(Velocity in ft/sec)

Soil—Unconsolidated Material		Rock—Consolidated Material	
Most unconsolidated materials	Below 3000	Most hard rocks	Above 8000
Soil—normal	800 to 1500	Shale—soft	4000 to 7000
—hard packed	1500 to 2000	—hard	6000 to 10,000
Water	5000	Sandstone—soft	5000 to 7000
Loose sand—above water table	800 to 2000	—hard	6000 to 10,000
—below water table	1500 to 4000	Limestone—weathered	As low as 4000
Loose mixed sand and gravel, wet	1500 to 3500	—hard	8000 to 18,000
Loose gravel, wet	1500 to 3000	Basalt	8000 to 13,000
Hard clay	2000 to 4000	Granite and unweathered gneiss	10,000 to 20,000
		Compacted glacial tills,	
		hardpan, cemented gravels	4000 to 7000
		Frozen soil	4000 to 7000
		Pure ice	10,000 to 12,000

Note: Occasional formations may yield velocities that lie outside of these ranges)   (Source, Soiltest, Inc.)

Where a two-layer condition exists, the thickness or depth of the upper layer can be determined from

$$H_1 = \frac{X_1}{2} \sqrt{\frac{V_2 - V_1}{V_2 + V_1}} \qquad (11\text{-}1)$$

where H_1 is the depth or thickness of the upper layer.

X_1 is the distance, taken from the travel-time graph, where the two plotted slopes intersect.

V_1, V_2 are seismic velocities in the upper and lower layer, respectively.

For the condition of three successively harder layers existing in an area, the travel-time graph will show three different slopes. The seismic velocity for each of the materials is the slope of the respective segment of the plot. The thickness of the upper layer can be calculated from Eq. (11-1). The thickness of the intermediate layer, H_2, can be determined from

$$H_2 = 0.85\, H_1 + \frac{X_2}{2} \sqrt{\frac{V_3 - V_2}{V_3 + V_2}} \qquad (11\text{-}2)$$

where H_1 is the thickness of the upper layer.

X_2 is the distance from the travel-time graph, where plotted segments 2 and 3 intersect.

V_2, V_3 are seismic velocities of layers 2 and 3 as determined from the travel-time graph.

There are certain significant limitations to the use of the seismic refraction method for determining subsurface conditions. These include:

1. The method should not be used where a hard layer overlies a softer layer, because there will be no measurable refraction from a deeper soft layer. Refraction seismic test data from such an area would tend to give a single-slope line on the travel-time graph, indicating a deep layer of uniform material.

2. The method should not be used on an area covered by concrete or asphalt pavement, because these materials will represent a condition of a hard surface over a softer stratum.

3. A frozen surface layer may give results similar to those obtained where a hard layer is over a soft layer, because of the velocity increase resulting from the better wave transmission through the more "solid" frozen material.

Further, some topographic and underground features will give seismic data that are difficult to interpret fully and correctly. Such situations include the condition of an irregular or dipping underground rock surface, the condition where discontinuities such as rock faults or earth cuts or banks exist, where layers having gradual changes in their velocity values occur, and the condition of thin layers of varying materials. Because of the possibility for misinterpretation of data with such occurrences, the seismic analysis should be performed by trained personnel, and as a minimum, spot checks should be made with borings or test pits.

   Where rock exists at or close to the surface at a construction site, predetermining the methods necessary for removal (to lower the surface elevation) or for excavation becomes important for estimating, bidding, and scheduling purposes. Seismic velocity data have been used to determine when the rock material is capable of being ripped with dozer rippers (or other similar equipment) and when drilling and blasting are required. Such seismic rippability information, as developed by one major construction equipment manufacturer, is shown in Fig. 11-34.

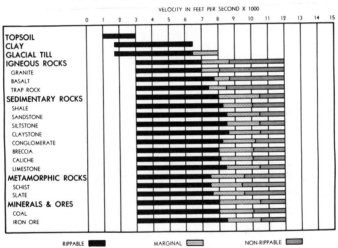

Figure 11-34. Seismic rippability chart developed by Caterpillar Tractor Co. for D9 with mounted No. 9 ripper. (Courtesy of Soiltest, Inc.)

ELECTRICAL RESISTIVITY / Resistivity is a property possessed by all materials. The electrical resistivity method for determining subsurface conditions utilizes the knowledge that in soil and rock materials the resistivity values differ sufficiently to permit that property to be used for identification purposes.

   To determine resistivity at a site, electrical currents are induced into the ground through the use of electrodes. Soil or rock resistivity can then be determined by measuring the change in electrical potential (voltage) between known horizontal distances within the electric field created by the current electrodes (see Fig. 11-35).

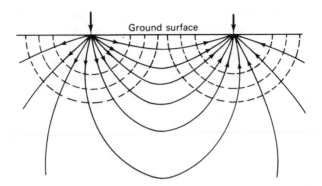

Figure 11-35. *Path of current flow (solid lines) through a soil in the electrical resistivity study (dotted lines are equipotential surfaces).*

A schematic diagram of equipment for resistivity testing and including electrode arrangements (the Wenner configuration) is shown in Fig. 11-36. With four electrodes equally spaced along a line as indicated, the resistivity is calculated from

$$\beta = 2\pi S \left(\frac{V}{I}\right) \tag{11-3}$$

where β is the resistivity of the earth material in ohm-feet.
 S is the electrode spacing.
 V is the difference in potential (volts) between the inner electrodes.
 I is the current flowing between the outer electrodes.

The calculated resistivity value is an apparent resistivity. This is a weighted average of all the earth material within the zone created by the electrodes' electric field. The depth of material included in the measurement (depth of penetration) is approximately the same as the spacing between electrodes.

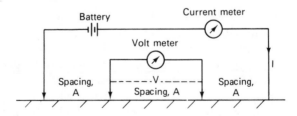

Figure 11-36. *Wenner configuration for arrangement of voltage and current electrodes.*

Two.different field procedures are in use for obtaining information on subsurface conditions. One method, "electric profiling," is well suited for establishing boundaries between different underground materials and has practical application in prospecting for sand and gravel deposits or ore deposits. The second method, "sounding," can provide information on the variation of subsurface conditions with depth and has practical application in indicating layered conditions and approximate thicknesses. It can also provide information on depth to a water table or water-bearing stratum. Fig. 11-37 shows field work for an electrical resistivity study being performed.

In the *electrical profiling* method, an electrode spacing is selected, and this same spacing is used in running different "profile" lines across

Figure 11-37. *Electrical resistivity field study being performed.* (Courtesy of Soiltest, Inc.)

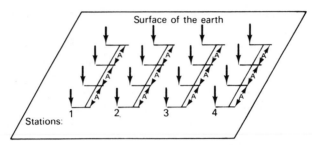

Figure 11-38. *Electrode arrangement in a profiling survey.*

an area; see Fig. 11-38. The information obtained applies for the particular location of the line and is sometimes referred to as an "electrical trench," implying that the subsurface data revealed are similar to the information that would be obtained by examining an open-trench excavation.

The information resulting from a profile line can most simply be plotted on arithmetic coordinates, as shown in Fig. 11-39. A change in the plotted curve indicates a change in the underground material. From the series of profile lines, boundaries of areas underlain by different materials can be established on a map of the area (areal map).

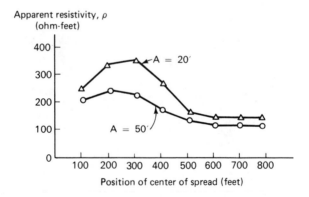

Figure 11-39. *Apparent resistivity versus position of the center of electrode spread for two values of electrode separation A.* (Source: Soiltest, Inc.)

In the *electrical sounding* method, a center location for the electrodes is selected and a series of resistivity readings is obtained by systematically increasing the electrode spacing, as indicated by Fig. 11-40. Since the *depth* of information recovered is directly related to electrode spacing, the series of resistivity data obtained from successively increased electrode spacings will indicate changes of resistivity with depth, and hence information on layering of materials. This method is capable of indicating subsurface variations where a hard layer underlies a soft layer, and also the condition of a soft layer underlying a hard layer.

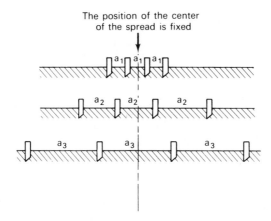

Figure 11-40. *Representative electrode position during a sequence of sounding measurements.*

Data can be presented on Cartesian (arithmetic scale) coordinates or logarithmic coordinates, as indicated in Fig. 11-41. Logarithmic coordinates are the most popular.

The spacing of electrodes is important. To obtain data that give a reliable indicating of conditions, the closest electrode spacing should be

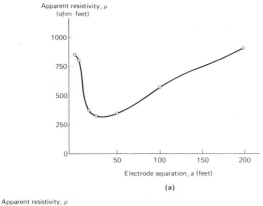

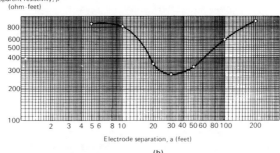

Figure 11-41. *Illustration of two methods for graphing sounding data: (a) Cartesian (arithmetic) coordinates; (b) logarithmic coordinates.*

no more than half the estimated thickness of the upper stratum. Three feet is the minimum recommended spacing, however. The largest spacing is between five and ten times the total depth of interest. The number of intermediate spacings selected should be adequate to provide sufficient points to plot a well-defined curve.

Representative sounding curves for some typical subsurface conditions appear in Fig. 11-42.

CORRELATION BETWEEN RESISTIVITY AND EARTH MATERIALS / In earth materials, resistivity decreases with increasing water content and increasing salt concentration. Increasing degrees of water and content and salinity make it easier for an electrical current to flow through the material. Consequently, nonporous materials (holding little water) will have high resistivity values. Such materials include most igneous and metamorphic rock, plus some dense sedimentary rock, such as dense limestone and sandstone. In soil materials, clean gravel and sand have a relatively high resistivity value. Silts, clays, and coarse-grained-fine-grained soil mixtures have comparatively low resistivity values. Soil formations in nonglaciated areas typically have lower resistivity values than soils in glacial areas.

Representative values of resistivity for commonly occurring earth materials are presented in Table 11-3.

TABLE 11-3 / REPRESENTATIVE RESISTIVITY VALUES

Types of Materials	Resistivity Ohm-ft
Wet to moist clayey soils	5 to 10
Wet to moist silty clay and and silty soils	10 to 50
Moist to dry silty and sandy soils	50 to 500
Well-fractured to slightly fractured bedrock with moist soil-filled cracks	500 to 1000
Sand and gravel with silt	1000
Slightly fractured bedrock with dry soil-filled cracks. Sand and gravel with layers of silt.	1000 to 8000
Massive bedded and hard bedrock. Coarse dry sand and gravel deposits.	8000 (plus)

(Source, Soiltest, Inc.)

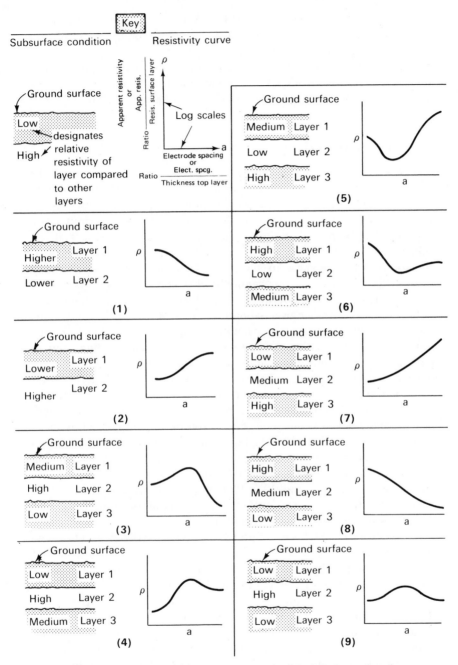

Figure 11-42. *Representative resistivity curves for differing subsurface conditions.*

THICKNESS OF LAYERS / A method of obtaining a reliable and accurate determination of the depth to, or thickness of, soil layers by using resistivity data is currently not available. Approximation methods exist, but they are cumbersome to handle and not particularly suited for use in field work. With present knowledge, depth or thickness information is best obtained from borings or by seismic methods.

In-place Testing

In-place testing, in the soil and foundation profession, refers to the procedure of determining soil properties or other subsurface conditions at the actual surface or subsurface location. In-place (in-situ) testing includes shear strength determinations, permeability determinations, in-place density determinations, plate-bearing and settlement determinations, lateral movement determinations, and pore pressure determinations. In-place shear strengths are determined where undisturbed samples for laboratory testing cannot be obtained, or where it is desired to eliminate the need to obtain samples. Field permeability determinations offer an advantage over laboratory tests, in that a larger volume of soil is being evaluated in its natural environment. Plate-bearing tests are used to simulate spread footing foundation loading, and to obtain load-settlement information on soil that will be supporting shallow foundations or structural slabs (floors, pavements). Lateral movement determinations are of interest where the stability of earth masses and slopes, or of embankments or retaining walls is being monitored, or if there is need to know about lateral soil movement caused by foundation or other loading. Underground lateral movements are measured with slope inclination indicator equipment. Methods for determining field permeability are described in Chapter 5. In-place densities are discussed in Chapters 4 and 12. Plate-bearing tests are discussed in Chapter 9. In-place shear testing, slope inclination equipment, and pore pressure devices are discussed below.

IN-PLACE SHEAR TESTS—VANE SHEAR / The vane shear test consists of inserting a vane (Fig. 11-43) into the soil and rotating by applying a torque. The torque is measured, and for a known-size vane is easily related to the shearing strength of the soil. In field explorations, the vane testing is typically performed at different depths in a boring as it is being drilled. The soil to be tested should be undisturbed by the boring operation. The vane shear test can be performed on any soil, provided that the vane can be inserted into the soil without causing significant disturbance. Practically, then, the method is useful primarily for fine-grained soils. For typical field or laboratory conditions, where little or no overburden pressure exists, the shear value being determined is primarily cohesion.

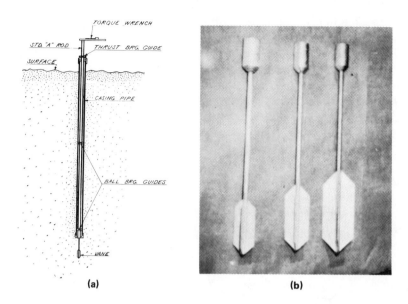

Figure 11-43. *In-place vane shear test apparatus: (a) schematic diagram of vane shear test procedure; (b) vanes for vane shear testing. (Courtesy of Acker Drill Co.)*

SLOPE INCLINATION INDICATORS / Slope inclination equipment provides information by indicating the slope or variation from a vertical axis at desired depths in a special casing installed in the ground; see Fig. 11-44. A device to determine the variation-from-vertical runs in special grooves or tracks on the interior of the casing.

Figure 11-44. *Slope indicator in use on a field project (equipment shown is a Soiltest slope meter probe). (Courtesy of Soiltest, Inc.)*

The pendulum principle is normally used within the device to establish the vertical axis and for reference. The slope-versus-depth data are recorded. If data collected at different times show a change in slope, this would indicate that lateral movement is occurring. Knowing the slope at different points from top to bottom of the casing permits the lateral position at any depth to be calculated.

PORE PRESSURES AND PIEZOMETERS / Pore pressure determinations refer to the procedure of measuring the water-pressure values developed in the void or pore spaces of an underground soil mass. Usually, the interest is in the "excess" pore water pressure, or the magnitude of pressure greater than a normal hydrostatic pressure resulting from the elevation of a groundwater table (phreatic surface). Excess pore water pressures can develop in fine-grained soil strata when a new structural loading is placed on the earth above the fine-grained soil, as discussed under "Consolidation" in Chapter 9. Excess pore pressures

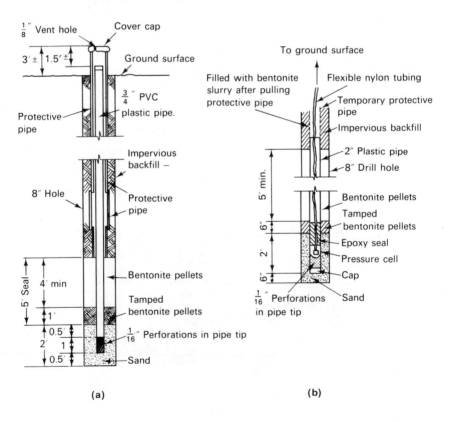

Figure 11-45. *Illustrations of piezometer types: (a) open tube type for use in permeable soil; (b) pressure cell for use in impermeable soil.*

provide information on the magnitude of stresses caused in the fine-grained soil by the new loading. On a typical construction project, pore pressure information is utilized to control the rate at which new loading occurs, to keep the stress resulting in the supporting soil from exceeding the strength of that soil.

The piezometer is a device to measure pore water pressure. In its simplest form, a piezometer would consist of an open tube or standpipe with its tip inserted into the soil layer to be checked. Before construction, the water level observed would be the phreatic surface level. As construction loading results in increased underground stresses, the water level in the open standpipe rises. The rise in elevation multiplied by the unit weight of water gives the magnitude of excess pore pressure. As the stressed soil consolidates, the pore pressure decreases, and the water level in the piezometer drops.

The simple piezometer just described would be limited to use for soil having a fairly high permeability. With fine-grained soil of low permeability, the time lag for changes in pore water pressures to be measured would be excessive. To indicate changes in pore pressures quickly, modifications (some extensive) have been made to the basic piezometer. The result is that some sophisticated types of equipment have evolved. Most types still include the piezometer unit, which has to be installed to the depth where readings are to apply, and pressure or gage lines which are carried to a monitor located on the ground surface. Some piezometer types have special conditions for which they are best suited. With some complex units, installation has to be accomplished by specially trained personnel. The diagrams in Fig. 11-45 show some piezometer types.

PROBLEMS

1. What publicly available maps serve as good references for obtaining information about surface conditions in an area?

2. List and describe some of the advantages of learning about land surface features from aerial photographs.

3. Outline the general procedure used to drill soil borings and obtain soil samples for classification and testing.

4. In soil borings, what is the standard penetration test?

5. Compare advantages and disadvantages of test pits versus soil borings for obtaining information on subsurface conditions.

6. Make a comparison between the static cone penetrometer and standard penetration test methods for determining subsurface soil conditions.

7. Prepare a boring log from the following information, as obtained from a $2\frac{1}{2}$-in. boring where standard split spoon soil samples were taken.

Sample Depth, in.	Blow Count N	Soil Classification
$2-3\frac{1}{2}$	3–4–4	Brown fine to medium sand
$7-9\frac{1}{2}$	7–8–7	Brown fine to medium sand
$12-13\frac{1}{2}$	8–9–10	Brown fine to medium sand in upper part of sample
		Red-brown clay in lower part of sample
$17-18\frac{1}{2}$	9–12–11	Red-brown clay
$22-23\frac{1}{2}$	9–25–27	Gray fine to coarse sand, silt and clay, occasionally gravel (compact glacial till)
$27-28\frac{1}{2}$	29–35–39	Gray fine to coarse sand, silt, clay with gravel (compact glacial till)
$31-31\frac{1}{2}$	100	Refusal (no sample recovered)

Water level encountered at 21 ft.

8. (a) Briefly describe the principles on which seismic refraction studies for subsurface explorations are based.
   (b) Briefly describe the principles on which electrical resistivity studies for subsurface explorations are based.

9. List and briefly describe the type of subsurface information that seismic refraction studies can provide, and also the limitations on information that can be obtained.

10. List and briefly describe the type of subsurface information that electrical resistivity studies can provide, and the limitations to data that can be obtained.

11. A seismic refraction study made for an area provides the following field data:

Distance from Impact Point to Geophone, feet	Time to Receive Sound Wave, seconds
50	0.025
100	0.05
200	0.10
300	0.11
400	0.12

(a) Graph the time-travel data, and determine the seismic velocity for the surface layer and underlying layer.

(b) Determine the thickness of the upper layer.

(c) Using the seismic velocity information, give the probable earth materials in the two layers.

12. List and briefly discuss the major advantages and disadvantages of in-place shear tests, such as those performed by the in-place vane shear.

CHAPTER 12
Earth Moving, Compaction, and Stabilization

In the natural location and condition, soil provides the foundation support for many of man's structures. But soil is also extensively used as a basic material of construction, as witnessed from the existence of earth structures such as dams, dikes, and embankments for roads and airfields. For situations where the natural topography needs to be changed to make the area more suitable for building development, soil is the material most used for filling low locations. The desirability of utilizing soil as a building material stems from its general availability, its durability, and its comparatively low cost.

When soil is used for construction purposes, it is typical for it to be placed in layers to develop a final elevation and shape. Each layer is compacted before being covered with a subsequent layer. Properly placed and compacted, the resulting soil mass has strength and support capabilities that are as good or better than many natural soil formations. In the case of earth structures such as dams, the compacted earth is capable of supporting itself and the forces to which it is subjected. With earth fills, it is possible to support buildings, highways, and parking areas on the compacted soil mass. Such soil is referred to as a *compacted earth fill* or a *structural earth fill*.

Whether soil is used as the foundation material to support buildings, roads, or other structures, or is used to build the structure itself (e.g., an earth dam), it is desirable that the in-place material possess certain properties. The soil should have adequate strength, should be relatively incompressible so that future settlement is not significant, should be stable against volume change as water content or other factors vary, should be durable and safe against deterioration, and should possess proper permeability. These desirable features can be achieved with a compacted fill by proper selection of the fill soil type and by proper placement. The important properties of a fill could be checked independently, but the more desirable characteristics such as high strength, low compressibility, and stability are normally associated with high-density (or unit weight) values and hence will result from good compaction.

Virtually any type of soil can be used for structural fill, provided that it does not contain organic or foreign material that would decompose or otherwise undergo change after it is in place. Granular soils are considered the easiest to work with on a construction site. This material is capable of developing high strength with little volume change expected after compaction. Permeabilities are high, which can be an advantage or disadvantage. Generally, compacted silts are stable, are capable of developing fairly good strength, and have limited tendency for volume change. Silty soils can be difficult to compact if wet, or if work is performed in wet periods. Permeability is low. Properly compacted clay soils will develop relatively high strengths. Their stability against shrinkage and expansion is related to the type of clay mineral; for example, the montmorillonite clays would have a greater tendency for volume change than the more stable kaolinite clays. Compacted clays have a very low permeability, a factor which can be used to advantage where movement of water needs to be restricted. Clay soils cannot be properly compacted when wet. Working with clay soil is difficult under wet conditions.

Field Procedures— General Considerations, Methods, and Techniques

The field procedure for constructing a compacted fill is simple in principle. The fill soil is transported to the area being worked, where it is spread in relatively thin layers, and each layer is then compacted to a predetermined density (unit weight). However, proper accomplishment of these requirements involves consideration of the soil type, its water content, and the type of equipment used for the compacting operation.

On big projects, where the soil fill is obtained from an earth cut or borrow pit on or near the construction site, self-propelled scrapers (Fig. 12-1) can be considered for hauling and spreading the fill. Self-propelled scrapers are widely used because of their excavating and self-loading capabilities, and the relatively fast haul speeds (see Figs. 12-2 and 12-3). With scraper operations, bulldozers or graders frequently

Figure 12-1. Self-propelled scraper spreading soil for an earth fill project. (Courtesy of Caterpillar Tractor Co.)

Figure 12-2. Scrapers self-loading in borrow pit. (Courtesy of Terex Division, General Motors Corp.)

Figure 12-3. Scraper being loaded in borrow area with assistance of push-dozer, a procedure to reduce time required for loading. (Courtesy of Terex Division, General Motors Corp.)

Figure 12-4. *Bulldozer utilized to spread soil fill.* (Courtesy of Terex Division, General Motors Corp.)

work in the fill area to help keep newly-placed uncompacted fill uniformly spread (Figs. 12-4 and 12-5). Compaction equipment follows the spreading equipment.

　　If the fill material is obtained from an off-site source, trucks are generally used for transportation (see Fig. 12-6). At the fill location, the soil is dumped and then spread with dozers and graders. Compaction follows the spreading operation.

　　Trucks are also commonly used for transporting the soil on cut and fill projects, particularly when the fill material is obtained from a borrow pit area where excavation can be efficiently handled by power shovel equipment or when the hauling distance is great; see Figs. 12-7, 12-8, and 12-9.

　　When the site being worked is of a limited area and fill is obtained

Figure 12-5. *Grader of type used on fill projects to spread soil and construct drainage ditches. Photo shows grader working on a roadside ditch.* (Courtesy of Galion Division of Dresser Industries, Inc.)

Figure 12-6. Front-end loader used in borrow area. Truck type shown is typically used for on-site or highway hauling. (Courtesy of Fiat-Allis Construction Machinery, Fiat-Allis Chalmers)

from an on-site cut, bulldozers may be used to perform excavation and to push the soil to the fill area, where it is spread and compacted.

The thickness of layers that can be properly compacted is known to relate to the soil type and method or equipment of compaction. Typically, granular soils can be adequately compacted in thicker layers than

Figure 12-7.   *Rear dump truck of type used for on-site hauling.*   (Courtesy of Caterpillar Tractor Co.)

Figure 12-8.   *Bottom drump truck, typically utilized for on-site hauling.* (Courtesy of Caterpillar Tractor Co.)

the fine-grained silt and clay soils. Generally, for a given soil type, heavy compaction equipment is capable of compacting thicker layers than light equipment.

The types of equipment commonly used for compaction include

Figure 12-9. *Power shovel loading operation.* (Courtesy of Terex Division of General Motors Corp.)

pneumatic or rubber tire rollers, drum-type rollers having projecting feet or lugs, such as the sheep's-foot roller, vibratory compactors, which impart vibrations into the soil, and smooth drum rollers. All types are available in a variety of sizes and weight. The vibratory compactors are most effective on cohesionless soils. The sheep's-foot and similar rollers having feet or other projections should be limited to cohesive soils. Pneumatic-tire rollers can be used with effectiveness on all soil types. The smooth drum rollers have a very limited effective depth of compaction, and hence their use should be restricted to situations where only thin layers or a surface zone needs to be compacted.

Hand-operated tamping and vibratory compactors are available for working in limited spaces and for compacting soil close to structures, where care is required to prevent damage.

The water content of the fill soil has an effect on its ability to be well compacted with reasonable effort. The fine-grained silt and clay soils, and granular soils containing fine materials, are particularly affected by variations in water content. Soil that is too dry is difficult to compact. Material that is too wet may be difficult to spread properly and to compact. When the soil is too wet at its source, measures should be taken to dry it before compaction is attempted at the fill area. Drying can be accomplished by scarifying and aerating the soil at the borrow pit or cut area, or by hauling it to an open area and loosely spreading it to permit aeration, or by letting it aerate after spreading across the fill area. These procedures presume that space and time are available. When the fill soil is too dry, water should be added and mixed throughout. Wetting can be achieved by spraying water at either the cut or fill area with a water truck, followed by a mixing operation performed with the excavating or earth-moving equipment. If the dry soil is spread across the fill area before wetting, dozers and graders can frequently perform some mixing after spraying by "rolling" the earth on the front of their blades as they work to even out the layer thickness. Where the fill source is a large borrow area, ponds can be created and flooded. This procedure permits the water to percolate through the soil before it is excavated and transported. Fine-grained soils will require a longer ponding period than granular soil.

When an area is being filled, the ability of the first fill layers to be properly compacted will depend on the condition of the natural material being covered. If weak or compressible soil exists, it may be very difficult to compact the fill properly. If poor material is left in place and covered over, it may compress over a long period under the weight of the earth fill, causing settlement cracks in the fill or any structure supported by the fill. Consequently, where the postconstruction settlement and deformation of an earth fill has to be limited, it is common practice to require that poor soil and nonsoil materials (vegetation, garbage, other waste material) be removed before filling commences.

To determine if the surface zone of the natural soil is adequate for supporting the compacted fill, the area can be "proofrolled." Proofrolling consists of utilizing a piece of heavy construction equipment (typically, heavy compaction equipment or hauling equipment) to roll across the fill site and watching for poor areas to be revealed. Poor areas will be indicated by the development of rutting or ground weaving. Where the height of a structural fill will be limited, even marginally poor soils should be removed. If the height of fill is to be great, marginal materials often can be left in place without future adverse effect.

Placement of a compacted fill may have to begin at an elevation that is below the water table. This requirement frequently develops where the excavation of poor material lowers the beginning working surface. Fill placed under water should consist of coarse, granular material. Fine-grained soil should not be used. The thickness of the initial fill layer may have to be considerably greater than the normal fill layer (several feet thick) in order to create a working pad capable of supporting equipment used for the subsequent placement and compaction of fill. A procedure that has been followed with success in situations where the disturbance to the in-place underwater soil must be minimized is to create a working base of very coarse material, such as cobble sizes. Clean sand and gravel materials are placed over this base. Close to and above the water surface, the granular material can be compacted with conventional equipment. Once above the water table, the placement and compaction of fill layers can proceed in the normal manner; such fill can consist of fine-grained soil as well as granular soil.

The working surface of earth fills made with fine-grained soils can deteriorate rapidly if exposed to accumulations of water from rain or other causes. These fills should be constructed so that rapid surface drainage can occur. This need can be satisfied by placing fill so as to achieve a cambered or turtle-backed surface. When the fill area is lower in elevation than the surrounding terrain, as is frequently the situation during early stages of construction, interceptor drainage ditches should be provided around the perimeter of the fill site. Accumulating surface water is normally not a problem in fills constructed of granular soils, because the material's high percolation rate permits rapid dispersing. The major detrimental effect of surface water on granular soil fills is the erosion possibility on slope areas.

Compaction of backfill in narrow trenches and against buildings normally cannot be accomplished by using the compaction equipment utilized for areal fills. Small or hand-operated equipment is required, and progress may be comparatively slow. Soil densification in such confined spaces may be attempted by using flowing water, such as from a hose. This method can be successful where the backfill consists of clean granular material and the material surrounding and underlying the fill is also a coarse soil, for the flowing water percolates rapidly through the soil and does not puddle. This rapid flow causes a rearrangement of soil particles, and densification results. The method should *not* be attempted where the backfill is a fine-grained soil or where the surrounding soils do not possess a high permeability. With such conditions, the water will not flow rapidly through the fill and will not densify the soil. The water may puddle in the excavation, causing a loose soil structure to remain after drying occurs. For fine-grained backfills, compaction should be achieved through the use of compaction equipment and by working with thin soil layers.

Field Compaction Equipment

The construction of a structural fill usually consists of two distinct operations—the placing and spreading in layers, and then the compaction process. The speed of the compaction operation is typically the more critical of the two steps and often controls the rate of a job's progress. The use of adequate and proper compaction equipment becomes a matter of economic necessity for contractors on almost all earth fill projects. Because of this need, various types of special compaction equipment have been developed for the construction industry. Some equipment has been designed to be specially effective for a particular soil type, whereas other equipment is for general or all-purpose use.

Soil compaction or densification can be achieved by different means—by tamping action, kneading action, by vibrating, or by impact. Compactors operating on the tamping, kneading, and impact principle are effective on cohesive soils. For cohesionless soils, equipment operating on the kneading, tamping, and vibratory principle are effective.

SURFACE COMPACTION EQUIPMENT / The *sheep's-foot roller* and similar rollers having projecting studs or feet are examples of equipment that compacts by a combination of tamping and kneading. Typically, these compactors consist of a steel drum manufactured with small projects; see Figs. 12-10 and 12-11. With most rollers, the drum can be filled with water or sand to increase the weight. As rolling occurs, most of the roller weight is imposed through the projecting feet. Contact pressures imposed by the projections can be fairly high. Pressures vary from about 100 psi (700 kN/m^2) for the lighter equipment to over 600 psi (4200 kN/m^2) for the heavier equipment in common use.

When a loose soil layer is initially rolled, the projections sink into the layer and compact the soil near the lowest portion of the layer. In subsequent passes with the roller, the zone being compacted continues to rise until the surface is reached. This continually rising effect experienced by the compactor is referred to as "walking-out" by the compactor. Such equipment is well suited for compacting clay and silt-clay soil.[1] The depth of layer that can be well compacted relates to the length of the projecting feet and the compactor weight. It is conventionally assumed that the larger, heavy units will properly compact layers on the order of 12 inches thick in three to five passes.

Figure 12-10. *Sheep's-foot rollers in use on a cohesive soil fill, being pushed-pulled by crawler tractors.* (Courtesy of Fiat-Allis Construction Machinery, Fiat-Allis Chalmers)

---

[1] Projection-type rollers are not recommended for cohesionless soil, because the studs (feet) continuously loosen the fill surface.

Figure 12-11.   *Closeup of projecting feet on a sheep's-foot roller.*

Small, light equipment is limited to working layers less than six inches thick where high soil densities are required.

Sheep's-foot-type compactors are available as both self-propelled units and as rollers only. A separate roller requires a tow tractor.

Pneumatic tire rollers compact primarily by kneading. This type of compaction equipment is available in a variety of designs, ranging from the conventional two-wheel-per-axle units to the multiwheel per axle unit (Fig. 12-12). Some equipment is provided with a "wobble-wheel" effect, a design in which a slightly weaving path is tracked by the traveling wheel. Another type is provided with an axle construction that permits individual wheels to follow the ground surface so that low spots are not skipped over.

Figure 12-12.   *Pneumatic roller being used to compact sand base for a highway. (Courtesy of Galion Division of Dresser Industries, Inc.)*

Pneumatic tire compactors are usually outfitted with a weight box or ballast box so that the total compaction load can be easily varied. Ground contact pressure can also be controlled somewhat by varying air pressure in the tires.

Pneumatic tire rollers are available as self-propelled units (Fig. 12-12) and as towed units (Fig. 12-13). These compactors are available in a wide range of load sizes, the heaviest having a capacity of about 200 tons (1800 kN). However, the heaviest units in common use are in the 50-ton (450-kN) range.

The pneumatic tire rollers are effective for compacting both cohesive and cohesionless soils, and are the best type of equipment for general compaction use. Light rollers (about 20 tons or 20,000 kg) are generally considered to be capable of properly compacting layers on the order of six inches (15 cm) thick with few passes. Equipment in the 40- to 50-ton (40,000- to 50,000-kg) category will usually compact layers on the order of 12 inches (30 cm) in thickness with three to five passes. The very heavy equipment is presumed to compact layers up to about one and one-half feet (45 cm) in thickness with a limited number of passes.

The effectiveness of pneumatic tire compaction is not limited to the specially made compaction equipment. Other heavy-tire equipment, such as trucks, graders, and scrapers is capable of providing an effective job of compaction (Fig. 12-14). In emergencies, such equipment can be pressed into service.

Vibratory compactors are available as vibrating drum, vibrating pneumatic-tire, and vibrating plate equipment. With the vibrating drum

Figure 12-13. *Fifty-ton pneumatic tire compactor showing soil-filled weight box.*

Figure 12-14. *Scrapers being utilized to haul and spread fill for a dam project. Note compaction effect achieved by rubber tire in photo foreground. (Courtesy of Fiat-Allis Construction Machinery, Fiat-Allis Chalmers)*

equipment, a separate motor drives an arrangement of eccentric weights so that a high-frequency, low-amplitude, up-and-down oscillation of the drum occurs. Smooth drums and sheep's-foot-type drums are available. On the pneumatic tire compactor, the separate vibrating unit is attached to the wheel axle. The ballast box is suspended separately from the axle, so that it does not vibrate. The vibrating plate equipment typically consists of a number of small plates, which are each operated by a separate vibrating unit. The drum and the pneumatic type equipment are available as either self-propelled or towed models.

On some vibratory compactors, the vibrating frequency can be varied by the equipment operator. Frequencies usually range between 1500 and 2500 cycles per minute, values that are within the natural frequency of most soils. The natural frequency is that value where the soil particles tend to oscillate in unison, giving maximum effect to the repeated impact imposed by the compactor, instead of having a random oscillation of particles occur, which would create a damping effect. Particles are thus "shaken" into a more dense arrangement. Usually, for granular soils the most effective results are achieved when the compactor travels at a slow speed, on the order of two to four mph (or 2 to 5 km/hr).

Smooth drum vibrators (Fig. 12-15) have proven very effective in compacting granular soils having little or no silt- and clay-sized material. Layers on the order of three feet deep (one meter) have been compacted to high densities (densities close to the maximum modified Proctor value). As the percentage of fine material increases, the thickness of layer that can be well compacted is reduced.

The *vibratory pneumatic* tire equipment also has been successful in compacting primarily granular soil. It is commonly presumed that at

Figure 12-15. *Vibratory smooth-drum compactor, typically used for cohesionless soil fills.*

least one-foot-thick (30-cm) layers of predominantly granular soil will be satisfactorily compacted by most vibratory smooth drum and pneumatic tire compactors after a few passes.

The effectiveness of vibratory equipment on cohesive soil (Fig 12-16) is not nearly so pronounced as the results achieved on granular soils.

The *vibrating plate* compactors generally have a limited depth of effectiveness. Their use has been primarily in compacting granular base courses for highway and airfield pavements.

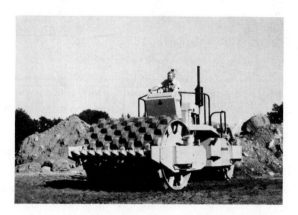

Figure 12-16. *Vibratory compactor with pads on drum, used where soil has cohesion.* (Courtesy of Dynapak, Stanhope, New Jersey)

Figure 12-17. *Smooth-drum roller compacting a base course for a roadway.* (Courtesy of Galion Division of Dresser Industries, Inc.)

Conventional smooth-drum rollers (Fig. 12-17) are not well suited for compacting earth fill. Because of the size of the drum and the large soil contact area, the resulting compaction pressures are relatively low. Smooth-drum rollers can be utilized for compacting limited thicknesses of material, such as granular base course on highway and airfield work. On earth-moving projects, the smooth-drum roller can be used advantageously to "seal" the surface of the fill at the end of each work day. Sealing provides a smooth surface so that rain water will quickly run off from the work area; it is not given the opportunity to percolate into the upper fill, where it might subsequently create a soft working surface.

Most self-propelled and towed equipment is large and cannot

Figure 12-18. *Hand-maneuvered vibratory compactor for working close to structures.* (Courtesy of Dynapak, Stanhope, New Jersey)

TABLE 12-1 / SOIL COMPACTION CHARACTERISTICS AND RECOMMENDED COMPACTION EQUIPMENT

General Soil Description	Unified Soil Classification	Compaction Characteristics	Recommended Compaction Equipment
Sand and sand-gravel mixtures (no silt or clay)	SW, SP, GW GP	Good	Vibratory drum roller, vibratory rubber-tire, pneumatic-tire equipment
Sand or sand-gravel with silt	SM, GM	Good	Vibratory drum roller, vibratory rubber-tire, pneumatic-tire equipment
Sand or sand-gravel with clay	SC, GC	Good to fair	Pneumatic-tire, vibratory rubber-tire, vibratory sheep's-foot
Silt	ML	Good to poor	Pneumatic-tire, vibratory rubber-tire, vibratory sheep's-foot
	MH	Fair to poor	Pneumatic-tire, vibratory rubber-tire, vibratory sheep's-foot, sheep's-foot-type
Clay	CL CH	Good to fair Fair to poor	Pneumatic-tire, sheep's-foot, vibratory sheep's-foot and rubber-tire
Organic soil	OL, OH, PT	Not recommended for structural earth fill	

manuever adequately to properly compact soil in confined areas and against structures. When structures are adjacent to a fill area, there should be a concern that heavy equipment will exert forces that could damage the structure. Consequently, small, portable compaction equipment should be utilized for working in areas of limited space and locations close to structures. Small vibratory drum and vibrating plate equipment is available; see Fig. 12-18. Pneumatic tampers and piston-type tampers are also available. It should be realized that with this small equipment, the thickness of the layer that can be compacted to a high density is frequently less than six inches.

VIBROFLOTATION / A method suited for compacting thick deposits of loose sandy soil is the vibroflotation process. Working on the surface with conventional compaction equipment limits the improvement in density to only the surficial zone of a deposit. The vibroflotation method, illustrated in Figs. 12-19 and 12-20, first compacts deep zones of soil and then works its way toward the surface. A cylindrical vibrator weighing about 2 tons (18 kN) and approximately six feet (2 m) long and 16 inches (40 cm) in diameter, called the *Vibroflot®*, is suspended from a crane and jetted to the depth where compaction is to begin. The jetting consists of a pressured stream of water directed from the tip of the vibroflot into the earth. As the sand is displaced, the vibroflot simultaneously sinks into the soil. Depths of about 40 feet (12 m) can be reached. At a desired depth, the vibrator is activated. The vibrations cause the soil to compact in the horizontal direction. Typically, the material from four to 5 feet (1.5 m) outward from the vibroflot is densified. Vibration continues as the vibroflot is slowly raised to the surface. As the vibrating process occurs, additional sand is continually dropped into the space around the vibroflot to fill the created void. To improve an entire site, treatment locations at approximately 10-foot (3-m) spacings are usually necessary.

COMPACTING BY EXPLOSIVES / Under proper conditions, explosives can be used to densify loose, sandy deposits. The shock wave and vibrations induced by the explosives' blast produce results similar to those achieved by vibratory compaction equipment. The method is most effective in dry cohesionless or completely saturated cohesionless soil. Where partial saturation exists, compressive stresses from the presence of air-water menisci act to prevent the soil-particle movement necessary for densification. With this method, the depth that can be effectively improved economically is usually limited. Loosest soils experience the most improvement. If variations exist in the original deposit, the blasting would tend to produce a more uniform deposit. Where explosives are used, it should be planned to compact the uppermost three feet or so by using conventional compaction methods, for little densification of this zone results from the blasting.

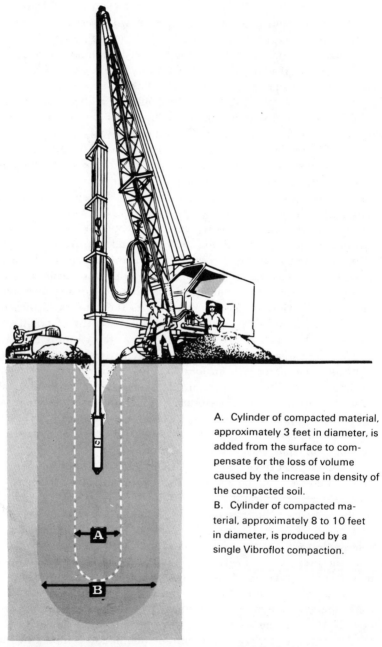

A. Cylinder of compacted material, approximately 3 feet in diameter, is added from the surface to compensate for the loss of volume caused by the increase in density of the compacted soil.

B. Cylinder of compacted material, approximately 8 to 10 feet in diameter, is produced by a single Vibroflot compaction.

Figure 12-19. *Vibroflotation equipment and process.* (Courtesy of Vibroflotation Foundation Co., Pittsburgh, Pennsylvania)

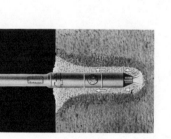

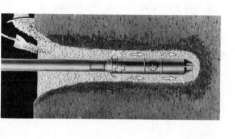

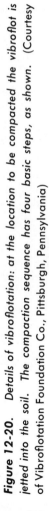

1. At start, lower jet is opened fully.

2. Water is introduced more rapidly than it can drain away. This creates a momentary "quick" condition ahead of the equipment, which permits the vibrating machine to settle of its own weight to the desired depth.

3. The water from the lower jet is transferred to the top jets and the pressure and volume are reduced just enough to carry the sand to the bottom of the hole.

4. Actual compaction takes place during the intervals between the one-foot lifts which are made in returning the Vibroflot to the surface. The vibrator is first allowed to operate at the bottom of the crater until the desired density around the lower part of the machine is attained. By raising the vibrator step by step and simultaneously backfilling, the entire depth of soil is compacted.

Figure 12-20. *Details of vibroflotation: at the location to be compacted the vibroflot is jetted into the soil. The compaction sequence has four basic steps, as shown.* (Courtesy of Vibroflotation Foundation Co., Pittsburgh, Pennsylvania)

313

Methods for Establishing
Required Soil Density

On structural earth fill projects, job specifications will indicate the soil density or degree of compaction that must be achieved in order for the fill to be considered satisfactory. The job specification requirements are typically based on the results of laboratory compaction tests (actually, moisture-density tests) performed on representative samples of soil to be used in the filling operation. The laboratory test determines the maximum density (or unit weight) for the soil, and the influence of moisture content on obtaining that density.

The most widely used procedure for moisture-density testing consists of compacting the soil in layers in a cylindrical mold by using a drop hammer (equipment illustrated in Fig. 12-21). For a particular method, the mold will have set dimensions, and the number of layers used to fill the mold will be specified, as will the weight and drop of the

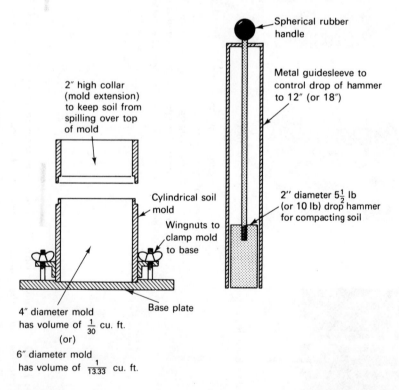

Figure 12-21. *Widely used cylindrical mold and drop hammer type of apparatus for performing moisture-density (compaction) test.*

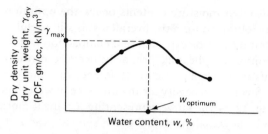

Figure 12-22. *Typical moisture-density curve obtained from laboratory compaction test trials.*

compacting hammer. To establish the moisture-density relationship for a soil, separate samples are each compacted at a different water content. Each sample is compacted in the same manner (same volume, same number of layers, same compaction energy). The compacted dry density and water content for each trial is then determined by weighing and drying the soil.

A comparison of results obtained from all the samples will reveal that the dry densities are different. This is caused by the variation in water content present during the compaction process. If the results obtained from all samples are plotted on dry density versus water content coordinates, a curve as indicated in Fig. 12-22 is developed. From such a plot, the maximum dry density is evident. The water content corresponding to the maximum density is termed the *optimum moisture content*. The optimum moisture is the best water content for achieving a high density for the given soil when a compaction energy corresponding to the particular laboratory test method is used.

With sands, the influence of moisture on the compacted density at low water contents is less well defined than for fine-grained soils; see Fig. 12-23. A scattering of dry density versus water content points

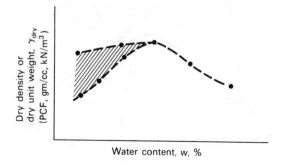

Figure 12-23. *Representative moisture-density curve expected for sands, indicating indefinite location of curve at low water contents.*

is rather usual at moisture contents below the optimum. However, the information obtained from the overall test is still useful.

It should never be expected that compaction will result in a no-void space condition for the soil, primarily because of the irregular shapes and various sizes of soil particles. Reference to a moisture-density curve should reaffirm the necessity for the presence of water and void spaces.

In Fig. 12-24 the curve representing a typical moisture-density relationship is shown, along with a curve showing the 100 percent saturation, or zero-air-voids, condition. The moisture-density curve approaches but does not overlap the zero-air-voids curve. This indicates that some air will always exist in void spaces during compaction.

The moisture-density curve for one method of compaction, as represented by Fig. 12-22, shows the maximum dry density for that test. If the results of a second test, which imparts more compaction energy, are included on the same coordinates, the new curve will be located upward and to the left, indicating that a greater dry density is attained at a lower optimum moisture content; see Fig. 12-25. Typically, a greater compaction energy results in a greater dry density coupled with a reduced dependency on water content in obtaining that maximum density.

The development of the aforementioned general method, which establishes the influence of water content on the ability of a soil to be compacted and provides a value of compacted density to use for field control, is credited to R. R. Proctor. Proctor developed the laboratory test procedure while working with compacted earth dam projects in the early 1930s. The laboratory procedure established by Proctor utilized a four-inch-diameter mold having a volume of $\frac{1}{30}$ cubic foot, a two-inch-diameter $5\frac{1}{2}$-pound hammer having a one-foot drop, and indicated that three layers be used to fill the mold, with each layer receiving 25

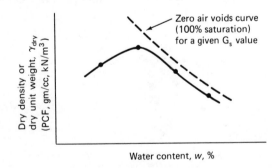

Figure 12-24. *Zero-air-voids curve related to moisture-density curve, indicating that soil does not become fully saturated by compaction.*

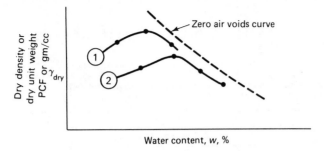

Figure 12-25. *Effect of compaction procedure on dry density. Curve 1 represents results from method using heavier compaction hammer and greater compaction energy than used to obtain curve 2.*

blows of the drop hammer.[2] This test is commonly referred to as the standard Proctor compaction test. The energy used to compact one cubic foot of soil is 12,400 foot-pounds. This energy compared favorably with the compaction energy transmitted by construction equipment in use at the time the method was developed.

With subsequent technological advances came the development of heavier vehicles and airplanes that would have to be supported on compacted earth bases, plus the desire to construct larger earth structures and the intent to support heavy building structures on compacted earth. Larger and heavier earth-moving and soil-compaction equipment was created. Consequently, laboratory test procedures were developed that would use a greater compactive energy and obtain higher values of dry density than the standard Proctor test. What has become known as the modified Proctor compaction test utilizes a four-inch-diameter mold having a volume of $\frac{1}{30}$ cubic foot as before, but requires a 10-pound hammer having a drop of 18 inches. Soil to fill the mold is compacted in five layers, with each layer receiving 25 blows.[3] The energy used to compact one cubic foot of soil in this test is 56,250 foot-pounds, a value comparing favorably with that provided with current construction methods. The presentation of laboratory data and the resulting moisture-density curve for a soil compacted in accord with the modified Proctor requirements are shown in Fig. 12-26.

If the compaction samples contain significant coarse material, a larger-size mold should be used, for the presence of coarse aggregates such as gravel in the $\frac{1}{30}$-cubic-foot mold can give dry densities that are too high to represent field conditions. A six-inch-diameter mold having a volume of 0.075 cubic foot is commonly used in place of the smaller mold. Layers are compacted with 56 blows of the drop hammer.

---

[2] ASTM Test Designation D-698.
[3] ASTM Test Designation D-1557.

Compaction Test Data Sheet

Compaction method

1. – 1/30 cu.ft. mold, 5–1/2 lb hammer, 12″ drops, 3 layers @ 25 blows/layer
2. – .075 cu.ft. mold, 5–1/2 lb hammer, 12″ drop, 3 layers @ 56 blows/layer
③. – 1/30 cu.ft. mold, 10 lb hammer, 18″ drop, 5 layers @ 25 blows/layer
4. – .075 cu.ft. mold, 10 lb hammer, 18″ drop, 5 layers @ 56 blows/layer

Trial	1	2	3	4	5	6	7
Wet density determinations							
Weight of mold and wet soil	13.84	14.30	14.00	13.89			
Weight of mold	9.32	9.32	9.32	9.32			
Weight of wet soil (W_T)	4.52	4.98	4.68	4.57			
Wet density	135.6	149.4	140.4	137.1			
Moisture determinations							
Cup identification	B-1	B-2	B-3	B-4			
Weight of cup plus wet soil	39.10	55.30	66.60	75.48			
Weight of cup plus dry soil	38.15	52.81	62.28	68.53			
Weight of cup	15.10	14.21	14.43	14.33			
Weight of dry soil	23.05	38.60	47.85	54.20			
Weight of water	0.95	2.49	4.32	6.95			
Water content – %	4.0	6.5	9.0	12.8			
Dry density – PCF	130.3	140.3	129.0	121.5			

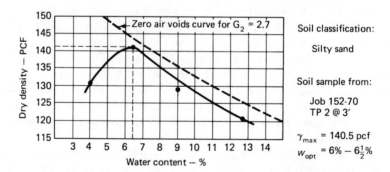

Soil classification:

Silty sand

Soil sample from:

Job 152-70
TP 2 @ 3′

γ_{max} = 140.5 pcf
w_{opt} = 6% – $6\frac{1}{2}$%

Figure 12-26.　*Laboratory moisture-density test results.*

Variations of the drop-hammer methods described above exist throughout the many organizations involved in developing earthwork specifications. Most commonly, the size of the drop hammer or mold is changed, or the number of layers or blows is varied as necessary to reflect local experiences or preferences. However, other types of laboratory test methods, such as those using kneading forces, have also been developed in attempts to reproduce more closely conditions resembling field compaction procedures. Their use has been limited because of the specialized techniques involved and the establishment of the method described previously. On a comparison basis, results obtained with other methods are not significantly different from the drop-hammer results.

A summary of the details for the laboratory compaction tests in common use is shown in Table 12-2.

Typical earthwork specifications for a structural fill project will require that the soil be compacted to a density equal to at least X percent of the maximum density obtainable with the (standard or modified Proctor) laboratory compaction test method. Currently, most compacted fills for buildings, roads, and dams use the modified Proctor as the reference test. For small dams, the standard Proctor may still be the reference. It is common to require that at least 95 percent of the maximum laboratory density be obtained by the field compaction (refer Fig. 12-27). Other typical requirements are 90 percent and 92 percent. The percentage selected is usually determined by the project designer on the basis of the project's requirements (e.g., fill for a parking area or fill beneath a building area) and his experience with various soil types. Occasionally, a requirement of 100 percent or greater is found, particularly for granular soils. Unless the project is unusual, such a high percentage should not be required. Values of 100 percent compaction referred to the modified Proctor test will be close to the maximum relative density for many sands.

When it is assumed that field compaction energy is similar to laboratory compaction energy, accepting a field compaction density less than the maximum laboratory density reduces the criticalness of water content in the fill soil. Reference to a moisture-density curve will reveal that it is possible to achieve a field density equal to, say, 95 percent compaction over a comparatively wide range of water content (Fig. 12-27).

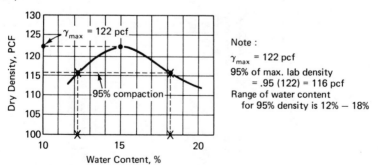

Figure 12-27. *Example of method to compute density for a specified percentage of compaction, and related range of water content.*

TABLE 12-2 / SUMMARY OF COMMON COMPACTION TEST EQUIPMENT AND PROCEDURES

ASTM Designation	AASHO Designation	Mold Size	Hammer Weight and Drop	Number of Layers at Blows per Layer	Upper Particle Size Limits (Sieve Size)
D-698 (A)*	T-99 (A)	4-in. dia, $\frac{1}{30}$ ft^3	5.5 lb @ 12 in.	3 at 25	-#4
(B)	(B)	6-in. dia, 0.075 ft^3	5.5 lb @ 12 in.	3 at 56	-#4
(C)	(C)	4-in. dia, $\frac{1}{30}$ ft^3	5.5 lb @ 12 in.	3 at 25	-$\frac{3}{4}$ in.
(D)	(D)	6-in. dia, 0.075 ft^3	5.5 lb @ 12 in.	3 at 56	-$\frac{3}{4}$ in.
D-1557 (A)†	T-180 (A)	4-in. dia, $\frac{1}{30}$ ft^3	10 lb @ 18 in.	5 at 25	-#4
(B)	(B)	6-in. dia, 0.075 ft^3	10 lb @ 18 in.	5 at 56	-#4
(C)	(C)	4-in. dia, $\frac{1}{30}$ ft^3	10 lb @ 18 in.	5 at 25	-$\frac{3}{4}$ in.
(D)	(D)	6-in. dia, 0.075 ft^3	10 lb @ 18 in.	5 at 56	-$\frac{3}{4}$ in.

*Standard Proctor compaction test.
†Modified Proctor compaction test.

Usually, the dry density of the soil being placed is the major item of concern on a compacted fill project. The water content of the compact soil is normally not specified, and is left to the choice of the contractor. An exception is on deep fills and earth dam projects involving cohesive soils, where the water content during placement and compaction is required to be slightly above optimum. At such water contents, the orientation of soil particles is considered to provide an earth mass that is more stable than a soil compacted dry.

Results of laboratory compaction tests are frequently included with the job specifications data assembled for a project. Experienced earth-moving contractors will pay attention to the moisture-density curve(s) and the natural moisture content(s) of the fill soil, for a comparison will indicate if the fill requires wetting or drying before compaction. If laboratory compaction results are not available, the approximate ranges of optimum water contents from modified Proctor tests presented in Table 12-3 can be used as a guide to evaluate the water content condition of a borrow soil (too wet or too dry).

TABLE 12-3 / APPROXIMATE RANGE OF OPTIMUM MOISTURE CONTENT VERSUS SOIL TYPE

Soil Type	Probable Value of Optimum Moisture, %, Modified Proctor Test
Sand	6 to 10
Sand-silt mixture	8 to 12
Silt	11 to 15
Clay	13 to 21

Field Control and Field Density Tests

On compacted earth fill projects it is usual practice to have in-place density tests performed on the compacted soil layers to ensure that the desired density is being achieved. Typically, each layer is tested at several random locations after it has been compacted. When tests indicate that satisfactory compaction has been obtained, the contractor can begin placement of the next layer. If the density tests indicate inadequate compaction, the contractor is notified, and more compaction rolling is performed. Normally, the personnel doing in-place density testing are directly responsible to the project owner or designer, and are independent of the earthwork contractor.

Several methods for determining or checking in-place density are in use. The simpler, more direct methods involve the "destructive testing"principle: a small hole is excavated in the compacted layer, and its volume is determined. The wet soil obtained from the hole is weighed and the water content is determined; these values permit the dry soil weight to be calculated. Knowing the dry soil weight and the in-place volume permits the dry density or unit weight to be determined. The *sand-cone method* and the *rubber-balloon method* are the most widely utilized application of this type testing.

A "nondestructive" type of testing is provided with nuclear moisture-density-determining apparatus. With such equipment, a nuclear source emits gamma rays, which are passed into the soil and reflected back to a detector. The gamma photon energy received at the detector is calibrated to indicate the wet density of the soil material. The amount of moisture is determined by calibration to a "thermal neutron" count moving through the soil from a fast neutron source. Moisture readouts are expressed as weight of water per unit of volume (i.e., pounds per cubic foot or grams per cubic centimeter).

Moisture-density determinations using the sand-cone and balloon methods are rather time-consuming. The minimum time required to complete all steps necessary to obtain a dry density and moisture determination at one location could be on the order of one-half hour. This is a drawn-out period in which construction progress awaits answers. When a series of in-place density determinations are made for a large area, a usual procedure, the time to obtain results is even greater. The time required for the performance of such testing is frequently cited as a major drawback. In contrast, density and moisture determinations are rapid with nuclear apparatus. Results for a test location are typically available in one to two minutes. At present, disadvantages associated with the use of nuclear moisture-density equipment include the relatively high purchase cost and the safety precautions necessary when one is dealing with a radioactive material.

On fill projects involving sufficient area, density testing can be planned to minimize interference with soil placement and compaction. If work operations are scheduled so that one area is receiving a fill layer while a different area is being compacted, the density testing is performed at locations that are believed to have received adequate passes of the compaction equipment. As soon as testing indicates that an area has been properly compacted, the compaction equipment moves over to begin work on a newly placed layer. Placement of a new layer begins in the area just checked.

When the source of a fill soil is from a natural deposit, it should be anticipated that the soil and its properties may vary within the deposit. Consequently, when density tests are performed, the tested soil type should be examined for correlation to the soil used to obtain the moisture-density curve. If a difference in soil exists, the laboratory-determined density value may not apply.

Field control of a compacted fill project should involve more than making in-place density tests. The thickness of newly placed soil layers should be watched to the greatest extent possible; proper depth and the possibility of undesirable soil's being mixed with the desirable soil should be checked. The behavior of the ground surface as earth hauling or compaction equipment moves over it should also be observed. Rutting, weaving, or other movement of the ground surface is typically an indication that poor soil exists at the surface or is buried below it. Where poor material is observed and it cannot be improved by additional compaction, it should be excavated and removed from the fill area. Restoring the area to grade involves placement and compaction of soil in layers, as is required at all other locations. With experience, field personnel will impose better control over a fill project by observing the performance of equipment travelling over fill areas than by continuously making in-place density tests. Density tests are performed with frequency at the beginning of the project to establish a familiarity with the soil types and related moisture. As the project progresses, the amount of density testing can be reduced. The running of in-place density determinations should not be discontinued completely, for they represent one of the permanent written records relating to earthwork performed for the project.

SAND-CONE METHOD / Determinations of in-place density using the sand-cone method involve the use of a sand-cone apparatus

Figure 12-28.   *Sand-cone density test being performed at a construction project.*

for obtaining only the volume of a density test hole. Field density will
be soil weight divided by total volume occupied. The soil weight is
determined directly by using the actual soil removed from the test hole.
The water content is calculated after obtaining the wet and dried weight
of the soil.

The sand-cone apparatus derives its name from the shape of the
cone-funnel and sand jar as shown in Fig. 12-28. The volume of a test
hole is determined by weighing the amount of sand originally in the jar
that is necessary to fill the test hole. The sand utilized with this equipment is a dry, free-flowing, uniformly graded sand whose bulk density is
known. Bulk density is the sand weight (voids included) per unit volume. Ottawa sand having a bulk density on the order of 100 pcf (16
kN/m^3) is commonly used. If the weight of jar sand to fill the test
hole is known, the volume of the hole is easily calculated. Test holes up
to about six inches (15 cm) in diameter and six inches deep can be
made with conventional equipment. Details of the sand-cone equipment
and test procedure are presented in ASTM Test Designation D-1556.

BALLOON METHOD / As with the sand cone, the balloon
apparatus is actually limited to determining the volume of a density test
hole. The weight of soil necessary to complete a density calculation is
determined by direct measurement of the material dug from the test
hole.

The typical balloon apparatus (Fig. 12-29) consists of a water-
filled vertical cylinder having a bottom opening over which a rubber
membrane or balloon is stretched. The cylinder is constructed so that
water levels in it are visible. Graduation marks indicating volumes are
etched onto the sides of the cylinder. A small hand pump (bulb or
piston type) is attached so that air can be forced into the top of the
cylinder. When the apparatus is placed over a density hole, the pumped
air forces the balloon and water into the hole. The volume is determined directly by noting the water level in the cylinder before and after
the balloon is forced into the hole. The water and balloon are retracted
from the test hole by reversing the air pump and evacuating air from the
cylinder. Outside atmospheric pressure forces the water and balloon
back into the cylinder. The apparatus is then ready for another test location.

Rubber balloon equipment is available in a range of sizes. The
smaller, more typical type is capable of measuring holes approximately
four inches in diameter and six inches deep (10 cm and 15 cm).
Larger equipment can measure larger holes. More information on the
balloon-type apparatus and test method is presented in ASTM Test
Procedure D-2167.

NUCLEAR MOISTURE-DENSITY METHODS / Surface-
type nuclear moisture-density equipment (Fig. 12-30) is coming into

1. Position base plate on flat test surface and set Volumeasure in place. Open valve for inflation of balloon. Pump down until water level in cylinder reaches lowest level. Take the initial base reading. Reverse bulb for vacuum. Pump balloon up into cylinder.

2. Remove Volumeasure, leaving base plate in place as template for hole. Dig down through opening in base to make hole to desired depth. Keep all removed soil in tight can for weight measurement.

3. Set Volumeasure on base plate over hole. Insert inflation end of bulb into holder hole.

4. Pump balloon down into hole. Record lowest point reached by water on the cylinder scale.

5. Invert the inflation-vacuum bulb to put it in vacuum position and pump balloon back up into cylinder. To determine volume of hole, subtract initial base reading from final reading.

Figure 12-29. Five-step procedure for using balloon apparatus to determine in-place density. Equipment demonstrated is the Soiltest Volumeasure. (Courtesy of Soiltest, Inc.)

Figure 12-30.   *Nuclear moisture-density apparatus in use.*   (Courtesy of Troxler Electronic Laboratories, Inc., North Carolina)

more prevalent use for performing density tests on compacted fill, replacing the sand-cone and rubber balloon methods, primarily because of the rapid results that can be obtained.

The principal elements in a nuclear density apparatus are the nuclear source, which emmits gamma rays, a detector to pick up the gamma rays or photons passing through the tested soil, and a counter or scaler for determining the rate at which the gamma rays reach the detector.

Commonly used nuclear sources include radium-beryllium and cesium-americium-beryllium combinations in equipment where moisture determinations as well as density information are to be provided. The detectors usually consist of gas-filled Geiger-Mueller tubes.

When the equipment is in use, the gamma rays penetrate into the soil, where some are absorbed, but some reach the detector by direct transmission or after reflecting off soil mineral electrons. The amount of gamma radiation reaching the detector is inversely proportional to the soil density. Densities are determined by obtaining a nuclear count rate received at the detector and relating such readings to calibration readings made on materials of known densities. Calibration curves are provided by the equipment manufacturer. The density determined by this method is a wet, or total, density.

Moisture determinations are obtained from a "thermal neutron" count. Alpha particles emitted from the americium or radium source bombard a beryllium target. This bombardment causes the beryllium to emit fast neutrons. These fast neutrons lose velocity if they strike the hydrogen atoms in water molecules. The resulting low-velocity neutrons are thermal neutrons. Moisture results are provided as weight of water per unit of volume. Dry densities are obtained by subtracting the moisture determination from the wet density determination. With this method for determining moisture, significant error can result if the soil contains iron, boron, or cadmium.

Several field procedures for making the moisture-density determinations are available. The direct transmission mode, as indicated in Fig. 12-31 (a), provides the most accurate results. In this method, the radioactive source is placed into the test material by utilizing a punched or drilled hole. Depths between 2 and 12 inches can be tested with conventional equipment.

The backscatter method operates by locating both the radioactive source and the detector on the surface of the test material; see Fig. 12-31 (b). The gamma rays are directed into the soil and some are reflected back to the detector. With this method, accuracy suffers if a gap exists anywhere between the bottom of the device and the soil surface.

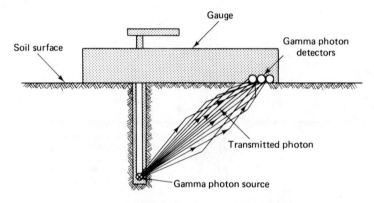

(a) Direct transmission density measurement

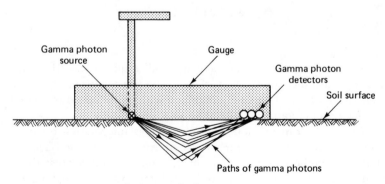

(b) Backscatter density measurement

Figure 12-31.   *Illustration of different modes for measuring soil density and moisture content by nuclear methods.* (Courtesy of Troxler Electronic Laboratories, Inc., North Carolina)

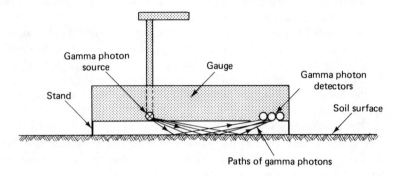

(c) Air-gap density measurement

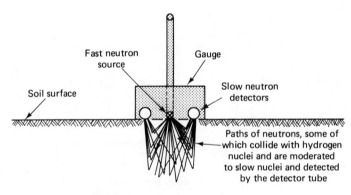

(d) Backscatter moisture measurement

Figure 12-31.   *(Continued.)*

A third procedure, the air-gap method, requires that the nuclear device sit a distance above the surface of the test material, as indicated in Fig. 12-31(c). The accuracy obtained with this method is related to attaining a proper gap height for thé equipment.

The direct transmission method provides information on the soil volume surrounding the gamma rays' source, whereas the backscatter and air gap methods provide data on the zone of soil nearest the surface at the test location. More detailed information for making moisture density determinations by nuclear methods is presented in ASTM Test Procedure D-2922.

Soil Stabilization

Soil stabilization refers to the procedure in which a special soil, a cementing material, or other chemical material is added to a natural soil material to improve one or more of its properties. One may achieve stabilization by mechanically mixing the natural soil and stabilizing material together so as to achieve a homogeneous mixture, or by adding the stabilizing material to an undisturbed soil deposit and obtaining interaction by letting it permeate through the soil voids. Where the soil and stabilizing agent are blended and worked together, the placement process usually includes compaction.

Many of the stabilization procedures of mixing and then compacting in place are for providing a limited thickness of treated material, such as in bases for road and airfield pavements, and beneath floor slabs.

One of the more common methods of stabilization includes the blending of natural coarse aggregate and fine-grained soil to obtain a mixture that possesses some internal friction plus cohesion, and provides a material that is workable during placement but will remain stable when in place and subject to a range of temperature and moisture changes. Improvement by the proportioning of coarse with fine soil is commonly referred to as *mechanical stabilization*. The proper combination of coarse- and fine-grained material (referred to as *binder*) is related to the size and shape of particles in the mixing materials. Generally, the best mixtures will include about one-quarter binder.

Sodium chloride and calcium chloride are additives noted for their water-holding properties. The presence of moisture in a soil creates capillary water in the voids, and the resulting compressive stresses imposed onto the soil particles increase the soil's internal shear strength. Spread on dirt roads, such materials help prevent dust. Calcium chloride is an especially preferable material on dirt roads, for it is capable of absorbing moisture from the air.

Portland cement and asphalt cement are additives intended to bond soil particles together. The result is a soil cement or asphalt-stabilized soil. In this respect, the cements are utilized as they are in the more familiar portland cement concrete and asphaltic concrete. In the stabilized soils the range of soil aggregates is not as select as the materials used for the concretes. The strength of a stabilized soil will be less than that of the concretes. Mixing can be in place on-site or in a mixing plant. In the construction of a proper soil cement, the soil needs to be at a water content close to the optimum percentage necessary for maximum compaction. After the cement and soil have been mixed, the mixture is compacted in place. Typically, the quantity of cement required ranges from about 7 percent by weight for sandy soils to about 15 percent for clay soils. Requiring only a limited wearing surface, soil cements have been satisfactory as bases for roads carrying low to medium volumes of traffic, provided that wheel loads were not heavy.

Asphalt-stabilized soils are capable of serving as the combination base and wearing surface for low-volume roads not subject to heavy wheel loads, or as a quality base material to support pavements of major roads.

Lime and calcium chloride have been used as additives to improve clay soils. A decrease in plasticity results as a base change, or exchanging of cations in the adsorbed water layers, occurs. The resulting material is more friable than the original clay. The lime stabilization process requires mixing, then curing for a period of a few days, followed by remixing and compaction. The stabilized soil will possess strength similar to low-grade concrete. When mixed with originally expansive clays, volume changes are prevented. Lime has been successfully used to stabilize road materials in areas where expansive clays have been a problem.

Various combinations of commercial and natural chemicals have been successful in acting as sealing or cementing agents for soil. The general categories are the silicate chemicals, the polymers, and chrome-lignin. As an illustration, sodium silicate mixed with calcim chloride reacts quickly to form calcim silicate, a hard and impervious material. The polymers represent those materials resulting from combining chemically complex organic manomers with a second chemical that acts as a catalyst to cause polymerization, or joining together of molecules in the manomer. One of the most widely known soil stabilizers, AM-9, is a polymer. Chrome-lignin is a slow-acting stabilizing gel formed from the reaction between the lignin in paper manufacture waste and sodium-dichromate. The chemicals are typically water soluble and produce the reaction necessary to obtain a bonding of particles after being placed in the soil. With most chemical additives, the reaction does not include the soil particle. An advantageous feature of most chemical stabilizers is that their setting or curing time can be varied.

Grouting and injection stabilization is a process wherein a bonding material is forced into (or through) a natural soil deposit in order to improve its strength, water resistance, or other properties. The method is utilized where it is necessary to improve soil that cannot be disturbed. Typically, this procedure is followed where treatment involves a considerable thickness or depth of earth, or where the treated area is close to existing structures or other facilities that cannot be disturbed. Chemical stabilizers are the materials commonly used in the process. Many stabilizers are marketed under special trade names. Injection and grouting are usually handled by specialty contractors who have proper equipment and have developed familiarity with one or more stabilization procedures. The injection or grouting is done under pressure. The stabilizing material moves through the void spaces in the soil. Consequently, the more viscous stabilizers are limited to treating soils having high coefficients of permeability. Some chemicals have viscosities comparable to water and can be used for finer soils, such as fine sand and silt deposits. At present, injection methods are not suitable for clay because of the very low permeability. With all injection-grouting methods, there is always some uncertainty in results obtained. The pressured stabilizer may follow paths through only the more permeable zones in a soil or along cracks and fissures, and may not be distributed uniformly through the soil mass as desired.

A summary of methods and materials applicable for stabilizing in-place soil of different types is presented in Fig. 12-32.

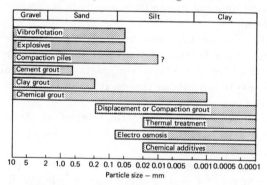

(a) Methods and particle-size range for in-place treatment of soils

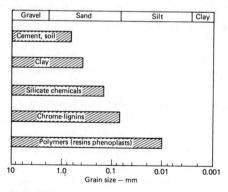

(b) Particle size ranges to which different grout materials apply

Figure 12-32. *Summary of various methods for stabilizing in-place soils.* (Ref. 101)

Alternate Methods
of Transporting Earth Fill

Most transporting of soil for earthwork projects is completed through the use of construction equipment and methods discussed in preceding sections. However, where large quantities of material are to be moved, alternate methods of transportation should also be considered because of possible savings in cost or time. Methods that fall into this category are conveyor-belt transportation and hydraulic transportation.

Belt-conveyor systems offer the possibility for economic advantage where large volumes of material are moved for distances of up to several miles. A belt-conveyor system is usually constructed to closely follow the terrain it crosses, and typically consists of a series of completely individual belt flights. Normally each flight has its own drive motor. The transported material passes from one flight to another in order to complete the trip.

The belts of a conveyor system are available in different widths; stock belts run to five feet wide. Conveyor speeds can vary, and rates of several hundred feet per minute are practical. When the wider belts are used, it is easily possible to transport several hundred cubic yards of material per hour.

On a compacted fill project, earth received at the deposit end of the conveyor is usually moved and spread by using conventional earth fill equipment and procedures.

A major disadvantage of a belt-conveyor system is the expense and time to construct it. Construction costs include the necessity for a maintenance road for servicing during operation, as well as the expense for erection and dismantling. Advantages include more independence from the effects of weather, easy round-the-clock operation, and reduced labor operating costs.

Hydraulic transportation refers to the method of mixing the borrow soil with large quantities of water and pumping the resulting slurry through a pipeline to a desired location. The method is highly dependent on an adequate source of water being available. Waterfront and dam-reservoir construction typify those projects that have the advantage of an available and economical source of water. For practical ease of handling at both the supply and deposit ends, cohesionless soil or soil possessing only slight cohesion is the most desirable, although all soil types can be transported hydraulically.

Where a sand supply is located below water (a popular situation for hydraulic transportation), suction dredging methods are used. The sand is sucked into a flexible pipe extending from a pump into the soil supply and then is passed into the transporting pipe. Underwater cohesive soils are cut into chunks by powered rotating cutters before entering the flexible pipe. The slurry passing through the pipe is usually 80 to 90 percent water and 10 to 20 percent solids.

Above the water table, the borrow soil can be excavated and washed into a ponding area by utilizing high-pressure streams of water, whence it is then piped to the desired location.

Land fills can be created with hydraulically transported soil. Such fill is referred to as *hydraulic fill*. At a free-flowing outlet, the coarse soil particles drop out close to the point of discharge, and the smaller particles are carried a farther distance by the fanning-out water. This is similar to the formation of a soil delta where a river enters a lake. If mixed soils are transported, periodic movement of the discharge end of the pipe to planned locations permits fill areas of the desirable soil type (coarse soil areas and fine soil areas) to be obtained.

The natural side slopes created by a flowing coarse soil slurry are usually in the range of between five and ten horizontal to one vertical. Hydraulically transported soft clay soils may develop slopes as shallow as 50 to 1. If it is necessary to contain the area receiving fill, dikes are constructed or sheetpiling is installed to form a ponding area. Soil particles will settle out of the slurry in this area. The water accumulating in the surface zone of the pond is drawn off. Sizable land areas have been reclaimed or created and relatively large earth dams have been constructed by using hydraulic fill methods.

Sand deposited by hydraulic methods will be in a relatively loose condition. Usually, it is necessary to densify the soil if it is intended to provide structural support. Normal compaction procedures can be followed, rolling equipment (preferably vibratory) being used to compact the fill in layers. Silt and clay hydraulic fills are difficult to compact after placement. These soil types would not be a choice for a structural fill but are used on land reclamation projects if economical sources of coarse material are not available. Where time permits, such fill is left to consolidate and stabilize naturally. Major structures are supported by piles or other special foundations. If necessary, the more involved and costly methods of densification, such as by pile driving or use of sand drains, could be performed, but results are somewhat unreliable.

PROBLEMS

1. Indicate the best type, and also other suitable types, of compaction equipment (such as sheep's-foot roller, vibratory roller) to use for easiest compaction of the following soil materials:
   Material A: well-graded fine to coarse sand.
   Material B: silty fine to coarse sand, trace clay.
   Material C: silt of low plasticity, trace clay and sand.
   Material D: silty clay of low plasticity.
   Material E: sand-silt-clay mixture, some gravel.

2. A contractor has bid on two different earthmoving-compacted fill
    projects. One project involves cohesionless soil, the other cohesive
    soil. The volume of soil to be excavated, transported, and com-
    pacted is similar for both projects. All working conditions are
    similar for both projects. The contractor's bid for compacted
    fill on the cohesive soil project is 50 percent greater than for
    compacted fill at the cohesionless soil project. Indicate if the con-
    tractor's bidding practice is proper or not, and why.

3. For an earthmoving project, the contractor is required to assume
    the responsibility for locating a source of off-site borrow to use for
    compacted fill. Two possible sources are found; the cost of pur-
    chase and hauling from either will be comparable. At location A,
    the soil is a silt-clay material and is found to have a natural
    moisture content in the 10-to-20-percent range. At location B, the
    soil is also cohesive and is found to have a natural water content
    in the 30-to-40-percent range. The contractor selects location A.
    Indicate if this represents a good choice, and why.

4. The following moisture-density data are results from laboratory
    compaction tests.

Water Content, %	Dry Unit Weight, pcf
8	111
11	113
14	115
17	114
20	109

    (a) Plot the moisture-density curve, and indicate maximum
         density and optimum moisture content.
    (b) What range of water content appears advisable in order to
         obtain 92 percent (or more) compaction?

5. A series of laboratory compaction tests on the same soil provides
    the moisture and density information shown. Plot the dry density-
    moisture curve, and determine the maximum dry density and
    optimum moisture content.

Water Content, %	Wet Density, pcf
7.0	118
9.5	129
13.5	136
15.0	132

6. Laboratory compaction test results for a soil are tabulated below. Review the data and determine if there are errors.

Water Content, %	Wet Density, pcf	Dry Density, pcf
9	122	112
12	129	117
15	137	119
18	107	109

7. Laboratory compaction tests have been performed on several different soils. The maximum dry density and optimum moisture content for each soil are as shown. Determine the probable type (classification) of soil for each test.

Soil	Max. Dry Density, pcf	Optimum Moisture, %
A	119	9
B	124	18
C	114	14

8. Sand-cone equipment is used to perform a field density test on a compacted earth fill. Ottowa sand is used in the cone, and is known to have a bulk density of 100 pcf.

    (a) From the information given below, determine the in-place dry density of the tested soil and the water content.

    Soil sample dug from test hole, wet weight   = 4.62 lb.
    Dried weight of soil sample   = 4.02 lb.
    Weight of Ottowa sand (sand cone) to fill test hole = 3.60 lb.

    (b) Determine the percentage of compaction of the tested soil if the laboratory moisture-density curve indicates a dry density of 115 pcf and an optimum moisture content of 13 percent.

9. A balloon-type apparatus is used to perform a field density test. The volume of the test hole determined by reading the water level graduations on the apparatus cylinder before and after digging the test hole is 0.025 ft.^{3} The wet weight of soil obtained from the test hole is 3.10 lb. The water content is determined to be 14 percent by drying a small sample on a field stove.

    (a) Determine the in-place dry density of the tested soil.

    (b) Determine the percentage of compaction if the tested soil is a compacted fill whose maximum density (from laboratory compaction tests) is 118 pcf.

10. Indicate the general advantages and disadvantages associated with hydraulic fill placement methods.

11. It is proposed to stop groundwater from seeping through a building's basement walls by pressure-injecting a bentonite clay slurry into the soil against the exterior of the leaking walls to seal the walls on the outside. The soil type is silt-clay. Will the method work in the indicated soil conditions?

12. Asphalt, cement, and lime are commonly used to stabilize soil for road bases. Which of these seem most suitable for stabilizing?

   (a)   Sand.
   (b)   Sand-silt mixture.
   (c)   Silt with trace of sand and clay.
   (d)   Clay with silt.

CHAPTER 13
Foundations:
Introductory Concepts

For many of man's structures, it is the earth underlying the structure that provides the ultimate support. The soil at a building location automatically becomes a material of construction affecting the structure's stability. Typically, soil is a material weaker than the other common materials of construction, such as steel, concrete, and wood. To satisfactorily carry a given loading, a greater area or volume of soil is necessarily involved. In order for loads carried by steel, concrete, or wood structural members to be imparted to the soil, load transfer devices— the structural foundations—are required. The major purpose of the structural foundation is the proper transmission of building loading to the earth in such a way that the supporting soil is not overstressed and does not undergo deformations that would cause serious building settlement. The type of structural foundation utilized is closely related to the properties of the supporting soils. A structural foundation performs properly only if the supporting soil behaves properly. Consequently, it is important to recognize that building support is actually being provided by a soil-foundation system, a combination that cannot be separated. Designers and constructors are aware of this relationship, but it has become common practice to consider the structural foundation separately, primarily because it is a cost item that is built or installed, while the supporting soil is usually the natural earth which is "there."

Since the soil-foundation system is responsible for providing support for the lifetime of a structure, it is important that all forces that may act over that time period be considered. For a building to last, its foundations should be designed for the worst conditions that may develop. Typically, the foundation design always includes the effect of the structure's dead plus live loads. It is important to also consider

load effects that may result from wind, ice, frost, heat, water, earthquake, and explosive blasts.

General Types of Foundations—
Foundation Categories

The various types of structural foundations can be grouped into
two broad catagories—shallow foundations and deep foundations. Generally, the classification indicates the depth of the foundation installation and the depth of the soil providing most of the support. Spread
footing and mat (or raft) foundations usually fall within the shallow
foundations category. Deep foundation types include piles, piers, and
caissons. The floating foundation, a special catagory of foundation, is
actually not a different type, but it does represent special application
of soil mechanics principles to a combination mat-caisson foundation.

SPREAD FOOTINGS / Spread footing foundations are typically of plain concrete or reinforced concrete, although masonry and
timber have also been used. The spread footing foundation is basically
a pad used to "spread out" building column and wall loads over a
sufficiently large soil area. Spread footings are constructed as close to
the ground surface as the building design permits (considering requirements such as basements or the need to resist lateral forces), and as
controlled by local conditions (considering factors such as frost penetration, soil shrinkage and expansion, the possibility of soil erosion, or

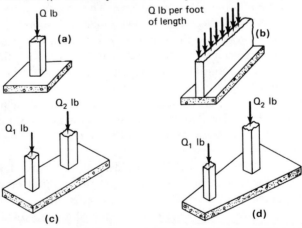

Figure 13-1. Types of shallow spread footing foundations: (a) square
spread footing to support column loading; (b) long (strip) footing to support
wall loading; (c) rectangular footing for two columns (combined footing) or
machine base; (d) trapezoidal footing for two columns (combined footing) or
machine base.

building code stipulations). Footings for permanent structures are rarely located directly on the ground surface. To be classified as a spread footing, the foundation does not have to be at a shallow depth; spread footings will be located deep in the ground if soil conditions or the building design requires. Spread footing foundations for building columns, walls, and equipment bases commonly have the shapes of squares, rectangles, trapezoids, and long strips; see Fig. 13-1. Usually, the shape and dimensions for a footing result from having the structural loading positioned so that theoretically a *uniform* bearing pressure on the soil beneath the foundation is achieved. For the support of walls and single columns, the loading is usually centered on the footing. For foundations supporting two or more column loads, or machinery, the positioning of the loading or weight often makes a rectangular or trapezoidal shape necessary.

MAT (OR RAFT) FOUNDATION / The mat (or raft) foundation can be considered as a large footing extending over a great area, frequently an entire building area (Fig. 13-2). All vertical structural loadings from columns and walls are supported on the common foundation. Typically, the mat is utilized for conditions where individual column footings, if used, would tend to be close together or would tend to overlap. The mat is frequently utilized as a method to reduce or distribute building loads in order to reduce differential settlement between adjacent areas.

PILE AND PIER FOUNDATIONS / Piles and piers are foundation types intended to transmit structural loads through upper zones of poor soil to a depth where the earth is capable of providing the desired support (Fig. 13-13a, b). In this respect, where loadings developed at one level are transmitted to a lower level, piles and piers are similar to structural columns. Though typically considered as being long slender structural members, such foundations obtain adequate lateral support from the embedding soil along their length so that there is no concern about buckling under axial load, as with conventional columns. These deep foundation types are also utilized in situations where

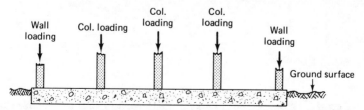

Figure 13-2. *Mat foundation for soils having low bearing capacity or where soil conditions are variable and erratic. Used to obtain low bearing pressure and reduce differential settlement.*

it is necessary to provide resistance to uplift, or where there is concern about possible loss of ground or erosion due to flowing water or other causes.

In years past, the pile or pier category was indicative of the method used to install the foundation or of its size. Piles were slender foundation units, usually driven into place. Piers, typically larger in area than piles, were units formed in place by excavating an opening to the desired depth and pouring concrete. Often, such foundations were large enough to permit a man to enter and inspect the exposed earth. Currently, a clear distinction between pile- and pier-type foundations is not always

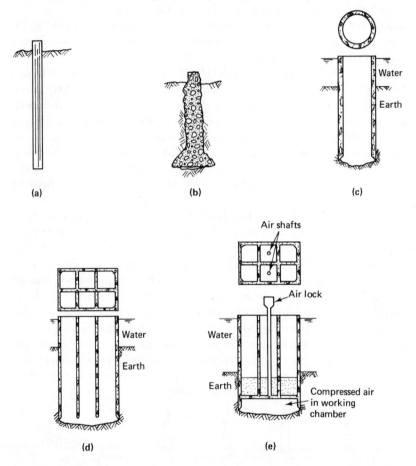

Figure 13-3.   *Representative types of deep foundations: (a) slender driven, drilled, or cast-in-place pile; (b) drilled or cast-in-place pier with enlarged base; (c) pile-type open caisson; (d) box-type open caisson; (e) pneumatic caisson.*

present because of occurring changes and innovations in construction or installation techniques. For example, there are some types of cast-in-place piles that are constructed by using the basic methods historically attributed to pier foundations. As a result, the developing practice is to classify all deep, slender foundation units simply as pile-type foundations, with terms such as *driven, bored,* or *drilled,* and *precast* or *cast-in-place,* to indicate the method of installation and construction.

CAISSONS / A caisson is a structural box or chamber that is sunk in place or built in place by systematically excavating below the bottom of the unit, which thereby descends to the final depth. Open caissons may be box-type or pile-type. Usually, the top and bottom are open during installation. When in place, the bottom may be sealed with concrete if necessary to keep out water, or the bottom may be socketed into rock to obtain a high bearing capacity. Pneumatic caissons have the top and sides sealed, and use compressed air to keep soil and water from entering the lower working chamber, where excavation to advance the caisson is occurring. Representative types of caissons are illustrated in Fig. 13-3c, d, and e.

FLOATING FOUNDATION / The floating foundation is a special type of foundation construction that is useful under proper conditions. Particularly, it has application in locations where deep deposits of compressible cohesive soils exist, and the use of piles is impractical. The floating foundation concept requires that a building's substructure (the below-ground structure) be assembled as a combination mat and caisson to create a rigid box, as shown in Fig. 13-4. This foundation is installed at a depth so selected that the total weight of the soil excavated for the rigid box equals the total weight of the planned building. In theory, the soil below the structure is, therefore, not subjected to any change in loading. For such an occurrence, there would be no settlement. Usually, however, some settlement does occur because the soil at the bottom of the excavation expands after excavation, and recompresses during and following construction.

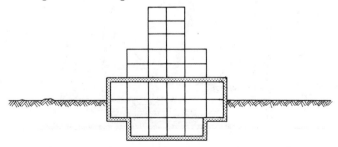

Figure 13-4. *Rigid box caisson foundation utilizing floating foundation concept.*

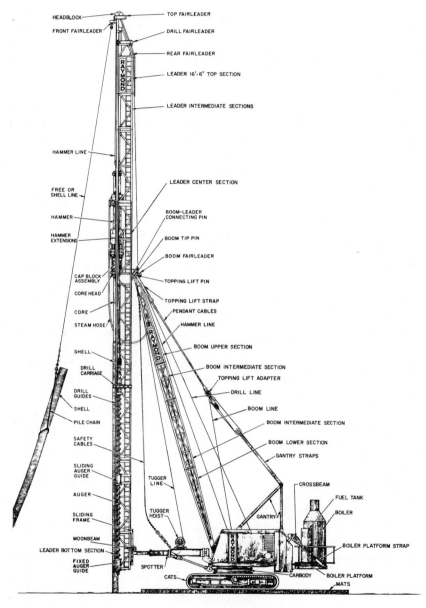

Figure 13-5. *Typical pile-driving equipment.* (Courtesy of Raymond International, Inc.)

Pile Foundation Types
and Installation Procedures

PILE DRIVERS FOR DRIVEN PILES / Driven piles are installed from the ground surface by hammering a ready-made unit or hollow shell, usually with special pile-driving equipment (see Fig. 13-5). Most driving is done with an impact type of hammer; a moving weight falls or is forced against the top of the pile. General categories of modern-day pile hammers include the single- and double-acting units (steam, compressed air, or hydraulic), and diesel hammers, illustrated in Fig. 13-6. The term *hammer* refers to the entire driving unit; usually the moving weight that strikes the pile is the *ram*.

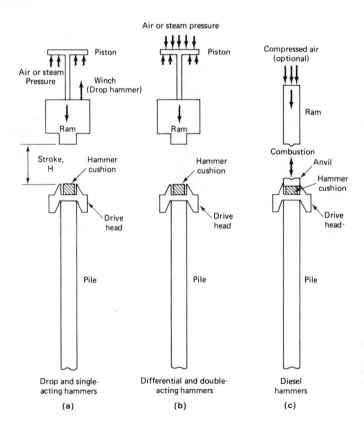

Figure 13-6. *Schematic diagram of ram-type pile-driving hammers.* (Ref. 110)

Single-acting hammers use steam or compressed air to raise the hammer ram to a ready-for-driving position. The ram is then released to enable it to drop on the top of the pile. A *double-acting hammer* uses steam or compressed air to raise the ram to a ready-for-driving position and also to accelerate the ram's downward thrust. *Differential acting hammers*, another category, are similar in operation to the double-acting hammers. Double-acting hammers actuated by hydraulic pressures are also available.

Diesel hammers are self-contained, self-activated units. The ram is located within an enclosed cylinder. Initially, the ram is mechanically raised to the top of the cylinder and released for its fall. A fuel mixture injected into the cylinder ahead of the falling ram is compressed due to the piston effect. Near the bottom of the stroke, the fuel is detonated and the force of the resulting explosion and the ram impact is delivered to the pile, driving it. Within the cylinder,

TABLE 13-1 / DATA ON REPRESENTATIVE PILE DRIVING HAMMERS

Hammer Make and Type		Weight: Ram or Moving Parts (pounds)	Stroke (inches)	Blows per Minute	Rated Eneroy Per Blow (foot-pounds)
Vulcan	2 (single-	3,000	29	70	7,260
	1 acting)	5,000	36	60	15,000
	0	7,500	39	50	24,375
	010	10,000	39	50	32,500
	016	16,250	36	60	48,750
	50C (differential acting)	5,000	15.5	120	15,100
	80C	8,000	16	111	24,450
	140C	14,000	15.5	103	36,000
MKT	S10 (single- acting)	10,000	39	55	32,500
	10B3 (double-	3,000	19	105	13,100
	11B3 acting)	5,000	19	95	19,150
	DE20 (diesel)	2,000	96	48	16,000
	DE40	4,000	96	48	32,000
Raymond	00 (single-	10,000	39	50	32,500
	000 acting)	12,500	39	50	40,600
	65C (differential acting)	6,500	16	110	19,500
Link Belt	440 (diesel)	4,000	37	86	18,200
	520	5,070	43	80	26,300

the force of the explosion raises the ram up to the top of the cylinder, where it is ready to begin another cycle.

Single-acting hammers need to rely on the weight of a heavy ram for driving piles. *Double-acting and diesel hammers* can develop high driving energy that is equal to or greater than the energy of the falling ram in a single-acting hammer, but they achieve this through greater impact velocity. One advantage of the double-acting and diesel hammers over the single-acting hammer is the greater operating speed. A greater number of blows will be delivered per unit of time. When long or heavy piles are driven, a hammer having a heavy ram should be used. For effective driving, the weight of the hammer ram needs to be approximately the same weight as the pile, or greater. Equipment manufacturers provide data on the important characteristics of their hammers, including the weight of the ram and the stroke, hammer efficiency, driving energy per hammer blow, and number of blows per minute (see Table 13-1).

Vibratory drivers rely on a principle different from the conventional hammers for installing a pile. Vibratory units are typically more contained and compact than the falling-ram type of hammer. Basically, the vibratory driver (Fig. 13-7) consists of a pair of counter-rotating weights that are synchronized so that the lateral components of thrust always counteract, or cancel, each other's effects. The vertical components of thrust are additive, however, and create up-and-down pulsations, or vibrations. For an installation, the driver is clamped to the pile. The

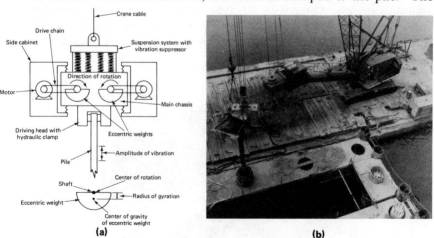

Figure 13-7. (a) Schematic diagram of vibratory pile driver; (b) vibratory hammer utilized to install steel casing cast-in-place concrete piles for a ship channel. Photo shows concete being tremied into a casing. The vibratory hammer is shown hanging from the crane boom.

pile is then vibrated into the earth under its own weight. Vibratory drivers may be low-frequency type (operating range between about 10 and 30 Hz or cps) or resonant type (capable of an operating frequency up to about 150 Hz or cps). The intent of a resonant driver is to create a condition of resonance between the driver, pile, and zone of affected soil so that penetration occurs very rapidly. Vibratory drivers are considered to be most effective when installing piles in sand and silty-sand soils.

An item of importance in pile installation is the factor of stresses created in the pile by the driving. For wood piles and concrete piles, excessive driving when high resistance is met (a high number of blows per inch of penetration) may result in structural damage or breakage in the pile. It is helpful for the foundation designer to impose an *upper limit* to the number of blows that can be safely imposed, for reference by field personnel. Breakage while driving can be difficult to detect, but is often indicated by a drop in blow count following a high blow count. Metal tips or end covers can be installed on piles to help protect them in hard driving and to aid penetration, as shown in Fig. 13-8.

PILE TYPES AND MATERIALS / The common types of pile that are *installed by driving* include *timber piles, steel H* (see Fig. 13-9) *or pipe, reinforced concrete,* and *prestressed concrete.* Generally with these types, the pile unit is complete and ready to be installed when it is delivered to the job site.

The *driven steel shell* (Fig. 13-10) is a different category of pile installed by driving. Often with this type, a hollow steel shell is installed to the desired depth by driving on a steel mandrel or steel core that fits inside the shell (Fig. 13-11). Much of the force of driving thus acts at

Figure 13-8. *Piles being provided with protective points.* (Courtesy of Associated Pile and Fitting Corp.)

(a)   (b)

Figure 13-9. *(a) H-piles being driven.   Photo shows battered piles.* (Courtesy of Associated Pile and Fitting Corp.) *(b) Installation of pipe piles.* (Courtesy of MKT Corp.)

the tip of the pile, so in effect the pile is being pulled into the ground. At the desired penetration the mandrel or core is withdrawn, and plain or reinforced concrete is placed in the shell. Corrugated or fluted steel is conventionally used for the shell so that it will possess adequate strength to withstand the stresses of driving. This type of pile is usually considered as a cast-in-place unit.

Timber piles (Fig. 13-12) can be treated or untreated "Treated" refers to the procedure by which the wood has been impregnated with a protective material or preservative such as creosote. Creosoting is a typical treatment for piles that are to be exposed to moisture but not permanently submerged in water. Untreated wood permanently below water will not decay and is considered to have unlimited life unless threatened by other effects. Preservative-protection treatment is required to protect wood piles from marine borers (if for a marine environment), or from wood-infesting insects such as termites, or from decay (wet rot) if the pile is embedded in soil above the water table.

Cast-in-place piles include the *steel-shell and concrete type* described previously, the *shell-less type*, which is formed by excavating to a desired depth and then filling the opening with concrete, and the *type where a concrete unit is formed in the ground without prior excavation.*

Where preexcavation is performed for a *shell-less pile* before the concrete is placed, the excavation may be unlined or provided with a temporary lining, such as a steel shell, whose purpose is to keep soil

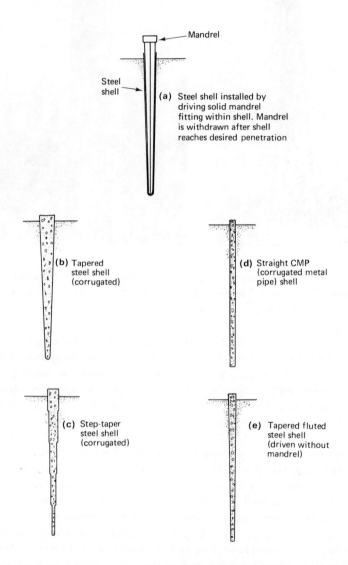

Figure 13-10.   *Types of driven steel shell, cast-in-place concrete piles.*

from caving into the open excavation and to seal off water from the soil walls.   Excavation for this type of pile is typically performed by augering or by applying a wash boring technique similar to that utilized when borings are drilled (Chapter 11).   With either of these methods, a bentonite clay slurry may be used in the excavation while it is being drilled (Chapter 11) instead of a temporary steel lining to keep the

Figure 13-11. *Installation of step-taper steel shell cast-in-place concrete piles.* (Courtesy of Raymond International, Inc.)

Figure 13-12. *Timber piles being installed with single-acting hammer.* (Courtesy of Associated Pile and Fitting Corp.)

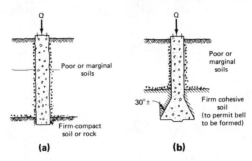

Figure 13-13. *Typical drilled-shaft foundations: (a) straight-sided; (b) belled or underreamed base.*

soil walls from caving in. The bentonite slurry is a heavy liquid whose purpose is to exert a lateral pressure sufficient to hold the soil walls in place. Concrete for the foundation unit is tremied (placed under water through a large-diameter flexible hose or tube in order to prevent contamination) to the bottom of the excavation. The pile is formed from the base up, and the slurry is displaced from the excavation. The shell-less, cast-in-place pile is often referred to as a bored pile or drilled-shaft foundation. Representative types are illustrated in Fig. 13-13.

In cohesive soils, the base of the excavation may be enlarged to provide a greater bearing area. The excavation for the shaft section and the belled or underreamed base is often performed by special auger-type drilling equipment, but can be accomplished manually. This type of foundation is often referred to as a drilled (drilled and belled) pier, drilled (drilled and belled) caisson, or drilled-shaft foundation (Fig. 13-14).

Figure 13-14. *(a) Truck-mounted rig for constructing drilled shaft foundations (Courtesy of Association of Drilled Shaft Contractors, Inc.); (b) crane-mounted equipment for constructing drilled shaft foundations. Photo shows underream tool for forming enlarged base. (Courtesy of Association of Drilled Shaft Contractors, Inc.)*

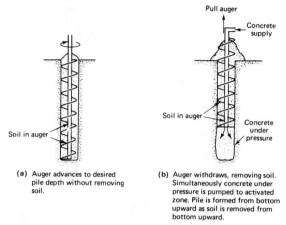

(a) Auger advances to desired pile depth without removing soil.

(b) Auger withdraws, removing soil. Simultaneously concrete under pressure is pumped to activated zone. Pile is formed from bottom upward as soil is removed from bottom upward.

Figure 13-15.   Augered cast-in-place pile.

A different type of cast-in-place pile is constructed by *augering into the earth without removing soil,* then pumping concrete to the bottom of the drilled zone through a hollow stem in the auger, and forming the pile from the bottom up as the auger withdraws and removes soil (Fig. 13-15).

Still another category of cast-in-place foundation is the *bulb-type uncased concrete pile* (Fig. 13-16). To construct this type, a mass of very dry concrete is placed in a steel casing standing on the ground surface and formed into a plug by a falling heavy ram. High friction develops between the concrete and inside casing wall, and continued blows of the ram force the concrete plug and casing into the ground. At the depth where it is desired to form a base for the pile, the casing is locked to the driving rig, and continued blows of the ram force the concrete plug from the bottom of the casing into the earth. An enlarged base forms as the driven concrete plug compacts the soil around this level. Additional dry concrete is then added and rammed while the casing is slowly lifted. A continuous pile shaft is formed as the rammed concrete continues to be driven from the bottom of the withdrawing casing.

The term *composite piles* usually refers to piles having different materials for their different sections of length. The most typical combinations include timber, or steel H pile or pipe pile for the lower section and cased cast-in-place concrete for the upper section, see Fig. 13-17. Composite piles are normally used for reasons of economy (timber is less costly than steel or concrete) or to take advantage of certain structural features that one material possesses (steel will fare better than timber or concrete in hard driving, or timber or cast-in-place concrete piles may not be available in the required length). Where composite piles are

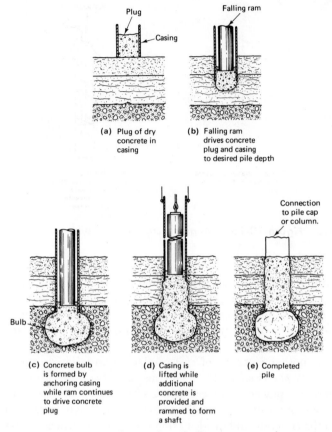

(a) Plug of dry concrete in casing

(b) Falling ram drives concrete plug and casing to desired pile depth

(c) Concrete bulb is formed by anchoring casing while ram continues to drive concrete plug

(d) Casing is lifted while additional concrete is provided and rammed to form a shaft

(e) Completed pile

Figure 13-16.   *Procedure to form bulb-type uncased concrete pile.*

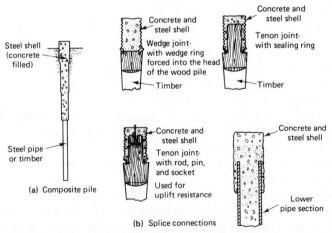

(a) Composite pile

(b) Splice connections

Figure 13-17.   *Composite pile showing splice connections: (a) typical makeup of composite pile; (b) typical splice connections.*

utilized, the splice joint between sections should be as strong as the pile materials, particularly if the pile will be subject to uplift forces. Composite pile is also the term applied to a pile cross section of more than one material, such as a concrete-filled pipe pile.

REPRESENTATIVE PILE LOAD CAPACITIES AND AVAILABLE LENGTHS / The pile type and cross section to be selected for a project will be influenced by soil conditions, required pile length, required structural capability of the pile, and consideration of the pile installation method. Information on the maximum lengths generally available for the various pile types is summarized in Table 13-2. Representative structural load ranges for various pile types are shown in Table 13-3.

METHODS TO AID PILE INSTALLATION / The installation of driven piles can be aided by the use of spudding or predrilling. These techniques are frequently utilized in situations where obstacles

TABLE 13-2 / AVAILABLE LENGTHS OF VARIOUS PILE TYPES

Pile Type	Comment, Available Maximum Length
Timber	Depends on wood (tree) type. Lengths in the 50- to 60-ft range (15 to 18 m) are usually available in most areas; lengths to about 75 ft are available but in limited quantity; lengths up to the 100-ft range are possible but very limited.
Steel H and pipe	Unlimited length; "short" sections are driven and additional sections are field-welded to obtain a desired total length.
Steel shell, cast-in-place	Typically to between 100 and 125 ft (30 to 40 m), depending on shell type and manufacturer-contractor.
Precast concrete	Solid, small cross-section piles usually extend into the 50–60 ft length (15 to 18 m), depending on cross-section shape, dimensions, and manufacturer. Large-diameter cylinder piles can extend to about 200 ft long (60 m).
Drilled shaft, cast-in-place concrete	Usually in the 50- to 75-ft range (15 to 25 m), depending on contractor equipment.
Bulb-type, cast-in-place concrete	Up to about 100 ft (30 m).
Composite	Related to available lengths of material in the different sections. If steel and thin-shell cast-in-place concrete are used, the length can be unlimited; if timber and thin-shell cast-in-place concrete are used, lengths can be on the order of 150 ft (45 m).

TABLE 13-3 / TYPICAL CAPACITIES FOR VARIOUS PILE TYPES

Pile Type	Typical Design Load Range (Tons)*
Wood	15 to 30
Concrete, cast-in-place, steel shell and uncased	30 to 75
Concrete, reinforced or prestressed (lower range for smaller cross section, upper range for the larger cross sections)	30 to 200
Concrete, bulb-type	75 to 1,000
Composite: wood and concrete	30 to 60
Steel pipe (lower range for smaller diameter and wall thickness, upper range for larger diameter and heavy wall thickness)	30 to 100
Steel HP (lower range for light sections, upper range for heavier sections	50 to 200

*1 ton = 8.9 kN.

that could damage the driven pile are buried in the soil to be penetrated or where compact or hard soil must be penetrated, or where driving vibrations may affect nearby structures. Another form of installation assistance is jetting, a procedure that eliminates some driving and hastens pile penetration.

Spudding refers to the procedure of driving a steel H or similar section into the earth to break up obstacles before installing the pile. When beyond the depth where the obstacles exist, the spud is withdrawn. The pile is then installed in the hole and driven to its final depth.

Predrilling consists of drilling a hole, approximately of the diameter of the pile, through very hard soils to eliminate the danger of pile damage that might result if driving were attempted. Predrilling is frequently utilized on projects where driving effects must be minimized, as protection against possible damage from driving vibrations to nearby facilities. The procedure is also utilized when piles are installed in clay soils to prevent ground heave, which can result from pile driving (Fig.

Figure 13-18. *Predrilling equipment for installing steel shell piles in clay soils.* (Courtesy of Raymond International, Inc.)

13-18). If ground heave occurs, previously driven piles may heave up also and have to be reseated.

Jetting is the technique of using a powerful stream of water directed below the tip of a long pile penetrating sandy soil, to wash ahead of the pile to assist it in advancing through the sand. The jetting nozzle may be temporarily attached directly to the pile. As sand is flushed from below the pile tip, the pile settles into the created void or is easily driven if the driving hammer is activated. Jetting causes the sand that eventually surrounds the pile to be loose. Typically, the jetting is stopped and the pile is driven for the last segment of the desired penetration, to develop high end friction and tip bearing. For short piles, jetting is also used to wash a hole at a pile location, before driving, to make installation easier.

OTHER   INSTALLATION   CONSIDERATIONS / Pile foundations may be required to resist lateral forces instead of or in addition to vertical loads. Driven piles and some types of formed-in-place piles can be installed at an angle to the vertical, to develop high resistance to lateral forces. Such piles are referred to as *batter piles*. Most vertical piles are capable of resisting lateral forces also, but usually of only slight magnitude.

Deep foundations may be required at locations where available headroom is limited. Driven and formed-in-place piles can be installed where space is limited, but special equipment is necessary, as illustrated in Fig. 13-19. If headroom is very limited, driven-type piles can be installed by working with very short sections and jacking. Sections are jacked and added to, until the desired penetration is reached.

The augered excavation cast-in-place piles can offer the advantage of relatively quiet, vibration-free installation compared to driven piles. Such factors may be important if construction is in a highly developed

Figure 13-19. *Pipe pile installation in area of limited headroom. Short leads are utilized along with short pile sections. (Courtesy of Associated Pile and Fitting Corp., and George W. Rogers Construction Corp.)*

area where there is concern about noise or the effect of driving vibrations on nearby structures.

Comparative installed costs for different foundation types will vary, depending on soil conditions to be penetrated, required pile length, desired load capacity, geographic area and labor costs, availability of pile materials, site accessibility and site conditions, and contractor availability. Although drilled-shaft or cast-in-place piles are often considered to be more economical than driven piles where short lengths are required, and vice versa, each project should be considered unique and costs should be determined accordingly.

Relating Soil Conditions and Foundation Types

A structural foundation serves as the intermediary element to transmit forces from a building's superstructure to the supporting soil. It is necessary to know soil conditions and soil properties underlying an area as well as the magnitude and type of building loading before selection of a proper foundation type can be made. After an appropriate foundation has been decided upon, each unit is sized for proper carrying capacity. Individual foundations should be analyzed as any other structural member is, to ensure that the element itself possesses adequate internal strength.

As a guide for developing an understanding of the intereffect of soil conditions and a required type of foundation, illustrations of different subsurface conditions and related foundation considerations are presented in Table 13-4. For the Design Comments, it was assumed that a multistory commercial structure, such as an office building, was to be supported.

TABLE 13-4
ILLUSTRATIONS RELATING SOIL CONDITIONS AND APPROPRIATE FOUNDATION TYPES

Soil conditions	Appropriate foundation type and location	Design comments

(1) El.0' , ground surface

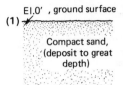

Compact sand, (deposit to great depth)

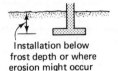

Installation below frost depth or where erosion might occur

Spread footings most appropriate for conventional foundation needs. A deep foundation such as piles could be required if uplift or other unusual forces were to act.

(2) El.0'

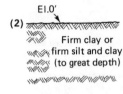

Firm clay or firm silt and clay (to great depth)

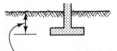

Installation depth below frost depth, or below zone where shrinkage and expansion due to change in water content could occur

Spread footings most appropriate for conventional foundation needs. Also see comment for (1) above.

(3) El.0'

Firm clay

El. −10'

Soft clay (to great depth)

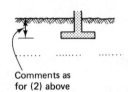

Comments as for (2) above

Spread footing would be appropriate for low to medium range of loads, if not installed too close to soft clay. If heavy loads are to be carried, deep foundations might be required.

(4) El.0'

Loose sand (to great depth)

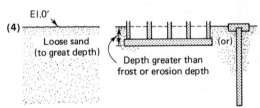

Depth greater than frost or erosion depth

Spread footing may settle excessively or require use of very low bearing pressures. Consider mat foundation, or consider compacting sand by vibroflotation or other method then use spread footings. Driven piles could be used and would densify the sand. Also consider augered cast-in-place piles.

(5) El.0'

(soft) Soft clay, but firmness
El. −25' increasing
with depth
(med. firm) (to very great
El. −50' depth)
(firmer)

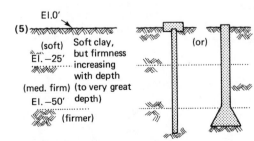

(or)

Spread footings probably not appropriate. Friction piles or piers would be satisfactory if some settlement could be tolerated. Long piles would reduce settlement problems. Should also consider mat foundation or floating foundation.

357

TABLE 13-4 cont'd.

Soil Conditions	Appropriate Foundation Type and Location	Design Comments

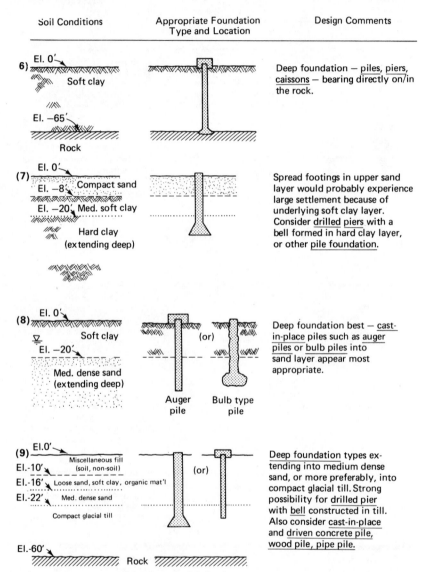

6) El. 0'
Soft clay
El. −65'
Rock

Deep foundation — piles, piers, caissons — bearing directly on/in the rock.

(7) El. 0'
El. −8' Compact sand
El. −20' Med. soft clay
Hard clay
(extending deep)

Spread footings in upper sand layer would probably experience large settlement because of underlying soft clay layer. Consider drilled piers with a bell formed in hard clay layer, or other pile foundation.

(8) El. 0'
Soft clay
El. −20'
Med. dense sand
(extending deep)

(or)

Auger pile   Bulb type pile

Deep foundation best — cast-in-place piles such as auger piles or bulb piles into sand layer appear most appropriate.

(9) El.0'
Miscellaneous fill (soil, non-soil)
El.-10'
El.-16' Loose sand, soft clay, organic mat'l
El.-22' Med. dense sand
Compact glacial till
El.-60'
Rock

(or)

Deep foundation types extending into medium dense sand, or more preferably, into compact glacial till. Strong possibility for drilled pier with bell constructed in till. Also consider cast-in-place and driven concrete pile, wood pile, pipe pile.

TABLE 13-4 cont'd.

Soil conditions	Appropriate foundation type and location	Design comments

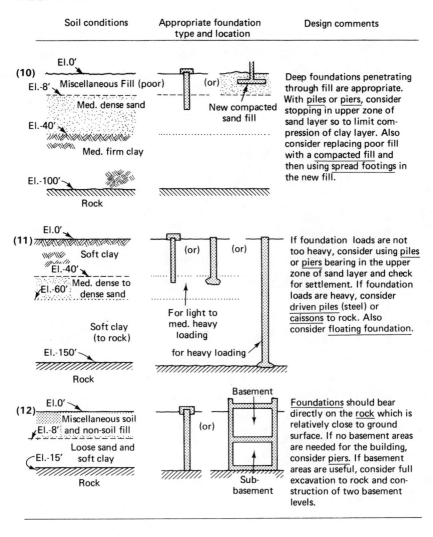

(10)

El.0'

El.-8' Miscellaneous Fill (poor)

Med. dense sand

El.-40'

Med. firm clay

El.-100'

Rock

(or)

New compacted sand fill

Deep foundations penetrating through fill are appropriate. With piles or piers, consider stopping in upper zone of sand layer so to limit compression of clay layer. Also consider replacing poor fill with a compacted fill and then using spread footings in the new fill.

(11)

El.0'

Soft clay

El.-40'

Med. dense to dense sand

El.-60'

Soft clay (to rock)

El.-150'

Rock

(or) (or)

For light to med. heavy loading

for heavy loading

If foundation loads are not too heavy, consider using piles or piers bearing in the upper zone of sand layer and check for settlement. If foundation loads are heavy, consider driven piles (steel) or caissons to rock. Also consider floating foundation.

(12)

El.0'

Miscellaneous soil and non-soil fill

El.-8'

Loose sand and soft clay

El.-15'

Rock

Basement

(or)

Sub-basement

Foundations should bear directly on the rock which is relatively close to ground surface. If no basement areas are needed for the building, consider piers. If basement areas are useful, consider full excavation to rock and construction of two basement levels.

CHAPTER 14
Foundations: Design Considerations and Methods

Building foundations need to be capable of carrying an imposed loading without undergoing movement that causes structural damage or affects the facility's planned usage. These considerations require that the soil responsible for supporting a foundation not be stressed beyond its strength limits. Simultaneously the deformations resulting within this soil because of loading and action of natural forces can not be excessive. The pressure that a foundation unit can impose onto the supporting earth mass without causing overstressing (or shear failure) is the soil's *bearing capacity*. Deformations occurring because of foundation loading usually cause settlement, but lateral movements may also be of concern. It is important also to consider the possibility of foundation movements due to natural phenomena, such as soil expansion and shrinkage if moisture changes or freezing occurs. The magnitude and type of loading (static, live, or repetitive), the foundation performance requirements (how much settlement is permissible), and properties of the supporting soil all have influence on the type and size of foundation that will be necessary and its resulting behavior.

Methods in widespread use for determining a soil's bearing capacity include the application of bearing capacity equations, the utilization of penetration resistance data obtained during soil explorations, and the practice of relating the soil type to a presumptive bearing capacity recommended by building codes. Permissible bearing capacities determined by the equation method or from building code tables typically do not consider effects of soil compressibility and possible influence of poorer soil layers underlying the bearing layer. Consequently, settlement determinations and other results of soil deformations must be analyzed separately (e.g., by using the methods discussed in Chapter 9).

Foundation design criteria developed from boring-penetration resistance data often relate a foundation bearing pressure to settlement. Design data in this form are convenient to use, but the methods available do not cover all foundation and soil types, and are felt to be less precise than the analytical methods for determining bearing capacity and settlement.

An additional method in use for determining the permissible or safe design loading, but applied primarily for pile foundations, is the field load test performed on an in-place foundation unit. Load tests relate carrying capacity and settlement together, an advantage. Disadvantages include the cost and time involved. Further, test results require care in their evaluation, for it is known that a load test on a small shallow foundation may not be representative of the behavior of a large shallow foundation. With piles, the load test results on a single unit may not be indicative of the behavior of a loaded pile group.

Of the common procedures for foundation design, the analytical method using soil mechanics principles (e.g., the use of bearing capacity equations with settlement analysis), and the use of penetration resistance data are preferred. Properly applied, these methods consider the effects of foundation type and size as well as the properties of soil to the depth that will have significant effect on the foundation performance. Load test data will provide reliable design information if properly related to the results of a subsurface exploration and a final foundation design. The use of presumptive bearing capacities is discouraged because of the heavy reliance on the soil-type description with little correlation to the soil's actual physical properties and no consideration of the possible existence of poor soil strata underlying the foundation bearing level.

The procedures for foundation design that follow represent currently accepted methods. These methods generally have a history of being conservative. As in the past, foundation design procedures are subject to improvement for greater precision as new techniques are developed to better determine soil properties and behavior, or as new analytical tools evolve. The foundations profession is still considered a "state of the art" profession. Scientific methods and principles are utilized, but exact answers are not always expected. Final decisions concerning best foundation type, design criteria, expected behavior, and methods for construction and field control are greatly influenced by experience and intuition.

Shallow Foundations

BEARING CAPACITY EQUATION—LONG FOOTINGS /

The ultimate soil-bearing capacity for foundations (the loading that will cause a shear failure in the supporting soils) is related to the properties

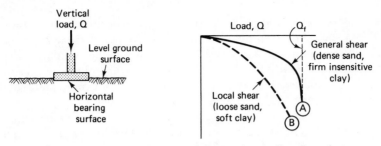

Figure 14-1. *Typical load-settlement relation for shallow foundations.*

of the soil, including the past stress history and proximity of the ground-water table, but is also affected by the characteristics of the foundation, including size, depth, shape, and roughness.

For the situation of an increasing load's being imposed onto a shallow, horizontal strip footing foundation resting on a homogeneous soil, characteristic load-settlement curves like those shown in Fig. 14-1 are obtained. For the condition where the foundation bears on soil that is relatively firm or dense (curve A of Fig. 14-1), a triangular zone forms beneath the footing when the foundation loading approaches the value of Q_f, as shown in Fig. 14-2. This soil wedge acts as part of the foundation. To each side there develops a zone of radial shear and plane (or mixed) shear. Some yielding occurs (plastic deformation) in the radial zones as foundation stresses cause the maximum shear strength of the soil to be mobilized. The effects are transmitted to the adjacent zones of plane (or mixed) shear, where high shear stresses and plastic deformation then also develop. The zones of radial shear can be considered as active zones where slippage associated with a bearing capacity failure

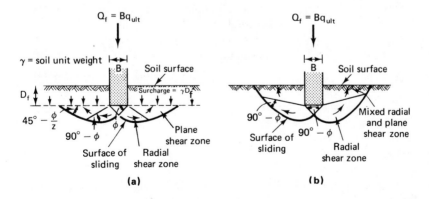

Figure 14-2. *(a) Terzaghi assumptions of soil failure zones for developing bearing capacity equation; (b) Meyerhof assumptions of soil failure zone in bearing capacity analysis. (Refs. 90, 92, 150)*

commences. The zones of plane (or mixed) shear are passive, reacting to the pushing force imposed by the soil in the radial shear zone. The affected soil zones increase in size with increases in the foundation depth and roughness of the base. Where complete yielding of the soil does occur, the foundation fails by experiencing downward movement or by tilting, depending on the symmetry of the failure zones.

The intensity of foundation pressure that will cause a bearing capacity failure q_{ult} is associated with the size of the failure zones extending below and above foundation level. In turn, the extents of these zones are functions of the weight and shear strength properties of the soil in these zones. The originating theoretical concepts for analyzing the behavior of shallow foundations are credited to Prandtl (1920)[1] and Reissner (1924).[2] Prandtl studied the effect of a long, narrow metal tool bearing against the surface of a smooth metal mass that possessed cohesion and internal friction but no weight. The results of Prandtl's work were extended by Reissner to include the condition where the bearing area is located below the surface of the resisting material and a surcharge weight acts on a plane that is level with the bearing area. Karl Terzaghi (1943)[3] applied the developments of Prandtl-Reissner to soil-foundation problems, extending the theory to consider rough foundation surfaces bearing on materials that possess weight. Terzaghi developed a general bearing capacity equation for *strip footings* which combines the effects of soil cohesion, internal friction, foundation size, soil weight, and surcharge effects, in order to simplify foundation design calculations. This equation introduces the concept of variable dimensionless *bearing capacity factors,* which are a function of the shear possessed by the supporting soils, Eq. (14-1).

$$q_{ult} = cN_c + 0.5B\gamma_1 N_\gamma + \gamma_2 D_f N_q \text{ (for strip footings)} \qquad (14\text{-}1)$$

where　N_c, N_γ, N_q = soil-bearing capacity factors whose values depend on the angle of internal friction.

　　　　c = cohesion of the soil below footing level.

　　　　γ_1 = effective unit weight of soil below footing level.

　　　　γ_2 = effective unit weight of soil above footing level.

　　　　B = footing width.

　　　　D_f = depth of footing below lowest adjacent soil surface.

　　　　q_{ult} = ultimate gross bearing capacity or soil-bearing pressure.

---

[1]Prandtl, "Uber die Harte, plastischer Korper" ("On the Hardness of Plastic Bodies"), *Nachr. Kgl. Ges. Wiss. Gottingen, Math.-Phys. Kl. 1920.*

[2]H. Reissner, "Zum Erddruckproblem" ("The Earth Pressure Problem"), *Proc. First Int'l. Congr. Appl. Mech.,* 1924.

[3]Reference 150.

Meyerhof (1951, 1953)[4] is credited with further evaluating the effects of shearing resistance within the soil above foundation level, the shape and roughness of the foundation unit, and the practical influence of a high water table. Through the years, as additional mathematical analyses were made to correct for assumptions and limitations included in the early work, and experiences were obtained on the actual performance of foundations, modifications to the Terzaghi bearing capacity factors have been made. However, the general form of the equation has been recognized for its simplicity of use, and retained. Appropriate bearing capacity factors can be obtained from the curves of Fig. 14-3.

For the condition where a foundation bears on a cohesionless soil ($c = 0$), the first term of the bearing capacity equation is zero, and the foundation pressure is determined by using only the last two terms. When a shallow strip footing bears on a purely cohesive soil ($\phi = 0$ deg), the bearing capacity factors for the second and third terms of the general equation become very small, and for practical purposes

$$q_{ult} = cN_c = 5.2c \quad \text{(for shallow strip footings on cohesive soil)} \quad (14\text{-}2)$$

Where a footing rests on loose sand or soft clay, a condition of local shear occurs within the supporting soils (Fig. 14-1, curve B) and the

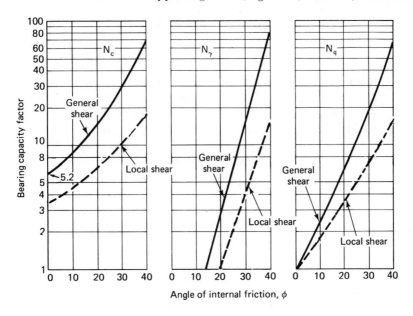

Figure 14-3. Bearing capacity factors for bearing capacity equation. (Ref. 96)

---

[4]References 90, 92.

failure zones do not extend as indicated in Fig. 14-2. For such conditions, the general bearing capacity equations can be applied, but reduced bearing capacity factors are required. Figure 14-3 includes curves
that provide the recommended values.

The values for soil unit weight to be used in the bearing capacity
equation are effective weights. When the soil in the yield zone is below
the water table, the submerged weight[5] must be used. For the condition where the water table is at the base of a footing, the submerged (or
buoyant) soil weight applies for the γ_1 term. Since the depth of the
failure zone is considered to be approximately equal to the width of the
footing, the full weight of the soil (not submerged weight) is used when
the water table is lower than a distance B below the footing. For in-
between depths of the water table, an interpolated value of effective soil
weight (between $\gamma_{wet\ soil}$ and γ_{sub}) can be used. Similarly, if the water
table extends above footing level, a submerged soil weight should be
used for γ_2 values.

When the bearing capacity equation is used for foundation design, it is conventional practice to apply a factor of safety between 2.5
and 3 to the value of q_{ult}.

$$q_{design} = \frac{q_{ult}}{3} \quad \text{or} \quad \frac{q_{ult}}{2.5} \tag{14-3}$$

Illustration 14-1:   A strip footing 3 ft. wide is supported in soil
having properties indicated by the sketch. What design loading
can be imposed onto the foundation per foot of length? Use a factor of safety of three.

Wall load, pounds per ft. of length

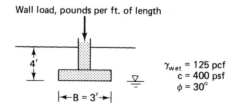

$\gamma_{wet} = 125$ pcf
c = 400 psf
$\phi = 30°$

$$q_{ult} = cN_c + 0.5B\gamma_1 N_\gamma + \gamma_2 D_f N_q$$

From bearing capacity factor chart obtain

$$N_c = 30, \qquad N_\gamma = 18, \qquad N_q = 20$$

---

[5]$\gamma_{sub} = \gamma_{sat\ soil} - \gamma_{water}$, or $\gamma_{sub} = \frac{1}{2}\gamma_{wet\ soil}$ (approximately).

$$q_{ult} = (400 \text{ psf})(30) + 0.5(3 \text{ ft})(\tfrac{125}{2} \text{ pcf})(18)$$
$$+ (125 \text{ pcf})(4 \text{ ft})(20)$$
$$= 23,700 \text{ psf}$$

$$q_{design} = \frac{23,700 \text{ psf}}{3} = 7900 \text{ psf}$$

Therefore, allowable wall loading

$$= (7900 \text{ psf})(3 \text{ ft wide})(1 \text{ ft long})$$
$$= 23,700 \text{ lb per foot of wall length}$$

Illustration 14-2: A wall footing is to be constructed on a clay soil, as indicated by the sketch. The footing is to support a wall that imposes a load of 10,000 lb per foot of wall length. What footing width should be provided to have a factor of safety of 3?

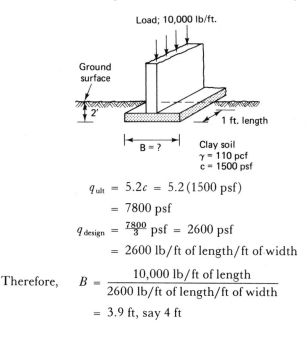

$$q_{ult} = 5.2c = 5.2(1500 \text{ psf})$$
$$= 7800 \text{ psf}$$
$$q_{design} = \tfrac{7800}{3} \text{ psf} = 2600 \text{ psf}$$
$$= 2600 \text{ lb/ft of length/ft of width}$$

Therefore, $$B = \frac{10,000 \text{ lb/ft of length}}{2600 \text{ lb/ft of length/ft of width}}$$
$$= 3.9 \text{ ft, say 4 ft}$$

SQUARE, RECTANGULAR, AND CIRCULAR FOOTINGS / The yield zones occurring in the soil beneath an infinitely long strip footing can for practical use be adequately analyzed in a two-dimensional study. Where foundations of a finite length are involved, such as with square, rectangular, and circular shapes, the yield zones are three-dimensional, and Eq. (14-1) does not directly apply. However, shape factors developed primarily on the basis of empirical data can be applied and permit the general form of the bearing capacity equation to be used for designing the common foundation shapes. The equations resulting for various shapes follow:

$$q_{ult} = 1.2cN_c + 0.4\gamma_1 BN_\gamma + \gamma_2 D_f N_q \quad \text{(for square footings)} \quad (14\text{-}4)$$

$$q_{ult} = a_1 cN_c + a_2 \gamma_1 BN_\gamma + \gamma_2 D_f N_q \quad \text{(for rectangular footings)} \quad (14\text{-}5)$$

where a_1 and a_2 are shape factors related by the length-to-width (L/B) ratio, as indicated in the following tabulation:

L/B	a_1	a_2
1	1.20	0.42
2	1.12	0.45
3	1.07	0.46
4	1.05	0.47
6	1.03	0.48
Strip	1.00	0.50

$$q_{ult} = 1.2cN_c + 0.6\gamma_1 rN_\gamma + \gamma_2 D_f N_q \quad (14\text{-}6)$$

(for circular footings with radius r)

Illustration 14-3:   A square foundation for a column is to carry a loading of 200 kips. Determine the footing dimensions if it bears on the surface of a sand that has a unit weight of 130 pcf and an angle of internal friction of 35 deg. Use a factor of safety of 3.

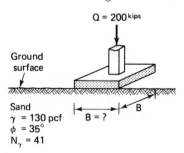

Q = 200 kips

Ground surface

Sand
γ = 130 pcf
φ = 35°
N_γ = 41

B = ?   B

For a footing on the surface of sand, $c = 0$, $D_f = 0$

$$\therefore q_{ult} = (0.4)\gamma_1 BN_\gamma$$
$$= (0.4)(0.130 \text{ kcf})(B \text{ ft})(41)$$
$$= 2.12 \, B \text{ ksf}$$

$$q_{design} = \tfrac{1}{3}(2.12\ B) = 0.71\ B\ \text{ksf}$$

$$Q_{design} = 200\ \text{kips}$$

$$q_{design} = \frac{Q}{A} = \frac{Q}{B^2} = 0.71\ B\ \text{ft}^3$$

$$B^3 = \frac{200}{0.71} = 283$$

$$B = 6.6\ \text{ft}$$

Illustration 14-4: In an industrial building, two columns are close together and will be supported on a common rectangular footing. The soil is a clay having a cohesion of 1700 psf. For the conditions indicated by the sketch, what footing dimensions should be provided? Apply a factor of safety of three.

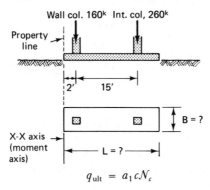

$$q_{ult} = a_1 c N_c$$

Assume $a_1 = 1.05$.

$$q_{ult} = (1.05)(1700\ \text{psf})(5.2)$$
$$= 9300\ \text{psf}$$
$$q_{design} = \tfrac{1}{3}(9300) = 3100\ \text{psf,}$$

but acceptable to use 3,000 psf or 3 ksf.

Two unknowns, B and L, require two independent equations: Use

$$\Sigma F_v = 0 \quad \text{and} \quad \Sigma M_{xx} = 0$$

(1) $\Sigma F_v = 0$

or downward forces equal upward forces

$$\text{Column loads} = (q_{design})(\text{footing area})$$
$$(160\text{k} + 260\text{k}) = (3\ \text{ksf})(BL)$$
$$BL = 140 \qquad (\text{weight of footing is neglected})$$

$$B = \frac{140}{L}$$

(2) $\Sigma M_{xx} = 0$

or moments due to column loads equal moments due to soil pressure reacting on footing. (Moment axis at wall end of footing.)

$$(160k \times 2 \text{ ft}) + (260k \times 17 \text{ ft.}) = 3 \text{ ksf } (BL) \frac{L}{2}$$

$$4740 = 1.5 \, BL^2 \text{ ft-kip}$$

$$BL^2 = 3150$$

From the calculations of step 1, substituting the value of $140/L$ for B, obtain

$$\left(\frac{140}{L}\right)(L^2) = 3150$$

$$L = \frac{3150}{140} = 22.5 \text{ ft}$$

$$B = \frac{140}{L} = \frac{140}{22.5} = 6.25 \text{ ft.}$$

There are some implications of practical importance in Eq. (14-1), as illustrated in Fig. 14-4. For a soil possessing both cohesion c and internal friction ϕ, the ultimate bearing capacity increases with increasing foundation width and foundation depth. In actual foundation design, the effect of using higher bearing pressures on larger foundations requires careful consideration where settlement is important. Because of the manner in which stresses from foundation loadings are transferred into the earth (Chapter 8), a large foundation tends to settle more than a small foundation when both impose the same intensity of bearing pres-

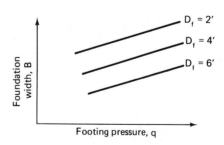

Figure 14-4. *Implications of general bearing capacity equation: allowable pressures increase with increases in footing width and depth.*

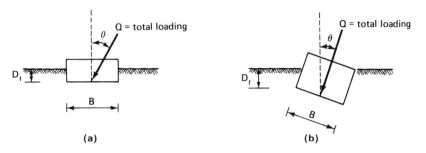

Figure 14-5. (a) Inclined loading on level foundation; (b) loading on inclined foundation.

sure, because the total load Q_t is greater. Using a greater bearing pressure with a large footing increases further the tendency for greater settlement. Consequently, if foundations for a structure are designed only on the basis of allowable bearing capacity, there is great possibility of differential settlement between the large and small footings.

FOOTINGS FOR INCLINED LOADS / If the loading imposed onto a horizontal foundation is not vertical [Fig. 14-5 (a)], or if the base of the foundation and loading is inclined [Fig. 14-5 (b)], the pattern of the soil yield zone beneath the foundation is different from the pattern that develops beneath a level footing carrying a vertical load (Fig. 14-2). Generally, the ultimate bearing capacity is lessened. For design purposes the reduced bearing capacity can be obtained from

$$q_{design} \text{ for inclined loading } = \tfrac{1}{3} q_{ult} R \qquad (14\text{-}7)$$

where q_{ult} = value obtained from the conventional bearing capacity equation which considers shape effects.

R = factor based upon the incline of the applied load or the foundation base, Table 14-1.

TABLE 14-1 / REDUCTION FACTOR
FOR INCLINED LOADING*

Inclination of Vertical Load or Foundation Base, θ Degrees Foundation as in Fig. 14-5	Inclined Load on Horizontal Foundation		Inclined Foundation	
	Cohesive Soil	Granular Soil	Cohesive Soil	Granular Soil
10 deg	0.80	0.60	0.92	0.75
20 deg	0.55	0.30	0.85	0.55
30 deg	0.40	0.15	0.80	0.40

*For condition where D_f is approximately $0.5B$.

FOOTINGS ON SLOPES / The bearing capacity for footings on slopes is less than that of those on level ground. The subsurface soil zones that provide resistance to foundation loading on the downhill side of the slope are smaller than they are where the ground is level. The safe loading is related to the soil failure zone involved. For footings of limited size, the soil failure zone is usually confined to the region near the vicinity of the foundation (regarding footings on level ground, see Fig. 14-2). For large footings, the failure zone may include all or most of the slope.

For continuous strip footings, the ultimate bearing capacity is evaluated from

$$q_{ult} = cN_{cs} + \tfrac{1}{2}\gamma BN_{\gamma s} \tag{14-8}$$

where N_{cs}, $N_{\gamma s}$ = bearing capacity factors for footings on slopes.

Values for the bearing capacity factors are shown in Fig. 14-6. For cohesive soil, N_{cs} is related to a slope stability factor, M_s, as defined in Fig. 14-6. The bearing capacity factor curves can also be used to indicate the footing setback distance where the slope inclination is no longer influential. For an adequate setback, the bearing capacity factors for level ground are used.

Illustration 14-5:   A bearing wall for a warehouse building is to be located close to a slope, as indicated by the sketch. For the given conditions, what size strip footing should be provided?

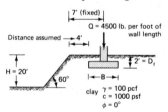

The allowable bearing pressure relates to the footing setback and the resulting b/B ratio, as well as the M_s factor. A trial and error procedure can be undertaken; it is initially assumed that $D_f/B = 0.5$ and $b/B = 1$. Since B is less than H, refer to the curves for $M_s = 0$. From Fig. 14-6(c); for $M_s = 0$, $b/B = 1$, $\beta = 60$ deg, get $N_{cs} = 4.9$. Therefore,

$$q_{ult} = cN_{cs} = 4.9c = 4.9\,(1000\text{ psf}) = 4900\text{ psf}$$

$$q_{design} = \frac{4900\text{ psf}}{FS} = \frac{4900\text{ psf}}{3\,(\text{assumed})} \simeq 1600\text{ psf}$$

Required footing width, $B = \dfrac{Q}{(q_{design})\,(1\text{ ft length})}$

$$= \frac{4500\text{ lb}}{(1600\text{ psf})\,(1\text{ ft})} \simeq 2.8\text{ ft}\qquad \text{Use } B = 3\text{ ft (conservative).}$$

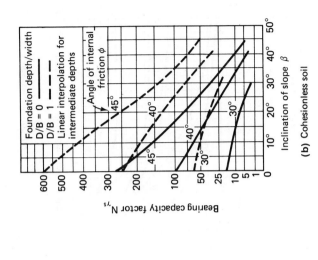

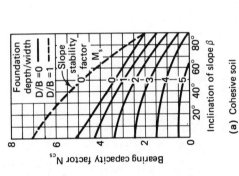

Figure 14-6 (a). *Bearing capacity factors for long footings on face of slope. (Ref. 94)*

Stability factor:
$M_s = \gamma H/c$
c = cohesion
γ = unit weight of soil
H = height of slope

Note: 1. To obtain a value of N_{cs} for footings where $B < H$, use the curves for $M_s = 0$. When necessary, interpolate for values of D/B between 0 and 1.
2. To obtain the value of N_{cs} when $B \geq H$, use the curve for the calculated stability factor M_s.

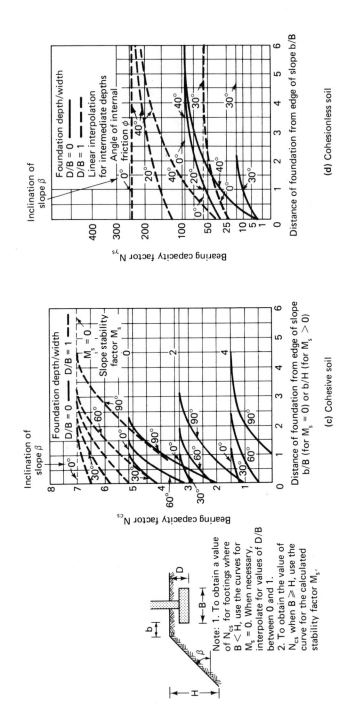

Figure 14-6 (b). Bearing capacity factors for footings on top of slope.

SPREAD FOOTING DESIGN USING PENETRATION RESISTANCE DATA / Obtaining foundation design criteria directly from the soil borings or soundings normally performed to explore conditions underlying a site is desirable from the view of cost and time savings. At locations where it is difficult to recover soil samples suitable for determining soil properties (a requirement when an analytical study for a foundation design is to be performed), obtaining foundation design criteria based directly on exploration data may be a necessity.

Typically, it is not a great problem to recover samples of soil possessing cohesion from borings, although care is required to minimize disturbance. On the other hand, obtaining undisturbed samples of cohesionless soil can be difficult, particularly when explorations extend below the groundwater table. Partly because of necessity, empirical relationships between the standard penetration test, SPT, (Chapter 11) and performance of spread footings supported on sands have evolved. Relationships have also been developed for the cone penetrometer (Chapter 11) and foundations on sand.

DESIGNING SPREAD FOOTINGS ON SAND BY USING STANDARD PENETRATION TEST / Early information relating standard penetration test results (the blow count N) in sand to spread footing size, bearing pressure, and settlement was presented by Terzaghi and Peck (1948)[6] and has been widely referred to for foundation design. This empirically determined design method was intended to provide foundations whose maximum settlement would not exceed one inch, with the expectation that the greatest differential settlement between different footings would not exceed one-half to three-fourths inch, a tolerable range for most structures. These early recommendations were based on limited data and were deliberately conservative.

Subsequent studies, which included performance evaluations of foundations designed in accord with the original recommendations, have indicated that modification of the original design criteria is necessary. Greater allowable bearing pressures can be permitted. Importantly, the value of N was found to be related to the soil overburden pressure (or depth) where the soil sample and blow count was taken. For a given soil density, the blow count is higher at deeper sampling depths because of the greater confinement and lateral pressure existing at deep, compared to shallow, locations. In contrast, blow-count values obtained close to the ground surface are lower than should be indicated because of lack of overburden pressure and confinement. Thus, when SPT values are used for foundation design, the field blow counts need to be

---

[6]Also in their subsequent publications through 1968, including Ref. 152.

corrected to reflect the effect of sample depth on actual soil properties
(see Table 14-2). The stress history of a sand deposit including effects
of compaction rolling or the presence of a high water table are factors
that also will affect a footing's performance, and these factors become
a design consideration.

TABLE 14-2 / CORRECTION FACTOR,

$$\frac{N_{design}}{N_{field}}, \text{ FOR SPT VALUES}$$

Effective Vertical Pressure ksf*	Approximate Depth for SPT Sample, Feet*	Correction Factor
0.25	2	1.7
0.50	4–5	1.4
0.75	7	1.2
1.0	9–10	1.1
1.5	14–15	0.95
2.0	18	0.90
2.5	22–24	0.85
3.0	26–28	0.80
4.0	35–38	0.75

*If water table is below sample depth. For correlation,
1 foot = 0.305 m;   1 ksf = 48 kN/m^2

Empirically developed foundation design methods are typically
based on the averaged or representative performance of case studies, but
will be conservative. Practically, deviations will always exist between
individual cases and the design recommendations that evolve. Often,
empirical design methods are recommended only for preliminary
studies. Realistically, designers frequently need also to apply the procedures to final designs because of lack of further information or
methods. When such methods are applied to future designs, it is necessary to expect a range in the predicted performance accuracy (e.g., an
expected maximum settlement of one inch could end up being somewhat
lesser or greater). Figure 14-7 presents design criteria for spread footings located at the shallow installation depths normally associated with
basementless buildings (three to five feet below grade). The bearing
pressures are intended to be values that produce a maximum settlement
on the order of one inch. The maximum differential settlement that
results between all foundation units on the site is expected to be less
than three-fourths inch.

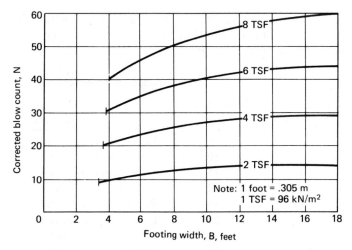

Figure 14-7. *Chart for spread footings on sand. Corrected standard penetration test results related to approximate soil bearing pressure for one-inch settlement (water table below depth B).*

The N values in Fig. 14-7 used for design are corrected values, which consider effects of depth or overburden pressure on the field SPT value of N. Table 14-2 provides a factor that should be applied to the field blow count in order to arrive at a modified N value.

The N values in Fig. 14-7 apply for dry or saturated sands. Where a condition of partial saturation exists, the sands possess some "apparent cohesion" because of the presence and effects of the air-water meniscus. Typically, this creates a greater strength. If the meniscus effect could disappear at some future time (through evaporation or submergence), a slight reduction should be applied to the N value used for design.

The procedure for designing foundations from Fig. 14-7 involves working initially with the most heavily loaded units. For each boring location, the average corrected N value (see Illustration 14-6) is determined for the soil that lies within the zone between the bottom of the footing and a depth of about $1.5B$. A value for B is based upon approximating a bearing pressure value from the design curves of Fig. 14-7. Since soil conditions are expected to vary somewhat at most construction sites, the N value selected for design is the lowest N_{avg}.

Illustration 14-6: Standard penetration test results from a soil boring located adjacent to a planned foundation for a proposed warehouse are shown below. If it is assumed that little site grading will be performed, and if spread footings for the project are to be founded 4 ft below surface grade, what foundation size should be provided to support a 400-kip column load? Assume that a one-inch settlement is tolerable.

Boring Log Summary

SPT Sample Depth, feet	Blow Count, N_{field}	Soil Classification
1	9	Fine to medium sand, trace silt
4	10	Fine to medium sand
8	15	Fine to medium sand
12	22	Fine to coarse sand
16	19	Fine to medium sand
20	29	Fine to coarse sand
25	33	Fine to coarse sand
30	27	Fine to coarse sand

Note: Water table encountered at 25 ft.

A solution requires that first the field SPT values be corrected. A tabular arrangement as shown below is practical. Correction factors are from Fig. 14-2.

SPT depth	N_{field}	Correction Factor	$N_{corrected}$	$N_{average}$ to depth shown
1 ft	9	1.70	15	—
4 ft	10	1.40	14	14
8 ft	15	1.15	17	15
12 ft	22	1.00	22	18
16 ft	19	0.95	18	18
20 ft	29	0.88	26	19
25 ft	33	0.83	27	20
30 ft	27	0.78	21	20

Reference to Fig. 14-7 indicates that for a corrected N value of between 15 and 18, the foundation design bearing pressure is approximately 2.5 tons/ft^2 or 5 ksf for footings 8 to 10 feet square.

$$B^2 = \frac{400 \text{ k}}{5 \text{ ksf}} = 80 \text{ ft}^2$$

Therefore use $B = 9$ ft.

The design bearing pressures shown in Fig. 14-7 presume the water table to be greater than a depth B below the bottom of the footing. If the water table is within the zone close to the bottom of the footing, a bearing pressure value obtained from the design curves should be reduced to keep settlement from exceeding the one-inch limit. For the condition where the water table is at the base of the foundation, a one-third reduction in the bearing pressure value should be applied. A linear interpolation can be assumed for water table depths intermediate between the foundation level and a distance B below it.

A modification to the blow count value obtained in the field should also be applied to the condition where fine sands and silty fine sands are submerged. This modification can be applied before correcting for the effects of overburden pressure. Letting N' be the modified field value, the following relationship is recommended:

$$N' = 0.6 N_{field} \quad \text{(for submerged fine sand and silty sand)} \quad (14\text{-}9)$$

The bearing pressure-settlement values indicated by Fig. 14-7 do not apply to narrow foundations. Where narrow footings are installed on cohesionless soil, bearing capacity values on the basis of shear failure of the supporting soil will control design. Angles of internal friction for the sand to permit analytical design can be estimated from the SPT blow count (Chapter 10).

Sites showing a blow count value of less than ten are indicating loose sand. The data of Fig. 14-7 should not be applied to such conditions. Consideration should be given to improving the density of the sands by vibratory compaction, vibroflotation, blasting, or other methods before construction. Where an entire area is improved, foundation design and performance will benefit, but other aspects of construction, such as building slabs, roads, parking areas, and buried utilities, are also affected. Foundation design should be made on the basis of the improved soil properties.

Different foundations usually carry different magnitudes of loading (e.g., interior column foundations generally are more heavily loaded than wall column footings). Foundations carrying different loads can have their allowable bearing pressures determined individually from the curves in Fig. 14-7, using the lowest corrected N_{avg} or an N value more appropriate for the actual footing area. Frequently, however, a bearing pressure is selected for the worst condition (lowest N value and heaviest foundation loading), and this pressure is used for proportioning all footing units. Although it is usually a safe procedure, this method tends to increase the range of differential settlement that results between light and heavily loaded foundations.

On building sites underlain by sand, many designers take the precaution of having the surface of the finish-graded area compacted by vibratory equipment to increase the density of the upper zone of soil and achieve a better degree of uniformity across the site before foundation construction takes place. Where sand fill projects are concerned, vibratory compactors are typically selected because of their effectiveness and efficiency. Densifying sand with vibratory methods causes high lateral pressures to develop in the compacted zone, a condition that produces high resistance to further volume changes and settlement. For the condition where the soil zone providing the foundation support has been compacted by vibratory methods, an increase in the bearing pressure values obtained from Fig. 14-7 can be made. Limited indications are that a one-fourth increase will be conservative.

Figure 14-7 is intended to apply only to sand. Where a subsoil contains gravel, the SPT blow count can be erroneously high. For gravelly soils, a large-diameter soil sampler should be used to recover samples for examination. Better, test pits can be excavated and the compactness of soil that will lie beneath footing level checked visually or with density testing equipment. Where deep soil zones need their density known, nuclear density meters capable of extending to deep depths can be used in the borings.

If a maximum settlement greater than one inch can be tolerated, or if it is desired to restrict the maximum settlement to less than one inch, bearing pressure values presented on Fig. 14-7 can be increased or reduced in the proportion of the final settlement desired, within reasonable limits (e.g., for half of one-inch settlement, assume half the bearing pressure).

The information of Fig. 14-7 can also be utilized for designing mat foundations. Although mat foundations typically impose comparatively low bearing pressures in accord with the purpose of their design, allowable pressures from the design can be doubled without expecting differential settlement of the mat to exceed three-fourths inch.

STANDARD PENETRATION TEST AND COHESIVE SOIL / Shallow foundations supported on clay are not designed directly from standard penetration test blow count data. A major reason for this situation has been the difficulty in relating blow count values to a cohesive soil's stress history accurately enough to permit reliable foundation settlement predictions. Since there exist practical procedures for separately determining the properties of a clay (vane shear tests can be performed in borings on the in-place soil, or laboratory strength and compressibility tests can be performed on undisturbed samples recovered from borings), some of the usual reasons for converting boring results directly into foundation design data do not apply. However, blow counts from the standard penetration test are useful for estimating the value of cohesion c (the unconfined shear strength):

$$c = \frac{N}{4} \text{ ksf} \quad \text{(for clay)} \tag{14-10a}$$

$$c = \frac{N}{5} \text{ ksf} \quad \text{(for silty clay)} \tag{14-10b}$$

The value of cohesion c can then be substituted into Eq. (14-2) to estimate the ultimate bearing capacity for the clay.

DESIGNING SPREAD FOOTINGS ON SAND BY USING STATIC CONE PENETROMETER / Static cone penetrometers are used extensively throughout Europe in the performance of investigations to determine subsoil stratification and related soil properties. In recent times, investigations utilizing penetrometer methods have received considerable attention in virtually all areas of the world. Interest has been generated by the developing awareness that direct relationships between penetrometer results and foundation design have been formulated.

For studying the relationship between static cone penetration results and foundation performance, much attention has been directed to the Delft (Dutch) cone penetrometer. This apparatus has a 60–deg cone point with a base area of 10 square cm. Many other static penetrometers in use have similar cone dimensions. To evaluate the soil in a boring or sounding, the cone is advanced a short distance to determine the point resistance q_c (kg/cm² or tons per square foot) of the undisturbed soil. European experience indicates that for spread footing foundations located at shallow depths (on the order of three feet or one meter) the allowable bearing pressure, q_{allow}, is about one-tenth of the point resistance obtained with the Dutch Cone or

$$q_{allow} = \frac{q_c}{10} \quad \text{(in kg/cm}^2 \text{ or ton/ft}^2\text{)} \tag{14-11}$$

More detailed design information, reflecting the effect of footing width and depth, has been developed, as shown in Fig. 14-8.

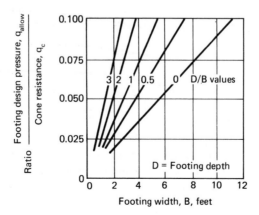

Figure 14-8. *Relationship among static cone resistance, footing size and depth, and allowable bearing pressure. (Ref. 128)*

Easily usable, refined relationships between static cone penetration, foundation size, bearing pressure and *settlement* have not been developed, although the problem has been receiving attention.[7]   For *order-of-magnitude information*, it is practical to use correlations between cone penetration resistance and the standard penetration test (SPT). For clean sands, and sands with little silt or gravel, the ratio of q_c to N usually ranges between 3.5 and 5.   A reasonable, practical approximation is

$$q_c = 4N \quad \text{(approximately)} \tag{14-12}$$

where q_c is in kg/cm^2 or ton/ft^2 and N is the blow count from a SPT.

With this value, the information of Fig. 14-7 relating shallow foundation design to settlement (an anticipated maximum settlement of one inch) has been replotted in terms of the static cone resistance q_c, see Fig. 14-9. In practical terms, the positioning of the bearing pressure curves are not highly sensitive to slight variations in the q_c-to N ratio.   Hence, the approximations of Fig. 14-9 are applicable for the sand categories indicated earlier.   The conditions assumed for developing Fig. 14-7 would also apply to Fig. 14-9 (i.e., the foundation embedment is on the order of three to five feet (1 to 1.5 m) below grade, the value of q_c is representative of the soil zone below the base of the footing to a depth extending approximately $1.5B$ further, and the water table is deep).   Reductions in bearing pressures recommended for the case of a high water table also apply to Fig. 14-9.

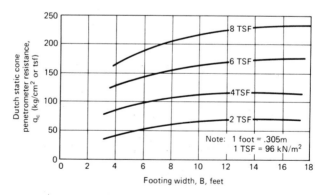

Figure 14-9.   *Chart for spread footings on sand.   Dutch static cone resistance related to approximate soil bearing pressure for one-inch settlement (water table below depth B).*

---

[7]Reference 130.

STATIC CONE RESISTANCE AND COHESIVE SOIL / In practical terms, good relationships have been established between q_c values and the cohesion and stress history of clay. The bearing capacity for the clay is subsequently calculated from the theoretical bearing capacity equations [Eq. (14-2)]. For the Delft (Dutch) cone:

$$c = \text{between } \frac{q_c}{15} \text{ and } \frac{q_c}{18} \qquad (14\text{-}13\text{a})$$

(for normally consolidated clay, $q_c < 20$)

$$c = \text{between } \frac{q_c}{10} \text{ and } \frac{q_c}{14} \qquad (14\text{-}13\text{b})$$

(for soft clays where a "local share" failure is expected

$$c = \text{between } \frac{q_c}{22} \text{ and } \frac{q_c}{26} \qquad (14\text{-}13\text{c})$$

(for overconsolidated clays, $q_c > 25$)

Settlement estimates can be obtained by using the methods described in Chapter 9. The value for C_c, the compression index, to apply when Eq. (9-3) is used can be determined from Table 14-3.

$$\Delta H = \frac{H_0}{1 + e_0} C_c (\log \sigma_{v_f} - \log \sigma_{v_0}) \qquad (9\text{-}3\text{b})$$

If soil samples are not recovered from borings or test pits, an appropriate value of e_0 can be obtained by estimating the water content from data in Table 14-3 while assuming full saturation and selecting a representative value for G_s.

TABLE 14-3 / CORRELATION OF CONE RESISTANCE, COMPRESSIBILITY INDEX, AND WATER CONTENT

Point resistance q_c (tsf, kg/cm^2)	Water content w (%)	Compression index C_c
$q_c > 12$	$w < 30$	$C_c < 0.2$
$q_c < 12$	$w < 25$	$C_c < 0.2$
	$25 < w < 40$	$0.2 < C_c < 0.3$
	$40 < w < 100$	$0.3 < C_c < 0.7$
$q_c < 7$	$100 < w < 130$	$0.7 < C_c < 1$
	$w > 130$	$C_c > 1$

(Ref. 128)

If the coefficient of volume compressibility m_v is defined as

$$m_v = \frac{\Delta e}{\Delta \sigma_v (1 + e_0)} \tag{14-14}$$

then Eq. (9-3b) can be rewritten as

$$\Delta H = H_0 (\Delta \sigma_v)(m_v) = \frac{H_0}{\eta q_c} \tag{14-15}$$

where $\Delta \sigma_v$ represents the stress increase in the clay layer, as for Eq. (9-3b) and η is obtained from Table 14-4.

TABLE 14-4 / VALUES OF THE η COEFFICIENT FOR CLAY AND SILT SOILS

	q_c in tsf or kg/cm^2	
CL–low-plasticity clay:	$q_c < 7$	$3 < \eta < 8$
	$7 < q_c < 20$	$2 < \eta < 5$
	$q_c > 20$	$1 < \eta < 2.5$
ML–low-plasticity silt	$q_c < 20$	$3 < \eta < 6$
	$q_c > 20$	$1 < \eta < 2$
CH–very plastic clay	$q_c < 20$	$2 < \eta < 6$
MH–OH–very plastic silt	$q_c > 20$	$1 < \eta < 2$

(Ref. 128)

Deep Foundations

The load-carrying capacity of deep foundations such as piles and piers are typically evaluated by analytical methods, by in-place field load tests on an installed unit, or through the application of pile-driving formulae.

The analytical method applies the principles of soil mechanics and requires knowing subsurface conditions and soil properties as well as having information on pile type, dimensions, and installation method. Field load tests consist of imposing loadings in increments onto the in-place pile or pier, and obtaining settlement data for each increment. The foundation design capacity is frequently taken where a predetermined tolerable settlement is exceeded or where the load-settlement plot is no longer proportional. Pile-driving formulae typically relate the dynamic energy imparted by the pile-driving hammer to the pile capacity. Most of the driving formulae in common use are not appropriate for determining the true capacity of a driven pile. Their lack of accuracy notwithstanding, driving formulae unfortunately are still in wide use as the only field control for pile installations. However, application of the driving formula for field control used in conjunction with load test results is a recommended procedure, for the driving record serves to relate each installed pile to the tested pile.

STATICAL ANALYSIS—GENERAL / For a pile or pier having adequate structural strength to carry an intended loading, the total downward capacity Q_t will be based on soil conditions (Fig. 14-10). The ultimate capacity of a pile is due to soil resistance developed by friction or adhesion between the soil and pile shaft, $Q_{friction}$ and the end bearing at the tip of the pile, Q_{tip}.

$$Q_{total} = Q_{friction} + Q_{tip} \qquad (14\text{-}16a)$$

$$Q_{total} = fA_{surface} + qA_{tip} \qquad (14\text{-}16b)$$

where $A_{surface}$ = total surface area of pile in contact with soil along the embedded shaft length.

f = average unit skin friction or adhesion between soil and pile surface.

q = bearing pressure of soil at pile tip.

A_{tip} = pile tip bearing area.

Methods in use to compute $Q_{friction}$ and Q_{tip} relate to the type of soil penetrated. The analyses differ for clay soil and for sand soil, as detailed below. Where strata of soil possessing different properties are penetrated, or the pile cross section and surface area vary along its length, the skin friction can be calculated by using segments of pile length and the appropriate soil value or pile area.

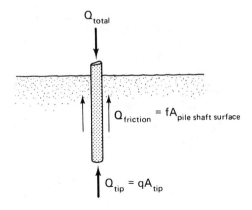

Figure 14-10.   *Basic concepts—pile capacity related to soil support.*

STATICAL METHOD—DRIVEN PILES IN SAND / The ultimate capacity of a single pile driven in sand is the sum of the point resistance and skin friction acting along the shaft of the pile:

$$Q_{total} = fA_{surface} + qA_{tip} \qquad (14\text{-}16b)$$

The unit skin friction for a straight-sided pile is a function of the soil pressure acting normal to the pile surface and the coefficient of friction between the soil and pile material (Fig. 14-11). Thus, skin friction on a pile shaft is determined in the manner used in engineering mechanics to compute frictional resistance developing between two solid materials with sliding impending.

Soil pressure acting normal to a vertical pile surface is assumed to be directly related to an effective vertical soil pressure (overburden pressure) and a lateral pressure coefficient K.

$$\sigma_h = K\bar{\sigma}_v \qquad (8\text{-}5)$$

where σ_h = horizontal soil pressure acting at any depth Z in a soil mass.

$\bar{\sigma}_v$ = effective vertical pressure (overburden) acting at same depth Z within a soil mass.

K = lateral pressure coefficient expressing the ratio of σ_h to $\bar{\sigma}_v$.

The unit value of skin friction f acting at any depth in a sand mass becomes

$$f_{sand} = \sigma_h \tan \delta = K\bar{\sigma}_v \tan \delta \qquad (14\text{-}17)$$

where $\tan \delta$ = coefficient of friction between sand and the pile surface.

Representative values for the coefficient of friction between silica sand

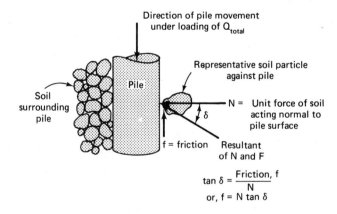

Figure 14-11. *Solid-to-solid friction developing against pile shaft.*

and pile materials are shown in Table 14-5. The total skin friction acting along the embedded length of pile is then

$$Q_{friction} = f_{sand} A_{surface} = (K \bar{\sigma}_v \tan \delta) A_{surface} \qquad (14\text{-}18)$$

TABLE 14-5 / COEFFICIENT OF FRICTION BETWEEN SAND AND PILE MATERIALS

Material	Tan δ
Concrete	0.45
Wood	0.4
Steel (smooth)	0.2
Steel (rough, rusted)	0.4
Steel (corrugated)	use tan ϕ of sand

In cohesionless deposits the vertical (overburden) stress σ_v is conventionally assumed to increase with depth, according to Eq. 8-1;

$$\sigma_v = \gamma_{soil} Z \qquad (8\text{-}1)$$

where γ_{soil} = unit weight of soil.
            Z = depth below ground surface.

However, *with driven piles* it has been found that the *effective* overburden stress of soil adjacent to the pile does *not* continue to increase without limit, as implied by Eq. (8-1). Adjacent to a pile, the effective vertical stress increases only until a certain distance of penetration, termed the critical depth D_c, is reached. Below this depth, the effective vertical pressure remains essentially constant, or changes at a low rate. The point where the critical depth is reached is influenced by the initial condition of the sand (loose or compact) and the dimension of the pile. Field and model tests indicate that the critical depth ranges from about ten pile diameters for loose sands to about twenty pile diameters for dense, compact sand. Figure 14-12 provides information on critical depth for different conditions to use in pile design.

The value of K that acts at a particular depth in a soil mass after the pile has been driven is affected by the initial condition of the sand (such as relative density), shape and size of the pile, and method of installation. Piles are driven into undisturbed ground or installed in predrilled holes or with the assistance of jetting. The latter two procedures are commonly utilized where it is anticipated that a pile cannot be driven to a desired depth without damage because of the presence of hard material or hazardous objects in upper zones of soil. Fig. 14-13 relates installation conditions with values of K that are commonly assumed in pile capacity analysis.

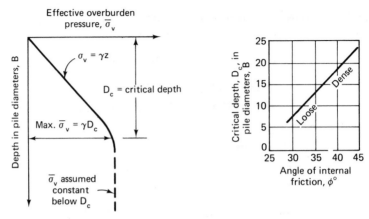

Figure 14-12. *Variation of effective overburden stress in sands adjacent to driven straight-sided piles. (Refs. 98, 160, 161)*

Driven piles causing large soil displacement (e.g., cylindrical piles) develop greater unit skin friction than piles with small volume displacement (e.g., H-piles). This is the effect of a greater K value resulting from the pile installation. Similarly, tapered piles develop greater capacity along the shaft than do straight piles because of the greater change in properties of the soil surrounding the pile, and the greater inclination in the angle of the soil force resultant acting against the pile shaft.

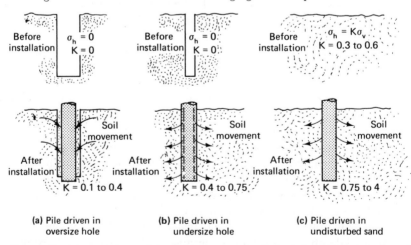

Figure 14-13. *Values of lateral pressure coefficient K used for design of piles in sand.*

Where information on soil conditions and properties along with the type and shape of the pile to be driven is available, the previously detailed methods should be applied to compute the skin friction. Because the condition of the sand surrounding the pile has significant influence on the skin friction that develops, it is possible to establish a probable range of skin friction values acting along the shaft of a pile based on the relative density of the sand. Such information is shown in Fig. 14-14 for driven piles having straight sides and limited diameter. Such data can be used to obtain *preliminary estimates* of skin friction during the absense of detailed information.

The end bearing pressure, q_{tip}, at a pile tip can be computed from the bearing capacity equation for deep foundations in cohesionless soil.

$$q_{tip} = 0.4\gamma BN_\gamma + \bar{\sigma}_v N_q \qquad (14\text{-}19a)$$

where N_γ, N_q = bearing capacity factors for deep foundations.
$\bar{\sigma}_v$ = effective vertical pressure acting at pile tip depth.
B = pile tip diameter or width.
γ = unit weight of soil in zone of pile tip.

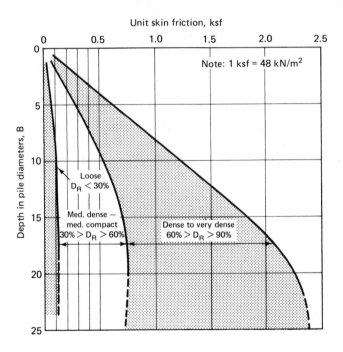

Figure 14-14.   *Range of values: skin friction versus depth for straight-sided piles driven in sand.*

This bearing capacity equation has the same form as the bearing capacity for shallow foundations. With most driven piles the first term of the equation is small compared to the second term because of the limited B dimension. Thus, for many design problems

$$q_{tip} = \bar{\sigma}_v N_q \qquad (14\text{-}19b)$$

and the total end bearing becomes, for practical purposes,

$$Q_{tip} = (\sigma_v N_q) A_{tip} \qquad (14\text{-}19c)$$

If the pile is a large-diameter pile, the tip dimension will be significant and the end bearing should be determined from Eq. (14-19a).

The value of the bearing capacity factor N_q is related to the angle of internal friction of the sand in the vicinity of the pile tip (several pile diameters above and below the tip) and the ratio of pile depth to pile width. Determining the value of N_q from field investigations is not easily accomplished because of the difficulty of separating it from other influential factors. Values of N_q presented by various investigators have been developed primarily from theoretical analysis. Results range widely (Fig. 14-15) because of the assumptions made in defining the shear zones near the pile tip. Values of N_q obtained from the curve attributed to Berezontzev in Fig. 14-15 are believed to be most applicable for commonly encountered soil conditions. The magnitude of effective vertical pressure $\bar{\sigma}_v$ which develops at the pile tip is limited below the critical depth. For design purposes, it should be assumed that the value of $\bar{\sigma}_v$ in Eq. (14-19) is equal to the effective overburden existing at the critical depth; see Fig. 14-12. The angle of internal friction for soil in the zone of the pile tip is determined from laboratory tests on recovered samples or from correlations with penetration resistance in borings (e.g., SPT test) or from soundings. If Dutch Cone data are available, the end bearing for a pile is directly related to the cone resistance q_c.

$$Q_{tip} = \frac{q_c}{2} A_{tip} \qquad (14\text{-}20)$$

If the pile analysis requires use of Eq. (14-19a), the value of N_γ for a deep foundation can be conservatively taken as twice the N_γ value used for shallow foundations.

Values of N_q shown in Fig. 14-15 assume that the soil above the pile tip is comparative to the soil below the pile tip. If the pile penetrates a compact layer only slightly, and loose material exists above the compact soil, an N_q value for a shallow foundation will be more appropriate than a value from Fig. 14-15.

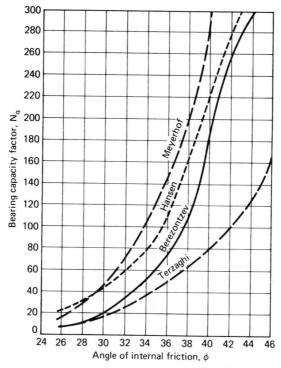

Figure 14-15. *Bearing capacity factor N_q for piles penetrating into sand.* (Refs. 87, 107)

Illustration 14-7: Prestressed concrete piles are planned for use as the foundation for a waterfront structure. Soil conditions are as indicated by the sketch. What is the approximate axial capacity for a single 12-inch square pile driven to a depth of 30 ft?

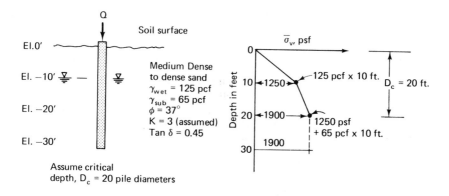

Total skin friction = (area of $\bar{\sigma}_v$ diagram) (K) (tan δ)

$\qquad\qquad\qquad\quad \times$ (pile circumference)

$$= \left[\left(1250 \times \frac{10}{2}\right) + \left(\frac{1900 + 1250}{2}\right) \times 10 \right.$$

$$\left. + (1900 \times 10)\right] (3 \times 0.45 \times 4)$$

$$= 220,000 \text{ lb} = 220 \text{ k}$$

End bearing = $\bar{\sigma}_v N_q A_{tip} = (1900)(80)(1) = 152,000 \text{ lb}$

$$= 152 \text{ k}$$

$$Q_{total} = 220 \text{ k} + 152 \text{ k} = 372 \text{ k};$$

$$Q_{design} = \frac{372 \ k}{FS}$$

STATICAL METHOD—DRIVEN PILES IN CLAY / The ultimate supporting capacity of a driven pile in clay is

$$Q_{total} = Q_{friction} + Q_{tip} \qquad\qquad (14\text{-}16a)$$

As driven piles penetrate into a saturated clay, the soil in the vicinity of the pile is remolded because of displacement and disturbance. The remolded strength of a clay is almost always less than the original strength (cohesion), but typically the strength improves with time (thixotropy). The rate of strength gain will be related to the consolidation characteristics of the clay, for dissipation of excess pore water pressures developed because of the pile installation must occur.

The skin friction, or adhesion f, that develops between the soil and pile shaft has been related to the undisturbed cohesive strength of the clay.

$$f_{clay} = \alpha c \qquad\qquad (14\text{-}21a)$$

where f_{clay} = unit adhesion or skin friction developed between clay and pile shaft.

$\qquad\quad c$ = cohesive strength of undisturbed clay.

$\qquad\quad \alpha$ = factor that relates adhesion to cohesion.

For *normally consolidated* clays in the soft-to-firm range (cohesion less than one ksf) the value of α can be taken as unity, and therefore

$f = c$ (for normally consolidated clays with $c \leq 1.0$ ksf)   (14-21b)

Equation (14-21b) may also be valid for normally consolidated clays having cohesive strengths greater than one ksf, but current data are too limited for confirmation.

Various investigators have reported values of α for clays that are

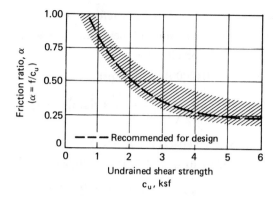

Figure 14-16. *Correlation of friction factor α and clay cohesion c_u.*

overconsolidated. Results vary, but fall within the band shown in Fig. 14-16.

The magnitude of Q_{friction} for clay to use in Eq. (14-16) then becomes

$$Q_{\text{friction}} = fA_{\text{surface}} = \alpha c A_{\text{surface}} \qquad (14\text{-}22)$$

It is recognized that the lateral pressure existing in the remolded soil adjacent to the pile shaft has an influence on the dissipation of excess pore water pressures in this zone. If sufficient pore water drainage can occur, the final strength of the clay conceivably may be greater than the undisturbed strength. Because of the phenomenon of strength gain with time, load tests on piles in clay should be performed some weeks after installation to obtain a reliable indication of actual capacity.

Illustration 14-8: A 14-inch square prestressed concrete pile is to be driven at a site where soil conditions are as indicated by the sketch. If a penetration of 40 ft is assumed, approximately what total skin friction is expected to develop along the embedded length of the pile?

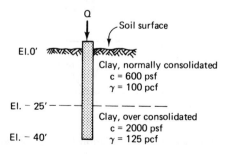

for depth 0 ft. to 25 ft. $\alpha = 1$, therefore

adhesion = 600 psi; cohesion   for depth 25 ft. to 40 ft.

$\quad\quad \alpha = 0.5\pm, f$ or adhesion = αc

adhesion = $(0.50)(2000) = 1000$ psf

$$Q_{\text{friction}} = (600\ \text{psf})\left(\frac{14\ \text{in.}}{12\ \text{in/ft}} \times 4\ \text{sides}\right)(25\ \text{ft})$$

$$+ (1000\ \text{psf})(\tfrac{14}{12}) \times 4)(15\ \text{ft})$$

$$= 70{,}000\ \text{lb} + 70{,}000\ \text{lb} = 140{,}000\ \text{lb} = 140\ \text{k}$$

Note: A value of Q_{total} would include Q_{tip} but then would be reduced by the factor of safety to yield Q_{design}.

Calculating the value of skin friction developing along the shaft of a pile by use of the method in Illustration 14-8 has proved to be fairly reliable, though not precise, for piles provided that the embedded length has not been too great (less than about 75 feet or 25 meters). For very long piles, the method has been found to be too conservative in the prediction of the frictional capacity. Reasons for the discrepancy are not yet clear. Disturbance and remolding may not be as severe at deep depths as for shallower depths, with the result that the α factors are too low. Or, as recent evidences imply, the manner in which skin friction in clay occurs may be basically similar to the development of friction in sands, where resistance is related to an effective lateral pressure.

In recognition of a probable relationship between effective overburden and skin friction, an empirical method for determining the frictional capacity of long cylindrical steel piles in normally consolidated clay has been developed (Lambda method). The method assumes that the displacement caused by the pile installation results in passive lateral pressures acting in the soil zone adjacent to the pile shaft. If the effective vertical pressure acting at a particular depth in a soil mass is $\bar{\sigma}_v$, the maximum horizontal pressure (passive pressure) is $\bar{\sigma}_v + 2c$, from the Mohr circle analysis for undrained shear strength; see Fig. 14-17.

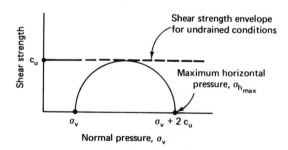

Figure 14-17. *Mohr's circle presentation indicating relation between vertical pressure and maximum horizontal pressure.*

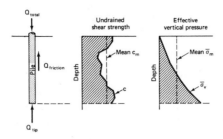

Figure 14-18. *Mean vertical pressure and mean shear strength used in designing long piles in clay. (Ref. 89)*

The unit skin friction is taken as

$$f = \lambda(\bar{\sigma}_v + 2c_u) \qquad (14\text{-}23a)$$

where λ = a friction capacity coefficient.

c_u = undrained shear strength (cohesion) of the clay.

The total friction capacity for a pile is then

$$Q_{friction} = \lambda(\bar{\sigma}_{v_m} + 2c_m) A_{surface} \qquad (14\text{-}23b)$$

where $\bar{\sigma}_{v_m}$ = mean effective vertical pressure for the embedded length.

c_m = mean undrained shear strength for depth of pile.

$A_{surface}$ = surface area of pile shaft.

The value of λ has been empirically determined for cylindrical steel piles embedded in clay. Values found to apply are presented in Fig. 14-19 for various pile lengths. Equation (14-23a) implies that the λ term

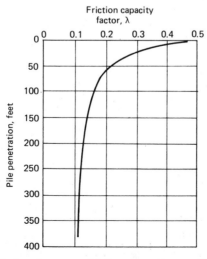

Figure 14-19. *Relation between friction capacity coefficient λ and pile length. (Ref. 89)*

is the empirical factor that serves to include a coefficient of friction between soil and pile surface, and accounts for the probability that the
lateral pressure against the pile is not fully passive.

Illustration 14-9.   A pipe pile with an outside diameter of 18 in.
and a flat end plate is driven 200 ft into clay soil having the
properties shown.   The water level is above the ground surface.
Applying the lambda (λ) method, calculate the skin friction expected to develop along the shaft of the pile.

Depth (feet)	Cohesion, c (psf)	Sub. Unit Wt. (pcf)
0 to 30	600	52
30 to 70	1000	58
70 to 90	1200	58
90 to 130	1600	55
130 to 200	3000	60

Note: Effective vertical pressure =
weight of soil overburden =
$\Sigma (\gamma_{soil} \times depth)$

$$c_m = \frac{\text{area of } c_u \text{ diagram}}{\text{pile depth}}$$

$$= \frac{\left[\begin{array}{c}(0.60 \text{ ksf} \times 30 \text{ ft}) + (1.0 \text{ ksf} \times 40 \text{ ft}) \\ + (1.20 \text{ ksf} \times 20 \text{ ft}) \\ + (1.60 \text{ ksf} \times 40 \text{ ft}) + (3.00 \text{ ksf} \times 70 \text{ ft})\end{array}\right]}{200 \text{ ft}}$$

$$= \frac{356 \text{ k/ft}}{200 \text{ ft}} = 1.78 \text{ ksf}$$

$$\bar{\sigma}_{v_m} = \frac{\text{area of } \bar{\sigma}_v \text{ diagram}}{\text{pile depth}} = \frac{1120 \text{ k/ft}}{200 \text{ ft}} = 5.6 \text{ ksf}$$

$$Q_{\text{friction}} = \lambda(\bar{\sigma}_{v_m} + 2c_m) A_{\text{shaft}}$$
$$= (0.12)[5.6 \text{ ksf} + 1.78 \text{ ksf}] \times (200 \times 3.14 \times \tfrac{18}{12})$$
$$= 840 \text{ kips}$$

Other studies have been performed to evaluate skin friction of piles in clay in terms of effective overburden pressure. Limited research suggests that the ultimate skin friction developed in normally consolidated clay may be expressed as

$$f = \beta\bar{\sigma}_v = K \tan\phi\bar{\sigma}_v = (1 - \sin\phi)(\tan\phi)\bar{\sigma}_v \qquad (14\text{-}24)$$

where β = skin friction factor.
  K = lateral earth pressure coefficient.
  ϕ = angle of internal friction for clay in drained shear.

For clays, the angle of internal friction in drained shear normally ranges between 15 and 30 deg. Accordingly, the value of β ranges between 0.2 and 0.3 The upper values apply to short piles (length less than 50 ft), while the lower limits apply to long piles. Though promising as an analytic tool, information on the method is currently too limited to recommend as a design procedure.

The values of *end bearing* for a pile in clay can be computed from

$$q_{\text{tip}} = cN_c \qquad (14\text{-}25)$$

where c = cohesion of the clay in the zone surrounding the pile tip.
  N_c = bearing capacity factor for deep foundations, ranging between about 6 and 10, depending on the stiffness of the clay. A value of 9 is conventionally used.

The above equation assumes that the lower section of pile penetrates at least five diameters into the clay whose cohesion is c.

If the subsurface investigation includes Dutch Cone data, the end bearing for a pile can be directly related to the cone resistance, q_c in kg/cm² or tons per square foot. For homogeneous conditions, where the pile penetrates at least eight diameters into the clay whose q_c is used in the design computation and no soft soil underlies the pile tip, the end bearing can be taken as

$$Q_{\text{tip}} = A_{\text{tip}} q_{\text{design}} = A_{\text{tip}} \frac{(q_c)}{1.5} \qquad (14\text{-}26)$$

FACTOR OF SAFETY / The value of Q_{total} obtained by using the analytical procedures discussed for piles in sand and clay represents an estimate of the maximum load that can be applied to a pile. Whenever possible, predicted capacities should be verified by a field load test before a final design loading is selected. Where load tests have not been performed, it is usual practice to apply a factor of safety of two during the analysis to determine the downward design load.

$$Q_{design} = \frac{Q_{total}}{2} \quad \text{(for downward loading)} \qquad (14\text{-}27a)$$

Piles subject to uplift develop resistance to pullout only from the skin friction developed along the embedded length. End bearing does not apply, but the weight of the pile can be included in uplift resistance. For design purposes, it is common to apply a factor of safety of *at least* two. If nonhomogeneous conditions exist or soil properties are not accurately known, a larger theoretical factor of safety is warranted. Tapered piles are commonly assigned a larger factor of safety than are straight-sided piles. Generally, a larger safety factor is applied to a design uplift capacity than to the downward capacity. The reason frequently stated is that an overrated capacity for downward loading probably results only in a greater-than-estimated settlement, whereas a pile whose uplift is overrated may experience significant upward movement with serious effects to the supported structure. The strength of the pile-to-pile cap connection becomes critical where uplift forces are imposed, for tensile failure at this location will negate any potential pullout resistance that a pile possesses.

$$Q_{design} = \frac{Q_{friction} + W_{pile}}{FS} \quad \text{(for uplift)} \qquad (14\text{-}27b)$$

where W_{pile} = weight of pile.
         FS = factor of safety, 2 or greater.

NEGATIVE SKIN FRICTION / When the properties of the soil through which a pile penetrates are evaluated, it is important to watch for and recognize the condition that will create "negative friction" or "downdrag." The phenomenon occurs when a soil layer surrounding a portion of the pile shaft settles more than the pile. This condition can develop where a soft or loose soil stratum located anywhere above the pile tip is subjected to new compressive loading. As an illustration, new loading to existing soil results when fill is placed on a construction site to raise the area. If the soft or loose layer settles after the pile has been installed, the skin friction-adhesion developing in this zone is in the direction of the soil pulling downward on the pile; see Fig. 14-20. Extra loading is thus imposed onto the pile. It is necessary to subtract negative skin friction values from the total load that the pile can support in order to know what building load can be carried. Values of negative skin friction are computed in the same manner as those of positive skin friction.

Where it is anticipated that an upper zone of soil would impose undesirable or intolerable downdrag on a pile, a common practice is to provide the pile with a protective sleeve or coating for the section that is embedded in the settling soil. Skin friction for this section of pile is eliminated, and downdrag is prevented.

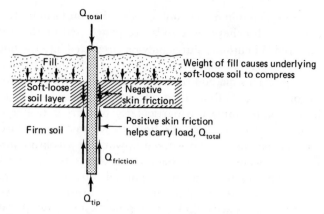

Figure 14-20. *Action of negative skin friction on pile shaft.*

OTHER DESIGN CONSIDERATIONS / The development of skin friction along the shaft of a pile does require some shear strain in the soil adjacent to the pile. If the tip and end section of a pile are embedded in a very firm or compact material, high end bearing can be developed with relatively little downward movement of the pile. Only limited deformation may occur in the upper zone of soil surrounding the pile shaft, and the maximum skin friction may not develop. The percentage of reduction in skin friction that should be attributed to comparatively soft or loose strata in contact with the pile shaft is generally a judgment factor, influenced by a comparison of relative values of cohesion or density in the various soil zones. Conversely, if a pile should penetrate through firm or dense soils and terminate in a soft or loose material, the maximum computed tip resistance may not be able to develop unless significant downward movement occurs.

STATICAL METHOD—BORED PIERS AND PILES / Bored piers and piles refer to the type of deep foundations that are constructed by drilling into the earth and subsequently placing concrete in the excavation, usually directly against the soil, to form the foundation unit. The concrete is often reinforced with steel. This type of foundation can be straight-sided for its full depth, or may be constructed with a belled or underreamed base if in cohesive soil, in order to increase the tip bearing. Bored foundations have gone under a variety of names in the past, such as drilled piers, piles, or caissons, and cast-in-place or cast-in-situ piles. When provided with an enlarged base, the foundation may be referred to as a drilled-and-belled type. The designation currently receiving general acceptance is *drilled-shaft foundation*.

Drilled-shaft foundations are constructed by utilizing mechanical auger drill equipment to excavate the hole in the earth. When subsurface conditions permit, such as a sufficiently strong cohesive soil and no groundwater, the hole can be drilled dry, after which the cast-in-

place concrete comes in direct contact with the soil forming the walls of
the excavation. If cohesionless soils are penetrated or the water table
is encountered, a bentonite slurry may be circulated into the hole as it is
being drilled to prevent soil cave-in and to assist in flushing soil cuttings
to the surface. (This method is similar to the procedure used for drilling
uncased soil borings.) Reusable protective casing may also be utilized
with the bentonite slurry if a particularly bad groundwater condition
is encountered. Concrete for the finished foundation is placed at the
bottom of the excavation and worked upward in a continuous pour so
as to force displacement of the bentonite slurry. If a casing has been
used, it is pulled as the concrete is poured, but in such a manner to
prevent the soil walls from falling into the excavation and mixing with
the concrete. This is accomplished by pouring high-slump concrete inside the casing and keeping the height of concrete above the bottom of
the withdrawing casing at all times. The earlier presence of a bentonite
slurry leaves a film on the soil surrounding the excavation, an effect that
will reduce the skin friction between the soil and hardened concrete.

Design criteria *for drilled shaft foundations in clay* have evolved from
instrumented studies performed on full-scale foundations. For foundations embedded in relatively homogeneous clay, the ultimate capacity
is due to the resistance provided from the end bearing and skin friction.
In calculating the foundation depth that effectively provides skin friction, the lower five-foot section and the belled section (if provided) are
neglected because of disturbance and loss of strength caused by construction. There is also a strong possibility of disturbance and loss of
strength occurring in the surface zone of soil. Skin friction should be
neglected in such a zone. A depth of five feet can be assumed unless
more accurate information is available.

$$Q_{\text{total}} = Q_{\text{base}} + Q_{\text{friction}} \tag{14-16a}$$

becomes
$$Q_{\text{total}} = cN_c A_{\text{base}} + fA'_{\text{shaft}} \tag{14-28a}$$

(for drilled shaft foundations)

where c = cohesion of the soil at base of foundation.

N_c = bearing capacity factor.

A_{base} = bearing area of foundation base.

f = unit skin friction developing between clay and concrete
shaft.

A'_{shaft} = surface area of shaft that is effective in developing skin
friction.

The skin friction that develops along the shaft is related to the
cohesion of the surrounding clay and the field procedure for drilling the
foundation. Values appropriate for design are shown in Table 14-6.

TABLE 14-6 / SKIN FRICTION VALUES
FOR DRILLED SHAFT FOUNDATIONS IN CLAY

Foundation Type and Drilling Method Utilized	Skin Friction f	Upper Limit On Skin Friction, ksf*
Straight shaft, excavation drilled dry	$0.5c$	1.8
Straight shaft, drilled with slurry	$0.3c$	0.8
Belled, drilled dry	$0.3c$	0.8
Belled, drilled with slurry	$0.15c$	0.5

Note: c is soil cohesion determined from triaxial testing, not in-situ vane shear tests.

$$*1 \text{ ksf} = 48 \text{ kN/m}^2.$$ (Ref. 118, 119)

If the base of the foundation is in very firm clay but the shaft is not, the supporting capacity of the unit is due primarily to end bearing. If the movement of the foundation base is very limited, the soil adjacent to the shaft does not experience the shear strain necessary for significant skin friction to develop, and thus

$$Q_{total} = Q_{base} = cN_c A_{base} \qquad (14\text{-}28b)$$

(for drilled shaft foundations bearing on very firm clay)

An N_c of 9 is assigned for all the conditions described previously.

When the methods presented above are applied to a foundation design, it is recommended that a factor of safety of 3 be applied to the bearing capacity of the base of the unit. Thus

$$Q_{design} = \tfrac{1}{3}cN_c A_{base} + fA'_s = 3cA_{base} + fA'_{shaft} \qquad (14\text{-}29)$$

(for drilled shaft foundations)

Illustration 14-10: A three-foot-diameter drilled-shaft foundation is constructed with a six-foot-diameter base in soil conditions shown by the sketch. The excavation is drilled dry. For the foundation length shown what maximum axial load (design load) should be planned?

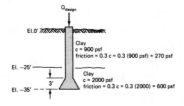

$$Q_{tip} = 9cA_{tip} = (9)(2000 \text{ psf})\left(\frac{\pi}{4} \times 6 \text{ ft} \times 6 \text{ ft}\right)$$

$$= 510,000 \text{ lb} = 510 \text{ k}$$

$$Q_{skin\ friction} = fA'_{shaft} = (270 \text{ psf})(\pi \times 3 \text{ ft})(25 \text{ ft} - 5 \text{ ft})$$

$$+ (600 \text{ psf})(\pi \times 3 \text{ ft})(10 \text{ ft} - 3 \text{ ft} - 5 \text{ ft})$$

$$= 51,000 \text{ lb} + 11,300 \text{ lb} = 62 \text{ k}$$

$$Q_{design} = \tfrac{1}{3}Q_{tip} + Q_{friction} = \tfrac{510}{3} \text{ kips} + 62 \text{ kips} = 232 \text{ kips}$$

The capacity of drilled-shaft foundations in *sand* can be analyzed by applying the procedures discussed for driven piles in sand.

$$Q_{total} = Q_{tip} + Q_{friction} \qquad\qquad (14\text{-}16)$$

$$= \bar\sigma_v N_q A_{tip} + (K\bar\sigma_v \tan \delta) A_{shaft}$$

where σ_v = effective vertical pressure considering the limits imposed by the concept of critical depth.

$\tan \delta$ = coefficient of friction between sand and concrete.

K = lateral pressure coefficient for a drilled foundation in sand.

The value of K for drilled-shaft or bored foundations ranges between about 0.3 for loose sand and 0.75 for compact sand. The value of $\tan \delta$ can be taken as equal to $\tan \phi$ for the sand when the excavation has been drilled dry, because of the roughness of concrete against soil. If a slurry has been used when drilling the excavation, some reduction should be applied. The design load should include a factor of safety at least equal to that for driven piles.

ARRANGEMENT FOR PILES IN A GROUP / Pile analysis discussed previously refers to the capacity of an individual unit. When used for foundation support, driven piles will almost always be used in a grouping. This requirement results from the desire to insure that the imposed structural load (e.g., column or wall load) falls within the support area provided by the foundation. Usually, driven piles are not used singularly beneath a column or wall because of the tendency for the pile to wander laterally during driving. If a single pile were to be used as the foundation, a designer could not be certain that the pile would be centered beneath the foundation. If an unplanned eccentric loading results, the connection between pile and column may be inadequate, or the pile may fail structurally because of bending stresses created within it. As a result, piles for walls are commonly installed in a staggered arrangement to both sides of the wall center line. For a column or isolated load, a minimum of three piles is used in a triangular pattern, even for light loads. Where more than three piles are required in order to obtain adequate capacity, the pile arrangement is

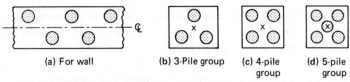

(a) For wall   (b) 3-Pile group   (c) 4-pile   (d) 5-pile
                                            group   group

Note: Pile spacings for groups typically range between
2½ and 4 pile diameters, center to center.

Figure 14-21. *Representative pile patterns for wall and column foundations.*

symmetrical about the point or area of load application. Representative patterns are illustrated in Fig. 14-21. Column and wall loads are usually transferred to the pile group through a pile cap. The pile cap is typically a reinforced concrete slab structurally tied to the pile butts (top) to help the group act as a unit (Fig. 14-22).

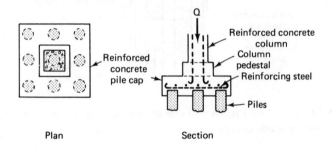

Plan   Section

Figure 14-22. *Pile cap of type provided for reinforced concrete column.*

The requirement for group arrangement of driven piles does not necessarily apply to bored piles (drilled-shaft foundations). Drilled shafts can be located and installed quite accurately. For building construction it is common to use a single, large-diameter, drilled-shaft foundation to support a column. For light loads, a single unit can usually be more properly matched with soil conditions to provide necessary capacity than can a driven pile foundation, where the three-pile minimum usually provides excessive carrying capacity. Economy can result. Additional cost savings will be realized with a single-pile foundation where the need for a pile cap is eliminated.

GROUP CAPACITY / The capacity of a pile group is not necessarily the capacity of the individual pile multiplied by the number of piles in the group. Soil disturbance because of pile installation, and overlap of stresses between adjacent piles, may cause the group capacity to be less than the sum of individual capacities. Conversely, soil between individual piles may become "locked-in" because of strong adhesion or densification from driving, and the group may tend to behave as an equivalent single large pile. Densification and improvement of soil surrounding the group can also occur. These latter factors tend to provide a group capacity greater than the sum of the individual piles. The capacity of the equivalent large pile is analyzed by determining the skin friction around the embedded perimeter of the group and calculating the end bearing by assuming a tip area formed by this same perimeter; see Fig. 14-23.

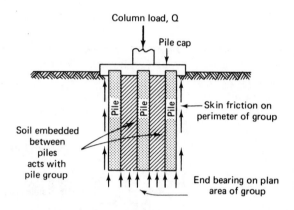

Figure 14-23. *Skin friction and end bearing acting on the single equivalent pile formed by a pile group.*

To determine the *design load* for a pile group, compare the sum of the individual pile capacities with the capacity of the single large equivalent (group) pile. With a proper factor of safety, the least capacity will be the design load. The capacity of the equivalent (group) pile is affected by soil type and properties, and pile spacing. Generally, there is a greater tendency for the group to act as a large single unit when pile spacings are close and the soil is firm or compact.

For driven piles embedded in cohesionless soil, the capacity of the large equivalent (group) pile will almost always be greater than the sum of the individual pile capacities. This is an indication that the soil in the immediate proximity of each individual pile controls the group capacity. Consequently, for design, the group capacity is typically based on the single-pile load and the number of piles in the group.

For bored piles in sand, the group capacity will be less than the sum of the individual pile capacities, in part because of limited densification of the soil zone surrounding the pile group (compared to the driven pile group), and the overlapping of stresses between adjacent units. The capacity of a bored-pile group is typically about two-thirds the capacity of the sum of the individual pile capacities.

The capacity of closely spaced piles embedded in clay is often limited by the behavior of the group acting as a single large unit. In determining the capacity of the single equivalent large pile, the skin friction and end bearing are calculated by using the general methods discussed for the individual pile. The unit value of perimeter skin friction is intermediate between adhesion or remolded shear strength and the undrained shear strength. For pile spacings greater than about three diameters, the group capacity is, for practical purposes, equal to the capacity of the piles acting individually. The capacity should be reduced by one-third if the pile cap for the group does not rest on the ground; this modification accounts for the condition that group load is controlled by the capacity of individual piles as influenced by stress overlap between adjacent units.

Illustration 14-11: A nine-pile group composed of 12-inch-diameter pipe piles is embedded 50 ft deep in soil conditions indicated by the sketch. A $2\frac{1}{2}$-diameter spacing is used. What design capacity should be estimated for the group?

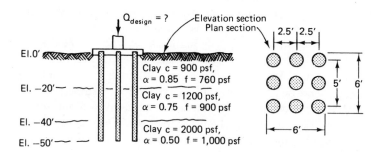

Note group
perimeter = 6' x 4 sides
= 24 ft.

Capacity, Single Pile

$$Q_{friction} = (0.85 \times 900 \text{ psf})(3.14 \text{ ft} \times 20 \text{ ft})$$
$$+ (0.75 \times 1200 \text{ psf})(3.14 \text{ ft} \times 20 \text{ ft})$$
$$+ (0.50 \times 2000 \text{ psf})(3.14 \times 10 \text{ ft}) = 136{,}000 \text{ lb}$$

$$Q_{tip} = 9cA_{tip} = (9 \times 2000 \text{ psf})\left(\frac{\pi}{4} \times 1 \text{ ft}^2\right) = 14{,}000 \text{ lf}$$

$$Q_{total} = 136 \text{ k} + 14 \text{ k} = 150 \text{ k/pile}$$

$$\text{Group total} = 150 \text{ k/pile} \times 9 \text{ piles} = 1350 \text{ k}$$

Capacity, Group

$$Q_{friction} = \left(\frac{900 \text{ psf} + 760 \text{ psf}}{2}\right)(24 \text{ ft} \times 20 \text{ ft})$$

$$+ \left(\frac{1200 \text{ psf} + 900 \text{ psf}}{2}\right)(24 \text{ ft} \times 20 \text{ ft})$$

$$+ \left(\frac{2000 \text{ psf} + 1000 \text{ psf}}{2}\right)(24 \text{ ft} \times 10 \text{ ft}) = 1,260,000 \text{ lb}$$

$$= 1260 \text{ k}$$

$$Q_{tip} = 9cA_{tip} = (9)(2 \text{ ksf})(6 \text{ ft} \times 6 \text{ ft}) = 650 \text{ k}$$

Group total $= 1260 \text{ k} + 650 \text{ k} = 1910 \text{ k}$

Use least group total for design. Therefore,

$$Q_{design} = \frac{1350 \text{ k}}{FS = 2} = 675 \text{ k}$$

PILE GROUP SETTLEMENT / The vertical movement that occurs at the pile cap level is the result of compressive shortening within the pile from the loading plus the settlement occurring in the soil supporting the pile. Analytical methods available to predict the settlement of pile groups provide only approximations. A widely used procedure assumes that the pile group acts as a single large deep foundation, such as a pier or mat. Where the piles are embedded in a uniform soil (friction plus end bearing piles), the total load is assumed to act at a depth equal to two-thirds the pile length. Conventional settlement analysis procedures assuming the Boussinesq or Westergaard stress distribution are then applied to compute compression of the soil beneath the pile tip. If the piles have their tip section embedded in a stratum firmer or more compact than the overlying soil (end bearing piles), the total load is assumed to act at a depth corresponding to pile tip elevation; see Fig. 14-24.

Compressive properties of soil below the pile tip are determined from laboratory tests on recovered soil samples (typical for cohesive soils) or from empirical correlations developed from soil exploration penetration tests (typical for cohesionless soil). For piles embedded in sand, the following expressions for approximating settlement can be applied.

$$\Delta H = \frac{q\sqrt{B}}{N} \qquad \text{(approximately)} \qquad (14\text{-}29a)$$

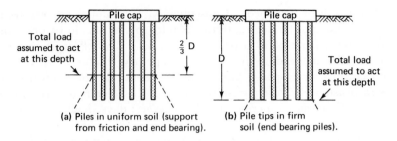

Figure 14-24.   *Assumed conditions for estimating settlement of pile groups.*

where ΔH = settlement in inches.
        q = net foundation bearing pressure, in tons/ft^2.
        B = width of pile group, feet.
        N = blow count from standard penetration test corrected for depth.

The N value applies to the sand in the zone most subject to compression by the pile load. A distance of one group width (or B feet) is recommended. If static cone penetration data (e.g., Dutch cone) is available for the soil below pile tip level, the settlement can be approximated from

$$\Delta H = \frac{qB}{q_c} \qquad \text{(approximately)} \qquad (14\text{-}29b)$$

where q_c = static cone resistance, in same units as q (the foundation load) and B (the pile group width).

If load testing is performed on a single pile, caution is in order when one is extrapolating settlement results to a group. Group settlements typically range from two to more than 10 times the settlement of the single pile. The greater settlement ratio (i.e., group settlement to single-pile settlement) occurs with the larger pile groups.

PILE LOAD TESTS / Applied to construction projects, on-site load tests are performed on test piles installed during the design stage to check estimated capacities (as predetermined by analytical or other methods) and to help develop criteria for the foundation installation contract, or as a check on contract piles installed by the builder to verify the carrying capacity (proof testing).

Both cohesive and cohesionless soils will have their properties altered by the installation of a driven pile. In clays, the disturbance causes remolding and loss of strength. With time, much of the original strength will be regained.   Consequently, to obtain load test results that are practical for design, testing of piles in clay should be performed only after a lapse of at least several weeks.   The effect of installing driven piles in sand is to create a temporary condition where extra resistance is developed.   Shortly after the installation, however, the extra resistance is lost (stress relaxation).   To obtain test results applicable for design, a period of at least several days should lapse before testing begins.

Loading applied to test piles frequently is obtained by hydraulically jacking against a supported weight box or platform, Fig. 14-25(a), or against a reaction girder secured to anchor piles, Fig. 14-25(b).   If the jack is outfitted with a pressure gauge, the magnitude of test load imposed onto the pile is read directly.   However, some designers prefer that a proving ring or pressure capsule be used for measuring the load, feeling that it is more accurate.

Measurement for pile movement is related to a fixed reference mark.   The support for reference marks need to be located outside of the soil zone that could be affected by pile movements.

Several different methods for performing pile load tests are in use. Probably the most common is the *slow maintained load* (Slow ML) *test* (ASTM D-1143).   With this procedure, the test load is applied in eight equal increments until twice the intended design load is reached.   Time-settlement data are obtained for each load increment.   Each increment is maintained until the rate of settlement becomes less than 0.01 inch per hour (2.5 mm per hour), or for two hours, whichever occurs first. The final load (double the design load) is maintained for 24 hours. Unloading also occurs in increments.

Another common procedure is the constant-rate-of-penetration test.   In this method, the load is increased on the pile as necessary to force settlement (penetration) at a predetermined rate.   A rate of 0.02 inch (5 mm) per minute is typical.   The force required to achieve penetration is recorded, thus giving load-settlement data similar to the "slow-maintained" test.   This test is considerably faster than the slow-maintained test.

Other general test methods include cyclic loading, where each increment of load is repeatedly applied and removed (or reduced), and the "quick-maintained" test, where the load increments are imposed for short periods of time.   Pile movement for all loads, or cycles of loads, is recorded.

The load-settlement data are used to determine the design load for the pile.   Frequently, the design load is obtained from a "failure load" indicated by the load test with a factor of safety applied.   The definition of failure load is arbitrary; it may be taken when a predetermined

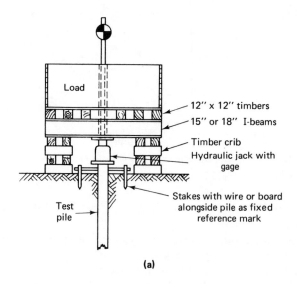

(a)

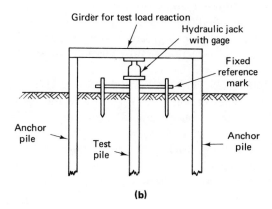

(b)

Figure 14-25. *Typical pile load test arrangements (a) weighted platform used as jacking reaction; (b) anchor piles and girder used to provide reaction.* (Source: ASTM D-1163)

amount of settlement has occurred (e.g., one-tenth of the pile diameter, or a certain number of inches) or where the slope of the load-settlement plot is no longer proportional. A method in use for the slow maintained load test is to plot both load and settlement values on logarithmic coordinates. The results typically plot as two straight lines; see Fig. 14-26. The intersection of the two lines is referred to as the "failure

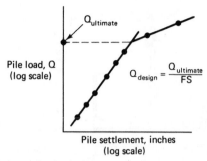

Figure 14-26. *One method of plotting pile load test data to determine pile design capacity.*

load." This is the failure load for design purposes and is not the actual failure or ultimate load.

The design factor of safety used for contract piles should relate to the extent of information known about subsurface conditions, and to the number and comparative results of pile load tests performed at different locations on the site. If soil conditions including properties are uniform and load tests at different areas compare well, a relatively low factor of safety could be justified (i.e., 1.5 to 2.0). If subsurface conditions and load test results are variable, a greater factor of safety is usually necessary (two to three) as protection against possible unexpected poorer soil conditions and lesser pile capacity.

PILE DRIVING FORMULAE / The axial capacity of piles driven in certain types of soil can be related to the resistance against penetration developed during driving. Relating the load capacity that a pile will develop to the driving resistance is appropriate for piles penetrating soils that will not develop high pore water pressures during the installation. Such soils include free-draining sands and hard clays. In saturated fine-grained soils, high pore water pressures develop because of soil displacement and vibration caused by the driving. The strength of the soil surrounding the pile is affected. A predicted capacity based on such soils' resistance to pile penetration is different from the capacity that develops after the excess pore pressures dissipate.

To use driving data to determine pile capacity requires knowledge of the effect that a hammer blow has on the pile and supporting soil. As the top of a pile is struck with the pile hammer, the force of the blow causes a stress wave to be transmitted through the length of the pile. Some of the force of the transmitted wave is absorbed by the soil surrounding the pile shaft, while some is imparted to the soil at the pile tip. What has become known as the *wave equation method* applies wave transmission theory to determine the carrying capacity developed by a pile and the maximum stresses that result within the pile during driving.

In a wave equation analysis, the method assumes that the pile and its behavior when embedded in soil can be represented by a series of individual spring-connected weights and spring damping resistances; see Fig. 14-27. The various weight values W correspond to the weight of

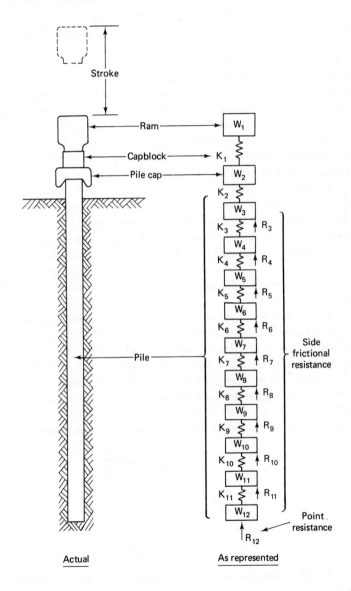

Figure 14-27. Method of representing pile for wave equation analysis. (Ref. 135)

incremental sections of pile. The spring constants K relate to the elasticity of the pile. The spring damping R represents the frictional resistance of soil surrounding the shaft of the pile and the soil resistance at the pile tip. Spring damping along the shaft of the pile accounts for a gradual diminishing of the longitudinal force (from the hammer blow) which travels along the length of the pile. Spring damping at the pile tip is necessary to account for the force which remains within the pile to be transmitted at the tip.

To obtain a solution for the wave equation, it is necessary to know approximate pile length, the pile weight, cross section, and elastic properties, and the pile hammer characteristics, including efficiency, ram weight, and impact velocity, to have data on the pile cap and capblock, and to assign values for soil damping and the spring constants. Determining the effect of a stress wave traveling through the pile is a dynamics problem. However, if the effect of a pile hammer blow at one particular instant in time is selected (the reaction of each weight and spring to the forces acting on overlying weights is determined), the analysis can be handled as a statics problem. By analyzing changing conditions for successive small increments of time, the effects of the force wave traveling through the pile to the tip will be simulated. This analysis requires a numerical integration, a task conveniently undertaken by computer. Results obtained will be only for a particular pile driven by a specified pile hammer. Separate analyses are required for different conditions.

For field use, it is convenient to have the pile capacity expressed in terms of inches of penetration per hammer blow, in order to know when driving can cease. Representative results for a solution are shown in Fig. 14-28. At present, a major drawback to greater application of the wave equation method lies in the practical requirement for computer use. In some cases, proper evaluation of factors such as the equivalent spring constant and soil damping values requires driving data for the pile under study. This means that results for field control are not always

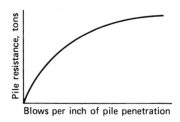

Figure 14-28. *Relation between pile capacity and driving resistance indicated by wave equation.*

available before the start of a project. This situation will change as more information on soil properties is accumulated and as more usable forms of the analysis evolve.

Wave transmission theory shows that the maximum force that can pass through a pile is related to the modulus of elasticity and mass density of the pile material, and the cross-sectional area, in a term referred to as *pile impedance.*

$$\text{Pile impedance} = \rho v A \qquad (14\text{-}30)$$

where impedance = ability of a pile to pass a longitudinal (axial) force by wave transmission resulting from a pile hammer blow.

ρ = mass density of the pile material, γ_{pile}/g (γ_{pile} is the pile materials unit weight; g is the acceleration of gravity).

v = velocity of longitudinal wave propagation, $\sqrt{E/\rho}$ (E is modulus of elasticity for the pile material).

A = cross-sectional area of pile.

If it is assumed that pile driving occurs with a hammer capable of delivering adequate energy, greater capacity is possible with piles having greater impedance values. Other comparative effects of pile impedance are shown by Fig. 14-29. The relative position of the curves indicate that high pile capacities can not be achieved when low-energy driving hammers or piles having low impedance values are used.

Wave transmission theory also explains other occurrences noted in pile driving. If the soil at the pile tip is very hard or compact, some of the stress wave may be reflected back along the pile. As a result, tensile stress (as well as compressive stresses) may develop within the pile

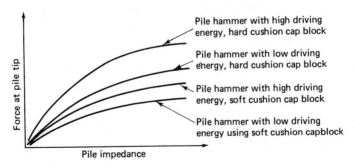

Figure 14-29. *Relationship among pile capacity, pile impedance, and equipment used for pile driving. (Ref. 110)*

during driving. If a pile material is weak in tension (e.g., poorly reinforced concrete), tension cracks may develop near the top of the pile. If the pile penetrates through firm soil into soft material, tensile stresses may develop near the tip and cause damage to the end section of the pile.

Expressions relating pile capacity to driving resistance have also been developed from work-energy theory and are simpler to apply than the wave equation. Such pile driving formulae have been in use since at least the mid-nineteenth century. A basic assumption for these driving formulae is that the kinetic energy delivered by the pile driving hammer equates to the work (work defined as the product of force and distance) done on the pile, or

$$(W_h)(H)(\text{Eff.}) = (\text{pile resistance})(\text{pile penetration per hammer blow})$$

$$(14\text{-}31)$$

where W_h = weight of pile hammer ram.
H = height of ram fall.
Eff. = efficiency of pile hammer.

For application, these formulae need to include corrections to account for energy losses during driving that are caused by factors such as elastic compression of the pile, soil, capblock, and cushion, and by heat generation. Variations in the many pile formulae that have been proposed result because of the different methods for handling these energy losses.

The assumption of work energy theory to develop pile driving formulae does not properly consider the effect of impact on a long member such as a pile, and therefore such formulae will not be theoretically correct. However, some expressions that include empirical energy loss correction factors have shown a reliability for predicting the axial capacity of piles, particularly those driven in cohesionless soil. Some of these formulae are lengthy and complicated (e.g., Hiley formula) and are not reported herein.[8] Load test studies on piles in sand indicate that the simple-to-use Danish formula is as reliable as any of the more complicated formulae.[9]

$$Q_{ult} = \frac{W_h H(\text{Eff.})}{S + \frac{1}{2}S_0} \qquad \text{(Danish formula)} \qquad (14\text{-}32a)$$

where Q_{ult} = axial capacity of the pile.
S = average penetration of the pile from the last few driving blows (pile "set").

[8]Reference 6.
[9]Reference 3.

S_0 = elastic compression of the pile

$$= \left[\frac{(2)\,(\text{Eff.})\,(W_h HL)}{AE} \right]^{1/2}$$

L = length of pile.
A = cross-sectional area of pile.
E = modulus of elasticity for pile material.

Where the formula is used, a factor of safety of three should be applied ($Q_{design} = \frac{1}{3} Q_{ult}$).

During pile installation, it is convenient to know when a desired capacity has been obtained and driving can cease. The pile formula above can be rearranged to indicate the final set, usually inches per blow, that provides the design capacity:

$$S = \left[\frac{(W_h H)\,(\text{Eff.})}{3\,Q_{design}} \right] - \frac{1}{2}\,S_0. \tag{14-32b}$$

If the set is to be in inches per blow, all lengths should be expressed in inches. Similarly, the units for the weight of ram and pile capacity have to be the same.

Illustration 14-12. Steel pipe piles intended to provide a design capacity of 50 tons are used at a construction site underlain by cohesionless soils. Preliminary analysis indicate that the required pile length will be between 45 and 50 ft. If the Danish formula will be used as a field control during driving to indicate when the desired capacity is obtained, what should be the set under the last driving blows? Pile installation will be accomplished with a Vulcan Model 0 hammer having a rated driving energy of 24,375 ft-lb.

Assume a pile length of 50 ft. Pile cross section = 10 in^2, E_{steel} = 29 × 10^6 psi

$$\text{Set } S = \frac{W_h H \times \text{Eff.}}{3\,Q_{design}} - \frac{1}{2}\,S_0.$$

where $S_0 = \sqrt{\dfrac{2 \times \text{Eff.} \times W_h HL}{AE}}$

$$= \left[\frac{(2)(24{,}375 \text{ ft-lb})(50 \text{ ft})(144 \text{ in}^2/\text{ft}^2)}{(10 \text{ in}^2)(29 \times 10^6 \text{ lb/in}^2)} \right]^{1/2}$$

$$= 1.1 \text{ in.}$$

Therefore,

$$\text{Set } S = \frac{(24.375 \text{ ft kip})(12 \text{ in./ft})}{(3 \times 50 \text{ tons} \times 2 \text{ k/ton})} - \frac{1}{2}(1.1 \text{ in.})$$

$$= 0.98 \text{ in.} - 0.55 \text{ in.} = 0.43 \text{ in.} \approx \tfrac{1}{2} \text{ in.}$$

One of the most widely known pile-driving formulae is the Engineering News (ENR) formula, presented in the late nineteenth century for determining the capacity of timber piles. The formula is not reported herein, because it does not have application for existing pile-drive methods. The use of the ENR formula should be discouraged. Though it has a theoretical factor of safety of six, various load test programs have indicated that the actual factor of safety can range from less than one (representing pile failure at less than the expected capacity) to above twenty.

Supervision of Foundation Construction

Since foundations frequently are covered or completely buried soon after their construction, there is limited opportunity for noting and correcting improper work or detecting conditions not in accord with design assumptions. To minimize the risk of problems resulting from improper or inadequate foundation construction, the practice of having such work guided (or supervised) by personnel with soil mechanics training is increasing.

The responsibilities of such personnel would include matching soil and groundwater conditions observed in excavations with the conditions indicated by the borings used for design. For projects where shallow foundations are being installed, the proper location, size, and depth should be verified. The bottoms of excavations should be examined prior to concrete placement to ensure that disturbed or muddy soil is not left in place; compaction or removal can be ordered if such conditions are observed.

Where drilled foundations are being installed, the bottom of the excavation and soil exposed in the walls should be examined whenever conditions permit. Depth to, or bearing in, a proper soil stratum should be verified. Proper location, dimensions, and installation of required reinforcing steel should also be verified.

With driven-pile foundations, driving records (the continuous record of blow count versus depth) should be kept for all piles. The size and length of piles should be verified. Detailed information on the equipment used for driving the piles should be obtained.

Personnel doing supervision work should keep a written record (daily logs or reports) of activities, observations, and decisions regarding the foundation work. Copies frequently are expected to be forwarded to the designer's office. Written records can serve as a diary of the construction progress, and frequently serve as a reference if disputes develop.

PROBLEMS

1. A long footing 3 ft wide is to be installed 4 ft below grade. The supporting soil has a cohesive strength of 1000 psf and the angle of internal friction is 20 deg. The unit weight is 115 pcf. Find the safe bearing capacity, using a factor of safety of three. The water table is deep.

2. Determine the wall loading that can be carried by a long footing 2.5 ft wide and located three feet below the ground surface. The supporting soil has a cohesion of 500 psf and an angle of internal friction of 25 deg. The soil unit weight is 120 pcf. The water table is expected to rise to the ground surface. Use a factor of safety of 3 with the general bearing capacity equation.

3. A square footing is to be designed to carry a column load of 250 kips. The footing will be installed at a shallow depth on a clay soil whose cohesion is 2000 psf. The water table is deep. If the factor of safety is three, what size footing should be provided?

4. A square footing is to be provided in order to support a 300-kip column loading. If the footing is to be located on the surface of a cohesionless soil deposit, what size footing is required? The soil has a unit weight of 125 pcf, and the angle of internal friction is 36 deg. It is expected that the water table could rise to the ground surface. For the design, apply a factor of safety of three to the bearing capacity equation.

5. A footing for an existing building is 6 ft sq., and three feet below the surface. A proposed building alteration would require this footing to carry a loading of 400 kips. Is the footing of adequate size? Soil investigation indicates that the foundation soils have an angle of internal friction of 35 deg and a unit weight of 120 pcf. Apply a factor of safety of three to the bearing capacity equation.

6. Design a square footing to carry a column load of 250 kips. The footing is to be installed 3 ft below the ground surface on a sand whose angle of internal friction is 33 deg and unit weight is 115 pcf. Use a factor of safety of three. Assume that the water table is deep.

7. What loading could a rectangular foundation 4 ft by 8 ft support if it is located on the surface of a soil that has a cohesion of 500 psf

and an angle of internal friction of 20 deg? The soil's unit weight is 110 pcf. Use a factor of safety of three.

8. A heavy machine is to be provided with a concrete pad foundation that needs to have a length equal to twice the width. The foundation will be installed close to the ground surface on a clay soil whose cohesion is 2500 psf. The machine weighs 1200 kips. To have a factor of safety of three, what size of foundation should be constructed?

9. Two columns (a wall column and an interior column) are to be supported on a common rectangular (combined) footing. The wall column imposes a loading of 120 tons, while the interior column carries 200 tons. The columns are 16 ft apart. The end of the footing near to the wall cannot project more than 3 ft beyond the wall column's center line. Using an allowable soil bearing pressure of 4000 psf, give the length and width of foundation that should be provided.

10. A long footing 2.5 ft wide has its base inclined 10 deg from the horizontal. The footing bears on a clay soil that has a cohesion of 2000 psf. What load acting normal to the footing can be imposed if a factor of safety of three is to be used?

11. A long footing is to be constructed in the slope of a clay hillside that makes an inclination of 30 deg with the horizontal. The cohesion of the clay is 1500 psf, and the unit weight is 105 pcf. If the slope stability factor is zero, and the D/B ratio is close to zero, what width footing is required to support a wall loading of 8000 pounds per foot of length? Use a factor of safety of three.

12. A long footing 3 ft wide is located on the slope of a 30-deg hill. The hill is sand which has an angle of internal friction of 35 deg and a unit weight of 115 pcf. If a D/B ratio of one is assumed, what wall loading can the footing carry? Use a factor of safety of three.

13. A four-foot-wide long footing is located at the top of a slope. The slope inclination is 30 deg. The hill is sand with an assumed angle of internal friction of 30 deg. If the D/B ratio is one and the b/B ratio is also one, what loading can be imposed on the footing? Use a factor of safety of three.

14. Soil borings indicate that a building site is underlain by clean fine to coarse sand. The following standard penetration test information is obtained from a boring log: At 2-ft depth, N is 12; at 6-ft depth, N is 14; at 10-ft depth, N is 18; at 14-ft depth, N is 20; at 20-ft depth, N is 25; at 25-ft depth, N is 26. No groundwater was

encountered. It is planned to construct a square footing 4 ft deep to carry a 300-kip column load near to this boring location. If a 1-in. settlement is tolerable, what size foundation should be provided? What size footing should be provided if the groundwater table was about at footing level?

15. Static cone penetrometer resistance data obtained from a building site underlain by sand show an average q_c value of 100 kg/cm^2. From this information, estimate the size of square footing necessary to support a column load of 250 kips. Assume that the footing will be placed approximately 4 feet deep and that a 1-in. settlement is tolerable.

16. Static cone penetrometer resistance data indicate that a building site is underlain by an overconsolidated clay. The average q_c value in the upper 20 ft of soil is 75. Using this information, estimate the cohesion of the soil and a safe bearing capacity for shallow square footings.

17. A 14-inch-diameter pipe pile is driven 50 ft into a dry loose sand whose unit weight is 105 pcf. The angle of internal friction is 33 deg. Compute the safe axial downward loading, using the statical analysis. Apply a factor of safety of two.

18. A 12-in. sq. prestressed concrete pile is to be installed in sandy soil for a marine structure. The pile will be 40 ft long. It will be jetted 30 ft, and driven the last 10 ft. The water table is at the ground surface. The saturated soil weight is 120 pcf. The angle of internal friction is 35 deg. For these conditions, assume that the critical depth for computing effective soil stresses is 20 pile diameters. Calculate the safe axial downward loading, using the statical analysis and a factor of safety of two.

19. Timber piles are to be used as friction piles at a site underlain by a deep deposit of clay soil whose cohesion is 1000 psf and unit weight is 105 pcf. For a pile embedded 30 ft, having a tip diameter of 8 in. and a 14-in. butt diameter, calculate the axial capacity, assuming skin friction only. Use a factor of safety of two.

20. A 12-inch steel H-Pile (HP) is driven 60 ft into clay soils as follows: From zero to 30 ft the cohesion is 800 psf and unit weight is 105 pcf; below 30 ft the cohesion is 2000 psf and unit weight is 115 pcf. Assume that the upper clay layer is normally consolidated and the deeper clay is overconsolidated. Determine the axial capacity of the pile, using a factor of safety of two. Compute the shaft capacity, assuming that skin friction is developed on the surface of the rectangle (12 in. sq.) that encloses the pile.

21. A 14-inch-diameter pipe pile is driven 50 ft into overconsolidated
     clays having the following properties: From zero to 20 ft the co-
     hesion is 1500 psf and the soil unit weight is 110 pcf; below 20 ft
     the cohesion is 2500 psf and the soil unit weight is 120 pcf. Cal-
     culate the downward and upward axial capacity of the pile, ap-
     plying a factor of safety of two.

22. A four-pile group consists of 12-in sq. prestressed concrete piles
     35 ft long. The piles are driven into an overconsolidated clay
     whose cohesion is 2000 psf and soil unit weight is 115 pcf. The pile
     spacing is $2\frac{1}{2}$ diameters. The water table is at the ground surface.
     Using a factor of safety of two, find the capacity of the pile group.

23. For an offshore project, a steel pipe pile with an outside diameter
     of 18 in. penetrates 150 ft through normally consolidated clays.
     Soil conditions are as follows: From zero to 50 ft the cohesion is
     800 psf and saturated soil unit weight is 105 pcf; from 50 to 100
     ft the cohesion is 1500 psf and the saturated unit weight is 115 pcf;
     below 100 ft the cohesion is 2400 psf and the saturated unit weight
     is 125 pcf. The water table is above the soil surface. Using the
     lambda method, determine the axial capacity of the pile, using a
     factor of safety of two.

24. A drilled-shaft foundation is to be constructed in clay soils having
     the following properties: From zero to 30 ft the cohesion is 1000 psf
     and the soil unit weight is 105 pcf; below 30 ft the cohesion is
     2500 psf and the unit weight is 115 pcf. The foundation is to ex-
     tend to a total depth of 36 ft. A three-foot-diameter shaft will be
     used with a belled (underreamed) bottom six ft in diameter. As-
     sume that the bell forms a 60-deg angle with the horizontal plane.
     Compute the design downward capacity if

     (a)   The excavation is drilled dry.
     (b)   It is necessary to use a bentonite slurry during drilling.

25. A 12-in. sq. prestressed concrete pile is driven 50 ft. Under the last
     blow of a Vulcan 010 hammer the pile penetrates 0.25 in. Using
     the Danish pile-driving formula, what safe design load is indi-
     cated? Is this size hammer proper for this type and length pile?

26. The Danish pile-driving formula is to be used for field control of a
     pile installation. Steel pipe piles penetrating approximately 40 ft
     are to provide a design capacity of 60 tons. Each pipe has a net
     cross-sectional area of 20 in^2. What should be the set under the
     last blow of a MKT-11B3 hammer to obtain the desired axial
     capacity?

27. The following data are obtained from a Slow ML pile load test.
Plot the load and settlement information on logarithmic coordinates and from the plot determine the "failure" load.

Load (tons)	Settlement (inches)
25	0.05
50	0.15
75	0.26
100	0.40
125	0.60
150	0.80
175	0.95
200	1.55
225	2:45

APPENDIX 1
Selected Soil
Testing Procedures

1 Grain Size Analysis: Particle-Size Distribution By Hydrometer Method[1]

Purpose and Use

Information on the particle-size distribution of fine soils, such as clays, silts, and very fine sands, is generally obtained by using the hydrometer method. The sieving method commonly used to obtain grain-size data for the coarser soils is not effective for fine soils.

In this method, a hydrometer is used to measure the specific gravity (or a related property) of a solution or suspension composed of demineralized water and the soil sample. With time, the solid particles settle out of suspension, and as this happens the specific gravity of the mixture decreases. The rate at which the specific gravity is decreasing is then used to calculate the quantity and size of soil particles in the solution.

The method is based upon the Stokes equation for spheres falling freely in a fluid of known properties. The application is not absolutely correct, since most soil particles are not round—in fact, many fine soils are elongated or flat. However, the method does provide a practical way to obtain a good approximation of the particle-size distribution of fine-grained soils, and is in wide use.

Test Procedure

Special equipment or materials needed to perform the hydrometer analysis described herein include a hydrometer, two 1000-ml graduated cylinders, a thermometer reading to 0.1 degree centigrade, distilled or

---

[1]Also refer to ASTM (American Society for Testing and Materials) Test D-422.

demineralized water, a watch or other timing device, and a mixing apparatus (a milkshake-type mixer works well). Any hydrometer that will read specific gravity of the solution or grams of soil solids per liter of suspension can be used. Hydrometers meeting specifications for ASTM Hydrometer 151H and 152H provide data in a form that is convenient to reduce.

1. Select and mix between 50 grams and 100 grams of soil sample with distilled water to create a heavy slurry. The smaller quantity is used for clay-silt soils, the larger weight for silt-fine sands. (If the sample contains a significant amount of coarse sand or gravel, it should be separated by screening on the #10 sieve and analyzed separately by sieving. Record the weights of the soil coarser than, and finer than, the #10 sieve.) Add a deflocculating agent to the mixture. The deflocculating agent is used to prevent individual soil particles from sticking together (or flocculating). A commonly used agent is 125 ml of a 4 percent solution of sodium hexametaphosphate. If the sample contains a large quantity of clay, let the mixture stand for several hours. When ready to perform the analysis, mix in the mixing machine until the soil particles are segregated (5 to 15 minutes).

2. Fill one of the 1000-ml cylinders with distilled water. This will be used to store the hydrometer between test readings and to determine corrections for hydrometer readings.

3. Place the soil mixture in the other 1000-ml cylinder and add distilled water to fill to the 1000-ml mark. At the start of the test, all soil particles should be in suspension. To achieve this, stop up the top of the cylinder with a rubber stopper or palm of the hand, and turn the cylinder end over end for 30 to 60 seconds. Observe to make sure that soil does not stick to the bottom of the cylinder during the last few turns before starting the test.

4. When all the soil is in suspension (the last invert of the cylinder), stand the cylinder on a table or bench at a location safe from temperature changes and wind drafts. This marks the beginning time for the test. Hydrometer readings of the solution will be taken at intervals measured from this beginning time.

5. As soon as the cylinder has been placed in the upright position (test beginning) on the table, carefully place the hydrometer in the suspension. It should be placed so that it does not bob or rotate. (It is suggested that a person practice placing the hydrometer in the clear water of the second gradu-

ated cylinder to develop a feel for the hydrometer behavior and also to make a few trial insertions in the soil suspension before making actual test readings.) For the test, the hydrometer is left in the suspension and readings are made at intervals of one-fourth, one-half, one, and two minutes measured from the start. Read the top of the meniscus that forms on the hydrometer stem. Record the temperature of the suspension after the two-minute reading.

6.   The suspension should be reshaken and the series of two-minute readings repeated. This procedure is repeated until a series of four readings that agree is obtained.

7.   After obtaining a reliable set of two-minute readings, reshake and start the test over again. This time, take no readings until 2 minutes have lapsed. Do not insert the hydrometer until about 10 seconds before the reading is to be made. After making the reading, carefully remove the hydrometer and store it in the graduated cylinder of distilled water. Obtain the temperature of the solution. Continue to obtain hydrometer and temperature readings at time intervals approximately twice the interval of the last reading, i.e., 2, 4, 8, 16, 32, 64, 128, 256, . . . minutes measured from the start of the test. Accurately record the time intervals between readings. Remove the hydrometer after each reading. Make readings until almost all of the soil particles have settled out of suspension or until enough readings have been obtained to determine the minimum-size soil particles of interest.

8.   For every hydrometer reading of the suspension, make a corresponding hydrometer and temperature reading of the distilled water. The cylinder of the distilled water should be located next to the test cylinder, so that the temperature of both cylinders of fluid will be the same. These hydrometer readings will be used to make a correction to each test reading, since a direct reading of the hydrometer placed in the suspension will include small errors. Errors result because of reading the top of the meniscus on the hydrometer stem (the hydrometer is calibrated to read at the bottom of the meniscus, but with the soil clouding the solution this would be difficult); because of temperature variations (the hydrometer is calibrated to read directly when the water temperature is 20°C); and because the addition of a deflocculating agent makes the specific gravity of the solution different from that of pure water. For hydrometer 151H, the correction to be applied is the difference between the reading made in the graduated cylinder of distilled water and one, provided that

the temperature of both cylinders is similar. For hydrometer 152 H, the correction to be applied is the difference between the reading in the plain water and zero.[2]

9. After the readings have been completed, carefully place the soil solution in a pan. Wash all soil out of the cylinder. A syringe is convenient to aid in this task. Place the pan in an oven to dry. Obtain the dry weight of soil used in the test.

DATA REDUCTION—HYDROMETER 151H / The percentage of the original soil sample that remains in suspension at the time of a particular hydrometer reading can be computed from

$$P = \left[\frac{100,000}{W_s} - \frac{G_s}{G_s - G_w} \right] (R - G_w)$$

where P = percentage of soil remaining in suspension (% finer).
W_s = dry weight of soil in grams.
G_s = specific gravity of soil.
G_w = specific gravity of water (use 1.0).
R = hydrometer reading with corrections applied.

DATA REDUCTION—HYDROMETER 152H / The percentage of the original soil sample that remains in suspension at the time of a given hydrometer reading can be calculated from

$$P = \frac{Ra}{W_s} \times 100$$

where P = percentage of soil remaining in suspension (% finer).
R = hydrometer reading with corrections applied.
W_s = dry weight of soil in grams.
a = factor to correct for specific gravity of soil
  = $1.00 - 0.2(G_s - 2.65)$ for specific gravities (G_s) between 2.50 and 2.95.

The soil particle diameter corresponding to the percentage remaining in suspension, as determined by either hydrometer 151H or 152H, can be calculated by using Stokes Law.

$$D = \sqrt{\frac{30N}{980(G_s - 1.00)} \times \frac{L}{T}} = K \sqrt{\frac{L}{T}}$$

where D = particle diameter in millimeters.
N = coefficient of viscosity of water in poises.
G_s = specific gravity of soil.

---

[2]If the meniscus on the hydrometer in plain water is above the one (reading less than one, hydrometer 151 H) or above the zero (reading less than zero, hydrometer 152 H), the correction is added to the hydrometer test readings, and vice versa. In determining the correction, read the bottom of the meniscus.

L = distance from water level in the cylinder to the level at which the density of the suspension is being read by the hydrometer. (Values of L for given hydrometer readings are provided in Table A1-1.)

T = time at which hydrometer reading was made, minutes. (Values of K for different temperatures and specific gravities are presented in Table A1-2.)

Obtaining particle diameters by the formula can be accomplished relatively quickly. Nomographs that solve the Stokes equation are also available.

Combined Sieve and Hydrometer Analysis

If the original soil sample contained coarse particles that were screened and analyzed by sieving, the percentages of the coarse particles finer than a given size in the total sample can be computed by dividing the total weight passing each sieve size by the original total sample weight, and multiplying by 100.

The total percentage finer (P'), for particle sizes determined by the hydrometer method, can be calculated from

$$P' = P \times \frac{W_h}{W_k}$$

where P = percentage of soil remaining in suspension in hydrometer test.

W_h = weight of soil used in hydrometer analysis.

W_t = original total weight of soil sample.

Summary

The hydrometer method is believed to indicate soil particle sizes somewhat larger than actual. The result is that sizes indicated from the sieving and hydrometer method may overlap. When one is plotting the results of a combined analysis as a grain-size distribution curve, it is convenient and sufficiently accurate to "shift" the hydrometer results towards the smaller particle range, to obtain a smooth continuous curve.

Record the data and results of calculations on the data sheet provided. The results should then be summarized as a grain-size curve where percentage finer by weight (ordinate) is plotted against particle diameter (abcissa) on semilog coordinates.

An illustration of test data for a hydrometer analysis is provided in Fig. A1-1.

TABLE A1-1 / L = VALUES OF EFFECTIVE DEPTH BASED ON HYDROMETER AND SEDIMENTATION CYLINDER OF SPECIFIED SIZES*

Hydrometer 151H		Hydrometer 152H			
Actual Hydrometer Reading	Effective Depth, L, cm	Actual Hydrometer Reading	Effective Depth, L, cm	Actual Hydrometer Reading	Effective Depth, L, cm
1.000	16.3	0	16.3		
1.001	16.0	1	16.1	31	11.2
1.002	15.8	2	16.0	32	11.1
1.003	15.5	3	15.8	33	10.9
1.004	15.2	4	15.6	34	10.7
1.005	15.0	5	15.5	35	10.6
1.006	14.7	6	15.3	36	10.4
1.007	14.4	7	15.2	37	10.2
1.008	14.2	8	15.0	38	10.1
1.009	13.9	9	14.8	39	9.9
1.010	13.7	10	14.7	40	9.7
1.011	13.4	11	14.5	41	9.6
1.012	13.1	12	14.3	42	9.4
1.013	12.9	13	14.2	43	9.2
1.014	12.6	14	14.0	44	9.1
1.015	12.3	15	13.8	45	8.9
1.016	12.1	16	13.7	46	8.8
1.017	11.8	17	13.5	47	8.6
1.018	11.5	18	13.3	48	8.4
1.019	11.3	19	13.2	49	8.3
1.020	11.0	20	13.0	50	8.1
1.021	10.7	21	12.9	51	7.9
1.022	10.5	22	12.7	52	7.8
1.023	10.2	23	12.5	53	7.6
1.024	10.0	24	12.4	54	7.4
1.025	9.7	25	12.2	55	7.3
1.026	9.4	26	12.0	56	7.1
1.027	9.2	27	11.9	57	7.0
1.028	8.9	28	11.7	58	6.8
1.029	8.6	29	11.5	59	6.6
1.030	8.4	30	11.4	60	6.5
1.031	8.1				
1.032	7.8				
1.033	7.6				
1.034	7.3				
1.035	7.0				
1.036	6.8				
1.037	6.5				
1.038	6.2				

Note: Reprinted by permission of ASTM.

TABLE A1-2 / VALUES OF K FOR USE IN FORMULA FOR COMPUTING DIAMETER OF PARTICLE IN HYDROMETER ANALYSIS

Temperature, deg C	Specific Gravity of Soil Particles								
	2.45	2.50	2.55	2.60	2.65	2.70	2.75	2.80	2.85
16	0.01530	0.01505	0.01481	0.01457	0.01435	0.01414	0.01394	0.01374	0.01356
17	0.01531	0.01486	0.01462	0.01439	0.01417	0.01396	0.01376	0.01356	0.01338
18	0.01492	0.01467	0.01443	0.01421	0.01399	0.01378	0.01359	0.01339	0.01321
19	0.01474	0.01449	0.01425	0.01403	0.01382	0.01361	0.01342	0.01323	0.01305
20	0.01456	0.01431	0.01408	0.01386	0.01365	0.01344	0.01325	0.01307	0.01289
21	0.01438	0.01414	0.01391	0.01369	0.01348	0.01328	0.01309	0.01291	0.01273
22	0.01421	0.01397	0.01374	0.01353	0.01332	0.01312	0.01294	0.01276	0.01258
23	0.01404	0.01381	0.01358	0.01337	0.01317	0.01297	0.01279	0.01261	0.01243
24	0.01388	0.01365	0.01342	0.01321	0.01301	0.01282	0.01264	0.01246	0.01229
25	0.01372	0.01349	0.01327	0.01306	0.01286	0.01267	0.01249	0.01232	0.01215
26	0.01357	0.01334	0.01312	0.01291	0.01272	0.01253	0.01235	0.01218	0.01201
27	0.01342	0.01319	0.01297	0.01277	0.01258	0.01239	0.01221	0.01204	0.01188
28	0.01327	0.01304	0.01283	0.01264	0.01244	0.01225	0.01208	0.01191	0.01175
29	0.01312	0.01290	0.01269	0.01249	0.01230	0.01212	0.01195	0.01178	0.01162
30	0.01298	0.01276	0.01256	0.01236	0.01217	0.01199	0.01182	0.01165	0.01149

Note: Reprinted by permission of ASTM.

Soil Sample From: 20B 202-69, B-3 @ 10'
Soil Classification _____
Hydrometer No. 6922 (152-H)
Specific Gravity of Soil 2.65, a = 1.0
Weight of Evaporating Dish No. 102 Plus Soil 334.5 gms
Weight of Evaporating Dish 292.5
Weight of Soil, W_s _____ = 42.0

Date	Time	Elapsed Time, Min.	Temp. °C	Hydrometer Test Reading	Hydrometer Calib. Reading	Hyd. Corr.	Correct Hyd. Reading for Cal.	Effective Hy. Depth L	K	P%	Particle Diameter mm
12/16		0.25	22.3	40	−1	+1	41	9.7	.01327	98	.083
	USE	0.50	"	38	−1	+1	39	10.1	.01327	95	.059
		1.00	"	36	−1	+1	37	10.4	.01327	90	.043
		2.00	"	32(+)	−1	+1	33(+)	10.8	.01327	80	.031
		0.25	22.4	41	−1	+1					
		0.50	22.4	39	−1	+1					
		1.00	22.4	36	−1	+1					
		2.00	22.4	32	−1	+1					
		0.25	22.4	40	−1	+1					
		0.50	22.4	38.5	−1	+1					
		1.00	22.4	36	−1	+1					
		2.00	22.4	32	−1	+1					
		0.25	22.4	40	−1	+1					
		0.50	22.4	38	−1	+1					
		1.00	22.4	35.5	−1	+1					
		2.00	22.4	32	−1	+1					
		2	22.5	32	−1	+1	33	11.1	.01325	80	.031
		5	22.5	22	−1	+1	23	12.7	.01325	56	.021
		8	22.5	20	−1	+1	21	13.0	.01325	51	.017
		16	22.6	14	−1	+1	15	14.0	.01325	34	.012
		32	22.6	10	−1	+1	11	14.7	.01325	27	.009
		64	22.6	7.5	−1	+1	8.5	15.1	.01325	20	.006
		120	22.6	5	−1	+1	6	15.5	.01325	14	.005

Comments, Calculations, etc.

Sample Calculations

$$P\% = \frac{Ra}{W_s} \times 100\%$$

$$= \frac{(41)(1.0)}{42} \times 100\% = 98\%$$

$$D = K \sqrt{\frac{L}{T}}$$

$$= .01327 \sqrt{9.7/0.25}$$

$$= .083 \text{ mm}$$

Figure A1-1. *Data sheet for hydrometer analysis.*

2 Liquid Limit, Plastic Limit, And Plasticity Index Of Soils

Purpose and Use

The liquid limit of a soil is the water content where the soil changes from the plastic state to the liquid state. The plastic limit is the water content where the soil passes from the plastic to semi-solid state. The plasticity index is the numerical difference between the liquid limit and plastic limit.

These properties of fine-grained soils (clays and silts), often referred to as the Atterberg limits, can be used to properly classify a soil. Through empirical correlations, these limits can be used to evaluate engineering properties of a soil such as strength and compressibility, suitability of a soil for road base and structural fill, and other behavior, such as susceptibility to frost-heave.[3] Testing to determine these limits is performed on remolded soil.

Test Procedure

LIQUID LIMIT—METHOD A / A special type of equipment required is a liquid limit device meeting the requirements indicated by ASTM Test Designation D-423-54 and **AASHO**[4] Test Designation T-89-60.

1. If the sample consists of silt and/or clay, prepare it by thoroughly remolding it using a spatula and pestle at the natural water content until a uniform consistency is obtained.

---

[3]These limits and tests are also commonly performed on mixed soils (soil containing both sand and "fines") and fine sands to determine if the soils will exhibit plastic behavior when exposed to water.

[4]American Association of State Highway Officials.

If dry, add small quantities of distilled water until a heavy paste-like consistency is obtained. Do not make the sample too wet. If the sample contains coarse particles, remove those larger than the #40 sieve. Prepare the sample as above. Make sure that no lumps remain and that an even consistency has been obtained.

2.   Before placing the soil in the cup of the liquid limit device, check to see that the cup will drop exactly one centimeter for each rotation of the device's handle. Adjust as necessary. Place a sufficient amount of soil in the cup to fill the bottom and smooth off the top of the soil with the spatula so that it is even with the lip of the cup. Using the grooving tool provided with the liquid limit device, cut a groove through the center of the soil in the cup. The axis of the groove passes from the cam on the liquid limit device to the front edge of the cup. It may be necessary to use several careful passes of the grooving tool to extend the groove to the full depth of the soil without tearing the soil.

3.   Lift and drop the cup by rotating the handle at a speed of two revolutions per second. Each drop is counted as one blow. Continue until the two sides of the soil groove flow together to close along a distance of one-half inch. Record the number of blows.

4.   Remove the portion of the soil that flowed together by using the spatula and cutting a one-half-inch strip perpendicular to the groove (from side to side of the cup). Determine the water content of this soil by determining its wet weight, placing it in an oven to dry, then getting the dry weight. For all weighings, use a scale having accuracy to 0.1 gram. The water content is the weight of water lost by drying divided by the dry weight of the soil, expressed as a percentage.

5.   Remove the remaining soil from the cup and mix it with the sample left from the earlier mixing. Clean the liquid limit cup. Add a small amount of distilled water to the soil to make it more fluid. Repeat steps 2, 3, and 4 at least twice more so that blow counts (to cause closure of a one-half-inch groove) between 10 and 25, 25 and 40, and 20 and 30 are obtained. Disregard results greater than 40 or less than 10.

6.   After obtaining water contents for each blow count, make a plot of the data on semilogarithmic coordinates where water content w is placed on the arithmetic scale and blow counts N on the logarithmic scale. Draw the best straight line through the three or more points obtained. This is referred to as "flow curve." The liquid limit is taken from this curve, and is the water content corresponding to a blow count of 25.

LIQUID LIMIT—METHOD B / A one-point method is also available whereby the liquid limit is determined by using the results of a single trial. In this method, the soil is prepared as described previously and the device and procedure referred to above are used.

The single trial must give a result where the blow count N is between 20 and 30. The blow count is recorded, and the trial is repeated to check the blow count. When the counts agree, the water content of the sample is determined. The liquid limit is then calculated from the single blow count value N_1 and the water content w_1 results as:

$$\text{Liquid limit} = w_1 \left[\frac{N_1}{25}\right]^{0.12}$$

PLASTIC LIMIT[5] / The plastic limit is taken as the water content at which a soil begins to crumble when it is rolled into a one-eighth-inch-diameter thread.

1. The sample should be thoroughly kneaded to obtain uniform consistency. The soil will be only damp at the plastic limit, so water should be added only to dry soil when making preparations for the test. Wet soil should be air-dried (not oven-dried) to reduce the water content. Reduction in water content can be accelerated by making a ball and rolling it on paper towels or other absorbent material.

2. The soil should be rolled into a thread on a flat surface of frosted glass or smooth unglazed paper. Take approximately 10 grams of soil and form it into a ball. Then roll the ball into a thread by keeping the hand flat as it moves back and forth in the direction that the fingers point. Rolling should occur at a rate of about 80 to 90 strokes per minute. If the sample can be rolled into a thread smaller than one-eighth inch it is too wet, and will require some drying. A satisfactory amount of drying can probably be obtained by rolling the soil on paper towels. If the thread crumbles before reaching one-eighth inch, it is too dry and a small amount of water should be added.

3. When a one-eighth-inch thread crumbles, the soil is at its plastic limit. Determine the water content at this consistency by determining the wet weight of the soil thread, oven-drying it and then obtaining the dry weight. This water content is the plastic limit. Use a scale sensitive to 0.1 gram for all weighings.

4. Repeat the procedure at least two more times so that three separate tests for the plastic limit are performed. The water

[5]Also refer to ASTM Test D-424.

contents of all the tests should agree, within practical limits. If agreement is not obtained, repeat tests should be performed.

Summary

The liquid limit is obtained from the flow curve plotted from results of the liquid limit test, and is the water content corresponding to where the flow curve crosses the 25-blow line. The plastic limit is the water content determined directly by the procedure described above. The plasticity index is the numerical difference between the water content at the liquid limit and at the plastic limit and indicates the range of water content through which the soil acts as a plastic.

If the soil is too coarse or too granular to be rolled into a thread, or if it does not "flow" in the liquid limit device, the plastic and/or liquid limit cannot be determined, and the plasticity index should be reported as "nonplastic" (NP).

An illustration of test data for a liquid limit and plastic limit determination, with finished computations and graphical plot, is provided in Fig. A1-2.

Liquid limit

Test no.	1	2	3	4	5
Cup. no.	101	102	105	109	
Wt. of wet soil and cup	26.50	21.53	23.44	25.88	
Wt. of dry soil and cup	20.87	18.15	18.93	20.20	
Wt. of cup	14.71	14.61	14.31	14.50	
Wt. of water	5.63	3.38	4.51	5.68	
Wt. of dry soil	6.16	3.54	4.62	5.70	
Water content %	91.4%	95.5%	97.7%	99.5%	
N-count	35	26	21	18	

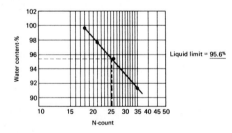

Liquid limit = 95.6%

Plastic limit

Trial no.	1	2	3	4	5	
Cup no.	C-8	C-6	C-1			
Wt. of wet soil and cup	15.15	16.79	15.57			Plastic limit = 36% (±)
Wt. of dry soil and cup	14.91	16.55	15.34			
Wt. of cup	14.21	15.90	14.70			Plasticity index = 60
Wt. of water	0.24	0.24	0.23			
Wt. of dry soil	0.70	0.65	0.64			Soil classification = CH
Water content %	34.2%	37.0%	35.9%			(clay of high plasticity)

Figure A1-2. *Data sheet for Atterberg limits.*

3 Moisture-Density Relationship of Soil And Compaction Test

Purpose and Use

Compaction is necessary when soil transported from one location to another is placed to support loads, such as building foundations and floor slabs or highways, or is to be part of an earth structure, such as a dike or dam. The compacted soil is referred to as a structural fill. Soil is compacted in order to make it as dense as practical. Obtaining a high density by compaction improves the engineering properties of soil by:

1. Increasing the shear strength.
2. Reducing the compressibility.
3. Decreasing the porosity so that it will be less susceptible to infiltrating water.

Soil placed for structural fill normally is spread and compacted in relatively thin layers. The thickness of the layer depends on the soil type and the ease with which it can be compacted, and the type of compaction or rolling equipment being used.

Laboratory tests and observations of field projects have established that a soil's water content affects the density that can be obtained and the effort required to obtain satisfactory compaction. Soil that is too dry is difficult to compact, as is soil that is too wet. There is a proper amount of water for soil, not too little or too great, that will enable the densest arrangement of soil grains, or maximum dry density, to be obtained. Such a water content is referred to as the *optimum moisture content*. This is the proper water content for obtaining the maximum density for a particular soil under application of a specific compaction force. Different soils will have different optimum moisture contents.

For a given soil, the maximum density and optimum moisture content will be affected by the compactive effort.

Knowledge of the moisture-density relationship of soil being used for a structural fill is necessary in order to determine if the fill is being properly constructed. When the maximum density of the soil is known, the density actually being obtained in the structural fill can be checked to determine if compaction is adequate.

Laboratory methods are the most commonly used way to determine the moisture-density relationship for soil. To be meaningful, laboratory methods should provide results that can also be obtained in the field with construction equipment. One widely applied method in use today utilizes a falling weight to compact soil that is confined in a cylindrical mold of known volume (Fig. A1-3). The soil is placed in layers in the mold, and each layer is compacted by a specified number of blows with the falling weight (or drop hammer). The energy to compact the soil is reported in foot-pounds per cubic foot.

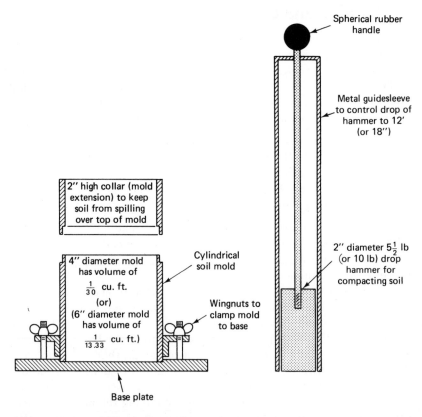

Figure A1-3. Widely used cylindrical mold and drop hammer type apparatus for performing moisture-density (compaction) test.

What is frequently referred to as the "standard compaction test" uses a four-inch-diameter mold[6] that provides a volume of $\frac{1}{30}$ cubic foot. The soil is compacted in three layers. Twenty-five blows of a 5.5-pound weight free-falling from a height of 12 inches is used to compact each layer. The energy used to compact a cubic foot of soil is, therefore,

$$\left[(5.5 \times 12 \text{ in.}) \times 25 \frac{\text{blows}}{\text{layer}} \times 3 \text{ layers}\right] \div \tfrac{1}{30} \text{ft}^3 = 12{,}400 \frac{\text{ft-lb}}{\text{ft}^3}$$

With the desire to increase loading which could be applied to structural fills and the development of heavier construction equipment, the need for greater compacted densities was recognized. The basic laboratory compaction test was modified. This "modified compaction test" uses a ten-pound weight falling 18 inches, and compacts the soil in the $\frac{1}{30}$-cubic-foot cylinder in five layers, using 25 blows per layer. The total compactive energy is then

$$\left[(10 \text{ lb} \times 1.5 \text{ ft}) \times 25 \frac{\text{blows}}{\text{layer}} \times 5 \text{ layers}\right] \div \tfrac{1}{30} \text{ft}^3 = 56{,}300 \frac{\text{ft-lb}}{\text{ft}^3}$$

Various modifications to the basic method described above have been made by different highway departments and governmental agencies concerned with soil compaction. Frequently, the variations are limited to differing mold size, hammer energy, or number of layers being compacted for the laboratory sample. Typically, for coarse soils or soil containing gravel, a six-inch-diameter mold having a volume of 0.075 cubic foot is used.

The determination of an optimum moisture content and maximum dry density is a trial-and-error procedure. Samples of the soil are compacted, the appropriate test method being used, at different water contents. A graphical plot of dry density (dry unit weight) obtained versus the related water content is made, and the optimum values are selected from the resulting curve; see Fig. A1-4.

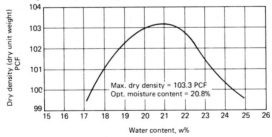

Figure A1-4. *Representative moisture-density curve resulting from laboratory compaction test.*

---

[6]The mold without collar is 4.0 in. in diameter and 4.585 in. high. When attached, the collar extends 2 in. above the top of the mold.

Test Procedure (Standard or Modified Compaction Test, Four-inch-diameter Mold)[7]

Special equipment needed includes the $\frac{1}{30}$ cubic foot compaction mold with collar attachment or, the 0.075 cubic foot mold may be used if the tested soil contains large coarse material, and the compaction hammer (either 5.5-pound weight having a 2 inch diameter face and a 12-inch drop, or a 10-pound weight having a 2-inch diameter face with an 18-inch drop). Scales for weighing in pounds and in grams, an oven, and hand-mixing tools are also needed.

1. The water content of the soil to be tested should be below the optimum moisture content. If the soil is received in a wet condition, it should be placed in an open pan and left to air-dry. A fan blowing over the pan will hasten drying. Granular soils will dry quickly, whereas silts and clays may require hours. The soil will be below optimum moisture content if a sample squeezed in the fist does not hold together or very easily falls apart. When dried, clumps of soil should be broken down to the individual particles.

2. The soil should be sieved through a 2-inch sieve, a $\frac{3}{4}$-inch sieve, and a #4 sieve. Material larger than the #4 sieve is normally discarded. However, if there is a significant portion of material coarser than the #4 sieve, all material finer than the $\frac{3}{4}$-inch sieve should be used. Replace material larger than 2 inches with an equivalent weight of soil in the $\frac{3}{4}$-inch to #4 sieve range. Select the test method to be used, standard or modified, and the size of mold (either $\frac{1}{30}$ cubic foot or 0.075 cubic foot). Record the empty weight of the mold.

3. Prepare the sample for the first trial. The soil should be thoroughly mixed so that it is at a uniform water content before it is placed in the mold and compacted. The soil layers placed in the mold should be slightly more than one-third or one-fifth of the height of the mold, depending on if the standard (three layers) or modified method (five layers) is used. Compact each layer with the required number of blows.[8] When compacted, the last layer should extend slightly above the top of the mold into the collar of the mold. After com-

---

[7]In accord with ASTM Test Designation D 698 and ASTM Test Designation D 1557.

[8]When using the $\frac{1}{30}$ cu. ft. mold, 25 blows per layer are used for compaction, with the 0.075 cu. ft. mold, 56 blows per layer are applied.

pacting, remove the collar and level the soil even with the top of the mold. Weigh the soil and mold. The wet weight of soil W_t is this total weight minus the weight of the mold.

4. Remove the soil from the mold by jacking or breaking it up, and obtain a representative sample (from the middle section of the compacted soil if possible) in order to determine the water content (about 100 grams of soil is sufficient). Place the sample in a cup, determine its wet weight, and then oven-dry to get the dry weight. From these data determine the water content (w = water content = weight of water lost by drying divided by dry weight of soil). Knowing the water content, determine the dry weight of the molded specimen from $W_s = W_t/(1 + w)$. The dry density will be $W_s \times 30$ if the 4-inch-diameter mold ($\frac{1}{30}$ cubic foot) is used, or $W_s \times 13.33$ if the 0.075-cubic-foot mold is used.

5. Prepare other trials at different water contents. It is desirable to have trials where the increment between water contents is about three percent. The soil from a previous trial can be re-used, but it should be thoroughly broken up and mixed so that the added water is evenly distributed through the soil. If a sufficient quantity of soil is available, it may be easier to prepare a separate mix at each desired water content. In the case of fine-grained soils, it is also desirable to let the mixed soil set for several hours before compacting. Determine water content, and wet and dry soil weight for each trial. At least four, preferably five, trials should be performed. Trials at increasing water content should continue until there is no change, or there is a decrease, in the wet weight of the sample, or until an excessive amount of water is observed squeezing out of the bottom of the mold during compaction.

6. Record all data as shown in Fig. A1-5. The results are then plotted to provide the moisture-density curve. The maximum density and the optimum moisture content are determined from the plotted curve.

Compaction method

1. — 1/30 cu. ft. mold, $5\frac{1}{2}$ lb hammer, 12″ drop, 3 layers @ 25 blows/layer (Standard test)

2. — 0.075 cu. ft. mold, $5\frac{1}{2}$ lb hammer, 12″ drop, 3 layers @ 56 blows/layer

3. — 1/30 cu. ft. mold, 10 lb hammer, 18″ drop, 5 layers @ 25 blows/layer ✓ (Modified test)

4. — 0.075 cu. ft. mold, 10 lb hammer, 18″ drop, 5 layers @ 56 blows/layer

Trial	1	2	3	4	5	6	7
Wet density determinations							
Weight of mold and wet soil	13.84	14.30	14.00	13.89			
Weight of mold	9.32	9.32	9.32	9.32			
Weight of wet soil (W_T)	4.52	4.98	4.68	4.57			
Wet density	135.8	144.4	140.4	137.1			
Moisture determinations							
Cup identification	B-1	B-2	B-3	B-4			
Weight of cup plus wet soil	39.10	55.30	66.60	75.48			
Weight of cup plus dry soil	38.15	52.81	62.28	68.53			
Weight of cup	15.10	14.21	14.43	14.33			
Weight of dry soil	23.05	38.60	47.85	54.20			
Weight of water	0.95	2.99	4.32	6.95			
Water content — %	4.0	6.5	9.0	12.8			
Dry density — PCF	130.3	140.3	129.0	121.5			

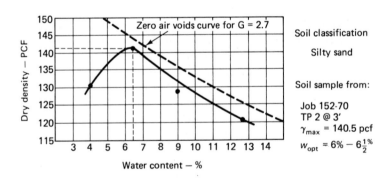

Soil classification

Silty sand

Soil sample from:

Job 152-70
TP 2 @ 3′
γ_{max} = 140.5 pcf
w_{opt} = 6% − $6\frac{1}{2}$%

Figure A1-5. *Data sheet for compaction test.*

4 Consolidation Test

Purpose and Use

This test is used to determine the compression properties of a soil, and the rate at which the compression occurs (consolidation). The consolidation test is a confined compression test; i.e., the soil being tested is confined within a ring to prevent lateral squeezing so that deformation under loading occurs in the vertical direction only. This is referred to as *one-dimensional* compression or consolidation. It is representative of conditions for many settlement problems, since the compressive soils are frequently located at some depth below the ground surface and confinement is provided by lateral pressures imposed by the surrounding soils.

Special equipment needed for the consolidation test includes a consolidometer, a device to apply compressive loading to the soil sample in the consolidometer, and an extensometer (or micrometer dial gage) to indicate change in thickness of the soil. Soil compression is due to a decrease in the voids that occur when loading is applied, and is not actually a decrease in volume of soil solids. Basically, the consolidometer is a ring to hold the soil being tested, with porous stones placed above and below the sample to permit pore water to drain from the sample as compression occurs. Schematic diagrams of two types of consolidometers are shown in Fig. A1-6. The loading device can consist of any equipment or method that permits a controlled compressive loading to be applied in increments.

When test results are to be used for settlement analysis, the soil to be tested should be undisturbed. It is preferable to start with an undisturbed sample slightly larger than the consolidometer ring will hold. The sample can then be trimmed with a wire, sharp-edged knife, or trimming lathe to remove the disturbed edges of the soil. The consolidation sample should have a diameter at least $2\frac{1}{2}$ times its thickness.

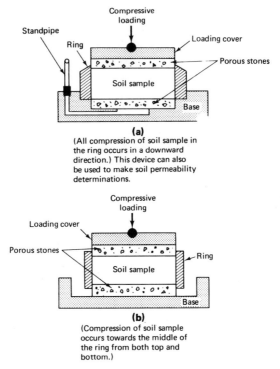

(a)

(All compression of soil sample in the ring occurs in a downward direction.) This device can also be used to make soil permeability determinations.

(b)

(Compression of soil sample occurs towards the middle of the ring from both top and bottom.)

Figure A1-6. (a) Fixed-ring consolidometer; (b) floating-ring consolidometer.

Test Procedure

1. Carefully trim the sample to fit into the consolidometer ring. Save trimmings to use for a water-content determination. Determine the weight (wet weight) of the consolidation sample before beginning the test, either before or after the sample is placed in the consolidometer ring. Also determine the thickness of the sample. These dimensions should be measured to an accuracy of 0.001 inch or 0.002 centimeter. If the diameter of the consolidometer ring is known, the total original volume of the soil can be determined.

2. Dampen the porous stones that will be placed on top and bottom of the soil sample. After the soil sample and consolidometer are assembled, place moist cotton, cellulose sponge, or other absorbent material around the soil sample so that it does not dry out during the test. (For soil samples

obtained from below the groundwater table, it may be desirable to keep the sample inundated during testing. For this situation, the sample is frequently kept moist as described above until the compressive pressure imposed on the sample in the test equals the overburden pressure that acted on the sample, when it is then inundated.)

3. Place the consolidometer in the loading device and impose a small pressure (equivalent to about 50 pounds per square foot) to "seat the load." Assemble the extensometer (or dial gage) to the consolidometer and record the reading as the starting reading.

4. Plan to impose compressive loads corresponding to 100, 200, 400, 800, 1600, 3200, 6400, 12,800 and 25,600 pounds per square foot (or any other similar magnitude loading where each successive pressure is twice the previous pressure). Each of these loadings should remain until the sample fully (or almost fully) consolidates under the imposed pressure before the next loading is applied. Generally, each loading is left on the sample for a period of 24 hours as a matter of convenience. It is frequently possible to impose lighter loadings for a shorter interval, however.

5. For each loading, record the decrease in thickness versus time (measured from when the new load was first applied). Readings should be taken frequently in the early stages of loading. Readings at intervals (all measured from the beginning) of $\frac{1}{4}$, $\frac{1}{2}$, 1, 2, 4, 9, 16, 25, 36, 64, and 128 minutes, and hourly thereafter, are suggested. Readings should be continued until it is assured that 100 percent primary consolidation has occurred. The method for determining this is discussed below.

6. If rebound or reloading information is desired, the compressive load is reduced, with time readings being made to determine the rate of expansion. For reloading data, compressive loads are then increased, and time readings are made as suggested above.

7. At the end of the test, when sufficient loading and unloading information has been obtained, the sample should be removed from the consolidometer, weighed, put into an oven to dry, and subsequently reweighed in order to determine the final water content of the sample and the weight of the soil solids. After this information has been obtained, the sample can be used for determining the specific gravity of the soil solids.

Calculations

1. From knowing the specific gravity of solids, the original thickness of the sample and its diameter, calculate the original void ratio of the sample; i.e.,

$$e_0 = \frac{V_v}{V_s} = \frac{V_t - V_s}{V_s} = \frac{h_0\pi d^2 - \dfrac{W_s}{G_s\gamma_w}}{W_s/G_s\gamma_w} = \frac{h_0\pi d^2}{W_s/G_s\gamma_w} - 1.0$$

where h_0 = original thickness of sample.
   d = diameter of consolidation sample.
   W_s = dry weight of soil.
   G_s = specific gravity of soil.
   γ_w = unit weight of water.

2. Calculate the void ratio at the end of each loading increment. Each void ratio can be calculated from

$$e = \frac{h\pi d^2}{W_s/G_s\gamma_w} - 1.0$$

where h is the sample height at the end of that loading period. (*Note:* $h = h_0 - \Delta h$, where Δh is the total decrease in sample thickness).

3. Plot the values of e versus the appropriate pressure on semilog coordinates, typified by Fig. A1-7, to give the compression test curve.

4. The compression index C_c is the slope of the steepest portion of the curve (virgin curve). The slope for the initial portion of the curve and the reloading cycle is C_r, the reload slope. These are shown in Fig. A1-7.

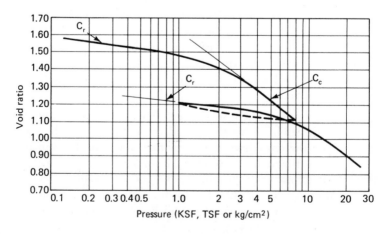

Figure A1-7.   *Consolidation test data, e versus log σ_v.*

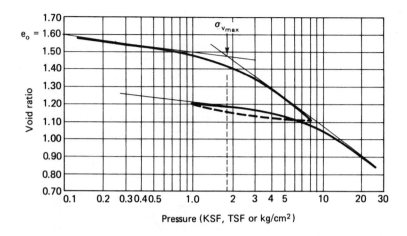

Figure A1-8. *Consolidation test data, estimating maximum pressure that acted on the soil in its past history.*

5.   The maximum pressure that has been imposed on the sample any time in its history can be estimated from the e vs. log σ_v (or strain vs. log σ_v) plot. Several methods are available, but the following method provides a reasonably good estimate, and is quick and simple:

    (a)   Through e_0, draw a line parallel to the slope determined for the reloading cycle.

    (b)   Extend the virgin slope backwards to where it intersects the line drawn for (a).

    (c)   The intersection of the two drawn lines occurs at a pressure close to the maximum pressure that has been imposed on the soil (precompression load). See Fig. A1-8.

6.   The coefficient of consolidation for any loading range (say for pressure increasing from 800 to 1600 psf) can be determined by one of two fitting methods—the log of time fitting method or the square root of time fitting method.

Log of time method / For the particular loading range, plot compression (dial readings will suffice) versus the log of time for the related compression to result; see Fig. A1-9. Draw tangents to the two straight-line portions of the resulting curve. The intersection of the tangents occurs at what is taken as the compression for 100 percent primary consolidation (d_{100}). Determine what is assumed as the point of zero percent consolidation (d_0) by taking the amount of compression

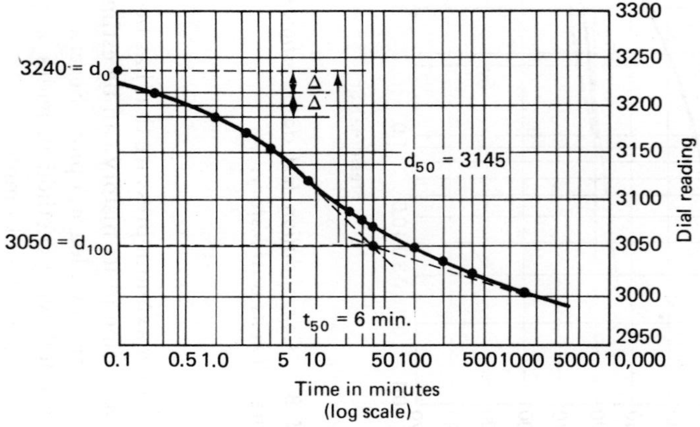

Figure A1-9. Log of time fitting method for determining coefficient of consolidation c_v.

Load increment 1 – 2 kg/cm²	
Elapsed time in minutes	Dial reading
0	3244
0.25	3214
1.0	3187
2.25	3179
4.0	3160
6.25	3142
9.0	3126
12.25	3112
16.0	3102
20.25	3090
25.0	3084
30.25	3075
36	3070
42.25	—
60.0	3058
100.0	3049
200.0	3034
400.0	3019
1440	2999

between 0.25 and 1.0 minute, and adding this value above the reading for 0.25 minute. The compression corresponding to 50 percent consolidation is midway between d_0 and d_{100}. The coefficient of consolidation c_v for the applied range of loading is

$$c_v = \frac{0.20 \left(\dfrac{h_{avg}}{2}\right)^2}{t_{50}}$$

where h_{avg} = average sample thickness for the range of loading considered, inches (or cm).

t_{50} = time in minutes for 50 percent consolidation to occur (time corresponding to d_{50} on the log of time plot).

c_v = coefficient of consolidation, inches^2/minute (or cm^2/min).

Square Root of Time Method / For the particular loading range, plot on arithmetic coordinates compression (dial readings) versus the square root of time for the indicated compression to occur (e.g., the dial reading for four minutes is plotted against the $\sqrt{4}$ or 2; the dial reading for nine minutes is plotted against 3, etc.). See Fig. A1-10.

Draw a tangent to the straight-line portion of the curve and extend it back to zero time. Through the point where the tangent crosses zero time, draw a line having a slope 1.15 greater (referenced to the ordinate). Where this new line crosses the curve is taken as the time for 90 percent consolidation to occur. (The time t_{90} is obtained by projecting the point down to the square root of time scale and squaring the value to convert from square root of time back to the true value.) The coefficient of consolidation c_v for the applied range of loading is

$$c_v = \frac{0.85 \left(\dfrac{h_{avg}}{2}\right)^2}{t_{90}}$$

where h_{avg} = average sample thickness for the range of loading considered, inches (or cm).

t_{90} = time in minutes for 90 percent consolidation to occur from the square root of time plot.

c_v = coefficient of consolidation, inches^2/minute (or cm^2/min).

The values of c_v will vary for each loading range. Further, c_v for 50 percent consolidation and c_v for 90 percent consolidation for the same loading range will not be equal.

An illustration of test data with finished computations and graphical plots for a completed consolidation test is provided in Fig. A1-11.

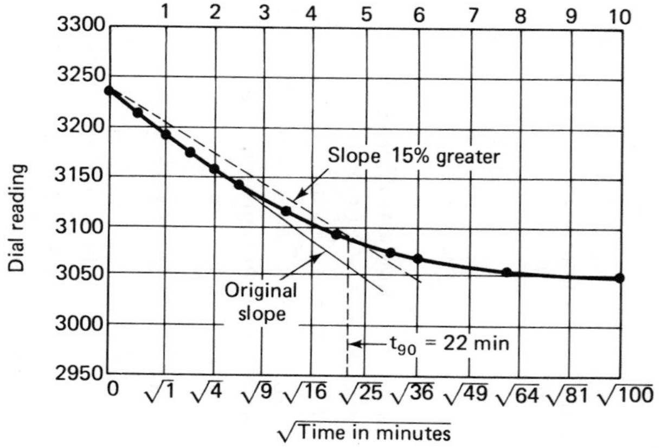

Figure A1-10. Square root of time fitting method for determining coefficient of consolidation c_v.

Load increment 1 – 2 kg/cm²		
Elapsed time in minutes	$\sqrt{\text{time}}$	Dial reading
0	0	3244
0.25	0.50	3214
1	1	3197
2.25	1.5	3179
4	2	3160
6.25	2.5	3142
9	3	3126
12.25	3.5	3112
16	4	3102
20.25	4.5	3090
25	5	3084
30.25	5.5	3075
36	6	3070
42.25	6.5	—
60	7.75	3058
100	10	3049
200	14.2	3034
400	20	3019
1440	38	2999

Pressure kg/cm²	Final dial reading	Dial change	Sample height inches	Void ratio	Fitting time		Coef. of consolidation, C_v	For T_{90}
					min.	min.	in²/min.	in²/min.
0	.7200	.007	1.000	1.90				
.125	7130	.010	0.993	1.79				
.250	7030	.025	0.983	1.76				
.50	6780	.022	0.958	1.69				
1.00	6560	.025	0.936	1.63				
2.00	6310	.062	0.911	1.56	2 min.	11 min.	.0214	.0165
4.00	5690	.080	0.849	1.38				
8.00	4890	+.025	0.769	1.16				
2.00	5140		0.794	1.23				
0.25	5860		0.966	1.43				

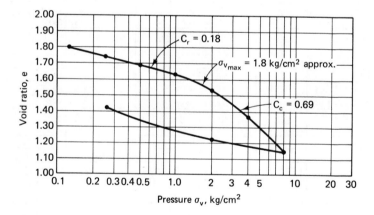

Figure A1-11. *Data sheet for consolidation test, and method of fitting for coefficient of consolidation c_v.*

Consolidation test data

Job 152-69

Soil type: _Gr. Br. Clay (CH)_
Boring: _Bor. 12 Sample 6_
Depth: _33' - 36½'_
Specific gravity, soil: _2.70_
Consol. ring diameter: _2.41" (A = 4.55m²)_
Starting sample thickness, H_0: _1.000"_
Initial void ration, e_0 = _1.80_

Total sample weight (start): _119.5 gm_
Total sample weight (finish): _109.5 gm_
Oven dry weight (finish): _71.5 gm_
Weight of water (start): _48.0 gm_
Water content (start): _67.0 %_
Weight of water (finish): _38.0 gm_
Water content (finish): _53.3 %_

Note: Dial Readings in inches

Time	Elapsed Time (Min.)	Dial Read'g	Compression	Time	Elapsed Time (Min.)	Dial Read'g	Compression	Time	Elapsed Time (Min.)	Dial Read'g	Compression
Loading Range:	0 – 0.125			Loading Range:	.125 - .25			Loading Range:	.25 - .50		
	O	7200	e_0 = 1.80		O	7130			O	7030	
	.25	7184			.25	7115			.25	6976	
	1.00	7190			1.00	7100			1.00	6939	
	2.25	7170			2.25	7092			2.25	6720	
	4.0	7163			4.0	7082			4.0	6901	
	6.25	7158			6.25	7076			6.25	6876	
	9.0	7150			9.0	7069			9.0	6860	
	12.25	7148			12.25	7061			12.25	6850	
	16.0	7143			16.0	7057			16.0	6841	
	20.25	7142			20.25	7051			20.25	6831	
	25.0	7141			25.0	7043			25.0	6822	
	30.25	7140			30.25	7041			30.25	6815	
	36.0	7139			36.0	7039			36.0	6908	
	42.25	7138			42.25	7017			42.25	6801	
	64.0	7137			64.0	7034			64.0	6795	
	100	7136			100	7033			100	6791	
	200	7135			200	7032			200	6788	
	400	7133			400	7031			400	6785	
	1440	7130	.007		1440	7070	.010		1440	6780	.025
			e 1.79				e 1.76				e 1.69

Figure A1-11.   *(Continued.)*

450

Time	Elapsed Time (Min.)	Dial Read'g	Compression
Loading Range: .50-1.0			
	0	6780	
	.25	6751	
	1.0	6719	
	2.25	6690	
	4.0	6662	
	6.25	6648	
	9.0	6629	
	12.25	6614	
	16.0	6602	
	20.25	6596	
	25.0	6589	
	30.25	6584	
	36.0	6580	
	42.25	6574	
	64.0	6568	
	100	6566	
	400	6564	$e = 1.69$
	1400	6560	.022

Time	Elapsed Time (Min.)	Dial Read'g	Compression
Loading Range: 1.0-2.0			
	0	6560	
	.25	6521	
	1.00	6472	
	2.25	6427	
	4.0	6413	
	6.25	6389	
	9.0	6373	
	12.25	6355	
	16.0	6343	
	20.25	6343	
	25.0	6335	
	30.25	6333	
	36.0	6330	
	42.25	6376	
	64.0	6319	
	100	6317	$e = 1.56$
	400	6315	
	1400	6310	.025

Time	Elapsed Time (Min.)	Dial Read'g	Compression
Loading Range: 2.0-4.0			
	0	6310	
	.25	6127	
	1.00	6035	
	2.25	5967	
	4.0	5915	
	6.25	5895	
	9.0	5862	
	12.25	5835	
	16.0	5809	
	20.25	5801	
	25.0	5795	
	30.25	5784	
	36.0	5771	
	42.25	5751	
	64.0	5738	
	100	5715	$e = 1.38$
	400	5701	
	1440	5690	.062

Time	Elapsed Time (Min.)	Dial Read'g	Compression
Loading Range: 4.0-8.0			
	0	5690	
	.25	5551	
	1.0	5466	
	2.25	5362	
	4.0	5283	
	6.25	5200	
	9.0	5160	
	12.25	5115	
	16.0	5091	
	20.25	5059	
	25.0	5022	
	30.25	5005	
	36.0	4988	
	42.25	4978	
	64.0	4960	
	100	4922	
	400	4908	$e = 1.16$
	1440	4890	.080

Time	Elapsed Time (Min.)	Dial Read'g	Compression
Loading Range: 8.0-2.0			
	0	4890	
	1440	5140	+.025
			$e = 1.23$

Time	Elapsed Time (Min.)	Dial Read'g	Compression
Loading Range: 2.0-0.25			
	0	5140	
	1440	5860	+.072
			$e = 1.43$

Figure A1-11.   *(Continued.)*

451

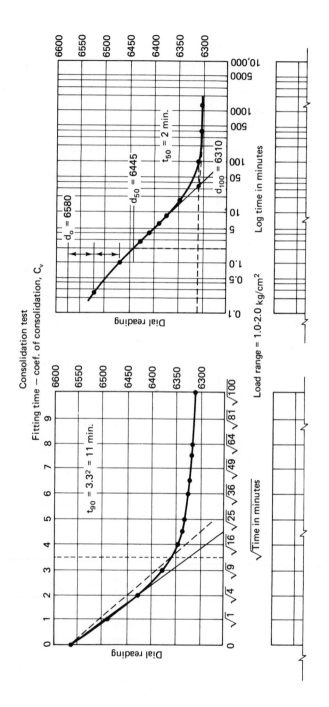

Figure A1-11. (Continued.)

452

Calculating c_v

c_v for t_{50}

$h_s = 0.936$ in.

$h_f = 0.911$ in.

$$c_v = \frac{T_v \left(\frac{h_{avg}}{2}\right)^2}{t_{50}}$$

$$= \frac{(0.20)\left[\frac{0.936 + 0.911}{2 \times 2}\right]^2 \text{in.}^2}{2 \text{ min}}$$

$$= 0.0214 \frac{\text{in.}^2}{\text{min}}$$

c_v for t_{90}

Sample height same as for t_{50}.

$$c_v = \frac{T_v \left(\frac{h\,avg}{2}\right)^2}{t_{90}}$$

$$= \frac{0.85 \left[\frac{0.936 + 0.911}{2 \times 2}\right]^2 \text{in.}^2}{11 \text{ min.}}$$

$$= 0.0165 \frac{\text{in.}^2}{\text{min}}$$

Note: Values of T_v for 50% and 90% consolidation are obtained from Fig. 9-16.

APPENDIX 2
Sample Specification
For Test Borings

SPECIFICATION FOR TEST BORINGS
(Project Name and Location)
As called for on the Subsurface Investigation Plan

1. *DEFINITION OF TERMS.* The word "Architect" as used herein refers to _____ _____. The words "Soils Engineer" refer to (*Soil Consultant's name and address*) or their authorized representative.

2. *QUANTITY OF WORK.* The Architect, Designer, or Engineer may increase or decrease the number of borings, change the location of the borings, increase or decrease the depth or the diameter of the borings before the start of the work, or during the progress of the work, or both, all of which shall not change the unit prices as submitted in the proposal.

3. *EXAMINATION OF SITE.* Each bidder shall visit the site of the proposed borings and fully familiarize himself with conditions as they exist. Failure to do so will not relieve the Contractor of his responsibility to accept the conditions at the site as they exist.

4. *PREVIOUS EXPLORATION.* (Four) preliminary borings were completed at the site during _____. These borings encountered _____ _____ _____. This

[1]Material in this appendix is courtesy Empire Soil Investigations—Thomsen Associates, Groton, New York.

information is provided as general information only. The Architect, Designer, or Engineer does not guarantee that conditions differing greatly from those disclosed by the preliminary borings will not be encountered at other points on the site.

5. *DEPTH AND SEQUENCE OF BORINGS.* Borings shall be made in the sequence as directed by the Soils Engineer and shall be carried to depths as indicated on Subsurface Investigation Plan or as otherwise directed by the Soils Engineer during the progress of the work. The Contractor shall notify the Soils Engineer by telephone two days before the rigs are mobilized at the site. The Soils Engineer will provide an Inspector at the site during the entire drilling period to supervise the drilling and sampling work. The Contractor shall concurrently operate at least two rigs during the progress of work.

6. *DRIVE SAMPLE BORING AND SAMPLING.* The borings shall be advanced through standard methods approved by the Soils Engineer by driving or drilling a pipe casing of not less than $2\frac{1}{2}$ inch nominal diameter into the ground. As the sinking of the casing proceeds, soil shall be removed from within it by jetting or blowing or by other approved methods ordinarily used by the Contractor. At every 5-foot interval, or as directed by the Soils Engineer, samples shall be obtained by driving a standard 2 inch O.D. split sample spoon of conventional design into material which has not been disturbed by the boring operation. After the spoon is lowered to the bottom of the hole, it shall be driven for three intervals of 6 inches each, or a total of 18 in. Driving shall be done with a 140-pound drive weight having a free fall of 30 in. The number of blows required to drive the sampler for each 6-in. interval shall be recorded.

In some of the borings the use of not less than 4 inch nominal diameter pipe casing will be required to facilitate recovery of $2\frac{1}{8}$-inch diameter rock core (NX). These borings shall be advanced in the same manner and split spoon samples shall be obtained as described directly above.

7. *CORE DRILLING.* For borings that are carried into solid bedrock, a double tube diamond drill shall be used to obtain a true representative sample of the rock not less than $2\frac{1}{8}$ inch in diameter or $1\frac{1}{8}$ inch in diameter as shown on the Subsurface Investigation Plan. The Contractor is to place the entire length of the core obtained in wooden core boxes, stating the percentage of core recovered as compared with

the actual drilling required to obtain the same. The time to core each one foot interval shall be recorded.

When core recoveries are poor or in soft rock, drilling shall be done with a Series M double-tube core barrel of such construction that the drilling water is fed to the bit without coming into contact with the core.

8. *GROUNDWATER OBSERVATION.* The groundwater level shall be observed at all boring holes in the following manner: after the hole has been completed to the required depth, the casing shall be pulled up one foot and left in the ground overnight. Water level readings shall be observed and recorded the following day. Overnight water level readings may be waived, if the Contractor can substantiate, through a sufficient number of observations, that the true water level has stabilized in the boring during or· shortly after termination.

9. *SAMPLES AND RECORDS.* Ordinary dry samples, immediately upon removal from the hole, shall be tightly sealed in eight ounce, large mouth, screw top, glass jars or bottles, having a diameter of approximately two inches. Each sample container shall be labeled to show clearly the number of the hole and the depth below the surface from which the sample came.

The Contractor shall at his own expense provide glass jars to receive ordinary dry samples, keeping a sufficient supply of these containers on the job to prevent any delay in the work.

During the progress of the boring operations, the Contractor shall keep a continuous and accurate log of the materials encountered during the advance of each hole. He shall also keep a complete record of the operations of advancing the casing and spoon. Blows on the sample spoon shall be recorded as specified in Paragraph 6. He shall include groundwater readings for all borings as specified in Paragraph 8. Coring time for each one-foot interval of rock coring shall be recorded as specified in Paragraph 7.

At the completion of the work, four white copies and one transparent copy of all records and logs, $8\frac{1}{2} \times 11$ inch in size, shall be delivered to the Soils Engineer. These records and logs shall be an exact unmodified copy of the driller's notes from the site. The samples, except Shelby Tube Samples, shall be packed in suitable boxes and be delivered to the office of the Soils Engineer. The Contractor shall give the Architect and the Soils Engineer every facility for obtaining his own records and determining any detail of the work as it progresses.

10. *DETERMINATION OF PAY QUANTITIES.* The amount of work to be paid for at the prices in the Proposal shall be measured as follows:

>   (a) *Mobilization and Demobilization.* The delivery to and removal from the site of all rigs and accessory equipment will be paid for at fixed lump sum price listed as "Mobilization and Demobilization" in the Proposal.

>   (b) *Drive Sample Boring and Sampling.* Payment will be made at the unit price quoted in the Proposal, for the actual linear feet of hole drilled, including split spoon sampling at five-foot intervals, measured from the surface of the ground to the depth reached in overburden. This unit price is to include any drilling required through boulders, obstructions, etc., in order to reach the depth ordered.

>   The unit price shall also include the readings of groundwater level in all borings, all sample containers, the transportation of samples as directed.

>   (c) *Additional Samples.* Split spoon samples in excess of the samples specified at five-foot intervals, recovered as directed by the Soils Engineer, or Shelby Tube Samples recovered as directed by the Soils Engineer, will be paid for at the fixed contract price per sample as stated in the Proposal.

>   (d) *Core Drilling.* Payment will be made at the unit price stated in the Proposal for the actual linear feet of drilling from solid bedrock surface to the bottom of the drilled hole.

>   (e) *Standby Time,* as required by the Engineer for permeability testing and other observations will be paid for at the fixed hourly contract price as stated in the Proposal.

11. *HOLES ABANDONED.* No payment will be made for holes abandoned before reaching final depth ordered unless the Soils Engineer decides that the boring so abandoned is acceptable.

12. *PAYMENTS.* No partial payments will be made. Payment in full will be made within 60 days after receipt by the Owner of the Contractor's invoice duly certified by the Architect or Designer and the Soils Engineer as being in order for payment. No invoice will be certified until all required items of work have been satisfactorily completed.

13. *COMPENSATION INSURANCE.* The Contractor shall take out and maintain during the life of the contract, Workmen's Compensation Insurance for all his workmen employed at the site and in case work is sublet, the Contractor shall require the Subcontractor to provide such insurance for all his employees.

14. *PUBLIC LIABILITY AND PROPERTY DAMAGE INSURANCE.* The Contractor shall take out and maintain during the life of the contract such Public Liability and Property Damage Insurance as shall protect the Owner and his agents against any claim arising from any injury to or death of persons or any damage to property resulting from the execution of this contract

This insurance shall be so written as to include Contingent Liability and Contingent Property Damage Insurance to protect the General Contractor against claims arising from the operations of the Subcontractors.

The Public Liability Insurance shall be in limits of ($100,000.00 to $300,000.00) on account of one accident.

The Property Damage Insurance shall be in limits of ($25,000.00 to $50,000.00) on account of one accident.

Certificates of such insurances shall be filed with the Owner.

The Contractor shall indemnify and save harmless the Owner, the Architect or designer, the Soils Engineer, and their agents and employees, against all loss or expense which the Owner, the Architect or designer, and the Soils Engineer may sustain, incur or become liable for on account of damage to or destruction of property, resulting from execution of work provided for in this contract, or due or arising in any manner from any act of negligence of the Contractor and any Subcontractor and their respective employees.

15. *LIQUIDATED DAMAGES.* The sum of one hundred and fifty dollars ($150.00) per day is to be agreed upon as liquidated damages for the work required to be finished within the time specified and shall be paid by the Contractor to the Architect for each and every calendar day in which any work of this contract is uncompleted after the time

specified for such completion. This shall not be regarded as a penalty to the Contractor for failure to complete the work on time, but as compensation to the Architect for the additional expense of maintaining inspection personnel on the project beyond the specified completion date.

16.   *SUBMISSION OF PROPOSAL.* Proposal for work described above shall be made in accordance with the form enclosed and submitted in *triplicate* in a sealed, opaque envelope marked with the name of the bidder and the name of the project bid upon.

      Bids shall be addressed to: (*Soil Consultant's name and address*).

      They shall be delivered to the above address no later than: ____(TIME)____ _____(DATE)_____ .

      If proposal is forwarded by mail, the sealed envelope shall be forwarded inside another envelope addressed to the Soils Engineer and forwarded by Registered Mail.

Form of Proposal

TEST BORINGS: (PROJECT)
                  (LOCATION)

DATE DUE: (_____)

TO: Soil Consultant's   OWNER: _____
        name and   _____
        address   _____

                   ARCHITECT/DESIGNER: _____
                                          _____

Gentlemen:

Pursuant to and in compliance with your request for bids dated _____ the undersigned hereby offers to furnish all plant, labor, materials, supplies, equipment and other facilities and things necessary, or proper for, or incidental to the test borings shown on the "Subsurface Investigation Plan" and called for in the "Specifications for Test Borings," Pages 1 to 7 inclusive, all addenda to the above, and having familiarized himself with the conditions for the sum of: ,

    1.   *Mobilization and Demobilization*   Lump Sum $_____
    2.   *Drive Sample Boring and Sampling* from surface to depth required including split spoon samples at five-foot intervals.

    (a)   2½-inch casing, _____ linear feet
          @ Unit Cost $_____ per Linear Feet
                                  Total $_____

    (b)   4 inch casing, _____ linear feet
          @ Unit Cost $_____ per Linear Feet
                                  Total $_____

3.   *Core Drilling* from surface of rock to bottom of completed hole.

    (a)   2⅛-inch diameter core (NX) _____ Linear Feet
          @ Unit Cost $_____ per Linear Feet
                                  Total $_____

    (b)   1⅛-inch diameter core (AX) _____ Linear Feet
          @ Unit Cost $_____ per Linear Feet
                                  Total $_____

4.   *Extra Samples* Split spoon samples obtained as directed in excess of samples specified at five-foot intervals.

    100 each @ Unit Cost $_____ each.
                                    Total $_____

5.   *Standby Time* For drilling equipment and crew standing by at Engineers request during field permeability tests or other observations.

    10 hours @ Unit Cost $_____ per hour.
                                    Total $_____

TOTAL SUM OF BID $_____

The bidder agrees to operate at least *two* rigs concurrently during the progress of the work.

If written notice of acceptance of this bid is mailed to the undersigned within thirty (30) days after the date of opening of bids, the undersigned agrees to commence work within ten (10) days and to complete the work within thirty (30) calendar days from date of notification of award. (Liquidated damages as specified in Paragraph 15 of the Specifications for Test Borings will be charged for each and every calendar day after the stipulated time of completion that is required to complete the work.)

Dated: _____   By: _____

APPENDIX 3
Typical Earthwork
Specifications

The following specification is suggested for use in contracts that involve cut and/or fill operations (particularly, the placement of load-bearing compacted fill), excavation for spread foundations, or trench excavations. It is intended that the specifications be used wholly or in part, or as a guideline, in preparing specifications for general contracts or subcontracts.

Specifications for
Site Preparation, Excavation,
Filling, and Grading

1. **Scope of Work** / The work included under this section includes:

    (a) Clearing and grubbing of site within contract limit lines shown on the plans.

    (b) Stripping of topsoil, removal of any existing unsuitable material, and stockpiling these materials where directed or removing them from the site.

    (c) Any other subgrade excavation required to bring the site to the lines and grades shown on the plans.

    (d) Drainage and/or dewatering of the site during construction, as required.

Material in this appendix is Courtesy Empire Soil Investigations—Thomsen Associates, Groton, New York.

463

(e) Compaction of subgrade prior to commencement of fill operation.

(f) Processing and supplying fill material meeting the requirements specified herein.

(g) Placement and compaction of fill material to the lines and grades shown on the plans as specified herein.

(h) Backfilling foundation excavations, utility trenches, and other areas of limited access.

2. **Soil Conditions** / Test borings have been made at the Owner's expense at the locations shown on the Plans. The approximate character of the encountered subsoils and the depths of various strata as disclosed by the borings are shown on the Boring Logs. This information is included as general information only and the Owner and/or Engineer does not guarantee that conditions differing greatly from those disclosed by the borings will not be encountered at other points on the site.

Additional borings and/or other means of soil exploration may be undertaken by the Contractor, but at no expense to the Owner or Engineer.

3. **Protection of Property** / Any existing roads and other property shall be suitably protected. If damaged by the Contractor, property shall be repaired by him to the satisfaction of the personal property owner and/or the public agency having jurisdiction over the damaged property.

Responsibility for cleaning public roads of any material carried onto these roads by trucks or other equipment shall be the Contractor's and his bid for the work shall include any costs for this work.

4. **Permits** / Any permits or licenses required to carry out this work shall be obtained and paid for by the Contractor and his bid price shall include any of these costs.

5. **Temporary Roads** / Any temporary haul roads which are not to be incorporated into the final roads and/or parking areas are the responsibility of the Contractor. Any haul roads made in areas which will become permanent roads shall be constructed in accordance with any portions of the specifications which apply to roads.

6. **Designation of Areas** / The Plans designate two (2) general areas within the contract limit lines:

(a) Building Areas.

(b)   Driveways, roadways, parking areas, and other areas outside building lines.

7.   **Clearing and Grubbing** / All trees and brush, including large roots, within the contract limit lines shall be cleared by the Contractor and suitably disposed of by burning (if allowed by local codes and agencies) or removal from the site.

8.   **Stripping** / Topsoil and sod shall be stripped from the site in all areas of excavation or fill. Topsoil shall be removed to its entire depth, and stockpiled in areas designated on the Plans, or removed from the site. The Owner, or his authorized representative, shall inspect the stripped surface and approve it prior to the commencement of excavation or compaction.

9.   **Compaction of Subgrade** / Following stripping, the subgrade in all fill areas shall be compacted sufficiently to develop to a depth of at least twelve (12) inches at least 90 percent of modified Proctor maximum density as determined in the laboratory in conformance with ASTM designation D-1557 or an equivalent test. The compaction will be checked by the Engineer and fill shall not be placed until the compaction of subgrade is approved by the Engineer.

In areas of excavation, the subgrade shall be compacted in accordance with the above, following the completion of the excavation to the required lines and grades.

10.   **Excavation** / Excavation shall be made to the lines and grades shown on the Plans, and any material which is suitable for either Type 1 or Type 2 fill shall be used as such. Any material which does not meet the specifications for fill material, as given below, or any material which is determined to be unsuitable by the Engineer shall be removed from the site, or used on the site as directed by the Owner or Engineer.

11.   **Backfilling** / Backfilling shall consist of placing and compacting fill material within excavated areas of limited access, where large compaction equipment cannot work. Backfill material, unless otherwise specified elsewhere on the Plans or in the Specifications, shall consist of inorganic run-of-bank gravel meeting the gradation and other requirements for Type 2 fill given in Section 12 below. The material shall be placed in nearly horizontal layers having a maximum thickness of six (6) inches. The moisture content shall be adjusted to facilitate compaction as deemed necessary by

the Engineer, and the material shall be compacted using approved, small compaction equipment. The excavations shall be of sufficient width to permit the use of approved small compaction equipment throughout the entire depth of the excavation around and adjacent to the constructed facility.

The backfill material within load-bearing areas shall be compacted to at least 95 percent of the modified Proctor maximum density, and backfill within non-load-bearing areas to at least 90 percent of the modified Proctor maximum density.

12.   **Fill Material** / Two types of fill material are required:

Type 1—Ordinary or general fill.

Type 2—Select run-of-bank gravel fill.

Type 1 fill shall be any type of fill material which consists of hard durable materials and soil binder, without excessive clay, organic matter, or any other deleterious material.

Type 2 fill shall be run-of-bank gravel consisting of hard durable pebbles, rock fragments, and soil binder. It shall be free of clay, organic matter, and other deleterious material, and shall conform to the following gradation requirements:

Sieve Designation	% by Weight Passing Square Mesh Sieves
3″	100%
$\frac{3}{4}$″	65% to 90%
#4	30% to 65%
#40	10% to 40%
#200	2% to 10%

If only occasional rocks or boulders are present in either Type 1 or Type 2 fill material as delivered to the site, a maximum stone size of six (6) inches is acceptable; however, no more than ten (10) percent by weight of the fill material shall consist of rocks or boulders larger than three (3) inches, and the gradation requirements above must be met for the portion of Type 2 fill that passes a three (3) inch square mesh sieve.

Acceptance of either Type 1 or Type 2 fill shall be based on the above requirements and final acceptance shall be made by the Engineer, and such acceptance or rejection of materials shall be binding upon the Contractor.

13.   **Placement and Compaction of Fill Material** / Prior to

placing fill, the subgrade shall be graded to provide adequate drainage, and shall be compacted as outlined in Section 9, above.

(a) Placement of Fill

The fill shall be spread evenly by mechanical or manual means, in approximately horizontal layers of six (6) to twelve (12) inches loose thickness, the thickness to be determined in the field by the Engineer. Thickness of lifts will depend on the type of material and the type of compaction equipment to be used.

(b) Moisture Control

At the time of compaction, the material in each layer of fill shall have a moisture content within ±2% of optimum moisture content for compaction, as determined by ASTM D-1557 procedure for determining the moisture-density relationship of the fill material. If, in the opinion of the Engineer, the fill material is too wet, it shall be dried by a method approved by the Engineer prior to commencing or continuing the compaction operation. Likewise, if, in the opinion of the Engineer, the fill material is too dry for proper compaction, the fill shall be moistened by a method approved by the Engineer prior to commencing or continuing the compaction operation. The Engineer, at his discretion, may permit a larger variation of moisture content than ±2% if the fill can be placed properly within the new limits, or may require a smaller variation of moisture content than ±2%, if it is necessary to properly control the fill.

The Engineer will, if deemed necessary, impose restrictions on routing of heavy construction equipment including loaded trucks over subgrade areas or compacted fill areas that show signs of adverse effects under the load of such equipment.

(c) Drainage of the Site

At all times the Contractor shall maintain and operate proper and adequate surface and subsurface drainage methods to the satisfaction of the Engineer in order to keep the construction site dry and in such condition that placement and compaction of fill may proceed unhindered by saturation of the area. During construction, the surface of the fill area shall be left in such condition that precipitation and/or surface water will run off without ponding.

(d) Compaction Equipment

It is the responsibility of the Contractor to select, furnish, and properly maintain equipment which will compact the fill uniformly to the required density; however, the Contractor's selection of equipment is subject to approval by the Engineer.

The Contractor shall have at the site at all times during the fill operation a smooth steel wheel roller, with a rated capacity of at least 10 tons, which shall be used to seal the surface of the fill at the close of each working day, and at other times when directed by the Engineer, to prevent the infiltration of precipitation and surface water into the fill material.

No fill shall be placed until approved compaction equipment is on the site and in working condition.

(e) Frost

No fill material shall be placed when either the fill material or the previous lift or subgrade on which it is to be placed is frozen. In the event that any fill which has been placed or the subgrade shall become frozen, it shall be scarified to break up all frozen material and recompacted, or removed, to the satisfaction of the Engineer before the next lift is placed. Any soft areas resulting from frost shall be removed or recompacted to the satisfaction of the Engineer before new fill is placed over the area.

(f) Compaction of Fill

Each lift within load-bearing areas shall be uniformly compacted to at least 95 percent of modified Proctor maximum density as determined in the laboratory by the Engineer in accordance with ASTM designation D-1557, and each lift within non-load-bearing areas shall be uniformly compacted to at least 90 percent of the modified

Proctor maximum density. Any lift, or portion thereof which is not compacted in accordance with the specifications, shall be recompacted or removed and replaced to the satisfaction of the Engineer. The degree of compaction of each lift shall be checked by the Engineer and each successive lift shall not be placed or compacted until the previous lift is inspected, tested, and approved by the Engineer. The fill is to be compacted to the lines and grades shown on the Plans, and the slopes shall at all times be maintained by the Contractor.

14. **Inspection, Testing, and Final Approval** / A commercial testing laboratory will be employed by the Owner with the approval of the Engineer, and this Testing Laboratory shall make such tests as deemed necessary by the Engineer, and shall be considered a representative of the Owner and the Engineer. The Owner's, Engineer's, and Testing Laboratory's personnel shall have access to any and all work areas at any and all times.

The acceptability of compaction will be established by tests made by the Testing Laboratory. The Testing Laboratory shall be directly responsible to the Owner and Engineer and will make the following tests:

(1) Sieve analysis.
(2) Maximum density and optimum moisture content (Proctor test).
(3) In-place density.

Tests (1) and (2) will be performed by the approved Testing Laboratory before acceptance and delivery of Type 2 fill material to the site. Any change in the source of material, or change of quality of the material, will require a new series of tests to determine acceptability.

Delivery and compaction of material within building roadway and parking areas shall be made during the presence of an inspector from the Testing Laboratory and shall be subject to approval by the inspector. This supervision by no means absolves the Contractor of responsibility of compaction as specified.

Suitable space shall be provided by the Contractor for the supplies and equipment of the Testing Laboratory.

The Testing Laboratory shall make reports to the Engineer as the work progresses, and at completion of the work, to aid the Engineer and Owner in giving their final approval to the work.

Upon completion and before final approval of the work, the Contractor shall remove all equipment, supplies, and the like from the work area, and leave the area in a neat and presentable condition.

Glossary

Adsorbed water. Water bound to soil particles because of the attraction between electrical charges existing on soil particle surfaces and (dipole) water molecules.

Atterberg limits. The liquid limit, plastic limit, and shrinkage limit for soil. The *water content* where the soil behavior changes from the liquid to the plastic state is the liquid limit; from the plastic to the semi-solid state is the plastic limit; and from the semi-solid to the solid state is the shrinkage limit.

Backfill. Soil material placed back into an area that has been excavated, such as against structures and in pipe trenches.

Bearing capacity. The pressure that can be imposed by a foundation onto the soil or rock supporting the foundation.

Boring. The method of investigating subsurface conditions by drilling into the earth. Frequently, soil or rock samples are also extracted from the boring for classification and testing.

Borrow. Soil or rock material obtained from an off-site source for use as fill on construction projects.

Caisson. Large structural chamber utilized to keep soil and water from entering into a deep excavation or construction area. Caissons may be installed by being sunk in place or by systematically excavating below the bottom of the unit to the desired depth.

Capillarity. The movement of water, due to effects other than gravity, through very small void spaces that exist in a soil mass. Water movement occurs in very small channels such as capillary-sized openings because of the affinity between soil and water which acts to increase the boundary of contact between the two materials and the surface tension property developed

471

by water in contact with air. Capillary flow can occur in a direction opposite to that of the pull of gravity.

Clays (clay minerals).   Very small soil particles having a crystalline (layered) structure, created as the result of the chemical alteration of primary rock minerals. Most clay particles, because of their mineralogical composition, are flat or plate-like in shape, with a large surface area to mass ratio. Clay particle dimensions are often smaller than 2μ.

Coarse-grained soil.   Those soil types having particles large enough to be seen without visual assistance. The coarse-grained materials include the sand and gravel (or larger) soil particles.

Cohesion.   The bonding or attraction between particles of fine-grained soil that creates shear strength.

Compaction.   The process of increasing the density or unit weight of a soil (frequently fill soil) by rolling, tamping, vibrating, or other mechanical means.

Compressibility.   The change, or tendency for change, that occurs in the thickness of a soil mass when it is subjected to compressive loading.

Consolidation.   The process by which compression of a newly stressed clay soil occurs simultaneously with the expulsion of water present in the soil void spaces. Initially, the newly imposed stress acting on the clay is imparted onto the water in the soil voids (pore water), and not onto the soil particles. Because of the increased pressure, the water is gradually forced out of the soil. As the pore water pressure is reduced, the magnitude of stress being imposed onto the soil particles is correspondingly increased. Compression of the clay layer occurs only as rapidly as pore water can drain from the soil, and thus is related to the permeability of the soil layer.

Density.   The weight or mass per unit of volume. In reference to soil, the term usually indicates weight per unit volume and is synonymous with unit weight.

Dewatering.   The procedure utilized to remove water from a construction area, such as pumping from an excavation or location where water covers the planned working surface; the procedure used to lower the ground water table in order to obtain a "dry" area in the vicinity of an excavation which would otherwise extend below water.

Drawdown.   The lowering of the level of the ground water table that occurs in the vicinity of a water well (on dewatering equipment) when it is pumped.

Earth pressure.   Normally used in reference to the lateral pressure or force imposed by a soil mass against an earth-supporting structure such as a retaining wall or basement wall, or on a fictitious vertical plane located within a soil mass. The *coefficient* of earth pressure refers to the ratio of lateral pressure to vertical pressure existing at a point in a soil mass.

Effective stress. The actual particle-to-particle contact stress (or pressure) existing between soil grains. This stress compensates for the possible buoyancy influence of water pressure. Effective stress relates directly to the shear strength possessed by a soil.

Fill. Earth placed in an excavation or other area to raise the surface elevation. Also referred to as earth fill or soil fill. *Structural earth fill* refers to material which is placed and compacted in layers in order to achieve a uniform and dense soil mass which is capable of supporting structural loading.

Fines or Fine-grained. Refers to silt- and clay-sized soil particles which exist in a soil mixture.

Flow line. The path of travel traced by moving water as it flows through a soil mass.

Flow net. A pictorial method used to study the flow of water through a soil. Utilized to indicate the paths of travel followed by moving water and the subsurface pressures resulting from the presence of the water.

Footing. Type of foundation typically installed at a shallow depth and constructed to provide a relatively large area of bearing onto the supporting soil.

Friction, internal. The particle (solid to solid) friction developed by cohesionless soils, and the property responsible for most of the shear strength which this type of soil can develop. The *angle* of internal friction, ϕ, references the shear strength to imposed normal stress acting at a point in a soil mass, i.e., $\tan \phi$ = shear strength/normal stress.

Ground water table. The surface of the underground supply of water. Also referred to as the *phreatic surface*.

Head. Shortened form of the phrase *pressure head*, referring to the pressure resulting from a column of water or elevated supply of water. Pressure would be computed from $\gamma_w h$, where γ_w is the unit weight of water, and h is the height or elevation of the water supply. The h term is the pressure *head*.

Heave. Upward movement of soil and foundations supported on soil, caused by expansion occurring in the soil as a result of such factors as freezing or swelling due to increased water content. *Frost heave* refers to the vertical soil movement that occurs in freezing temperatures as ice layers or lenses form within the freezing soil and cause the soil mass to expand.

Hydraulic gradient. Mathematical term indicating the difference in pressure head existing between two locations divided by the distance between these same locations. Given the designation i.

In-situ. Refers to soil when it is at its natural location in the earth and in its natural condition.

Liquefaction. Loss of strength occurring in saturated cohesionless soil exposed to shock or vibrations when the soil particles momentarily lose contact. The material then behaves as a fluid.

Penetration test.   Term generally applied to subsurface investigative methods for determining a strength–related property of a soil by measuring the resistance to advancement of penetration or boring equipment.

Permeability.   The ability of water (or other fluid) to flow through a soil by traveling through the void spaces. A high permeability indicates flow occurs rapidly, and vice versa.

Pier.   Category applied to column-like concrete foundations, similar to piles. The pier is generally considered the type of deep foundation which is constructed by placing concrete in a deep excavation large enough to permit manual inspection. Pier is also used frequently to indicate heavy masonry column units which are used for basement-level and substructural support.

Pile.   The relatively long, slender, column-like type of foundation which obtains supporting capacity from the soil or rock some distance below the ground surface.

Plasticity.   Term applied to fine-grained soils (particularly clays) to indicate the soils' (plus included water's) ability to flow or be remolded without raveling or breaking apart.

Pore pressure.   Water pressure developed in the voids of a soil mass. *Excess pore pressure* refers to pressure greater than the normal hydrostatic pressure expected as a result of position below the water table.

Relative density.   Term applied to sand deposits to indicate a relative state of compaction compared to the loosest and most dense conditions possible.

Rollers, compaction.   The category of construction equipment utilized to compact (or densify) soil by rolling it. The compaction force typically results from the heavy weight of the equipment and/or vibrations transmitted from the equipment into the soil.

Sand.   The category of coarse-grained soil whose particle sizes range between about 0.07 mm and 5 mm in diameter.

Seepage.   Generally refers to the quantity of water flowing through a soil deposit or soil structure such as an earth dam. Also may refer to the quantity of subsurface water leaking into a building's underground (basement) area.

Seismic exploration.   The method of determining subsurface soil and rock conditions (without excavation) by inducing a shock wave into the earth and measuring the velocity of the wave's travel through the earth material. This *seismic velocity* indicates the type of earth material.

Settlement.   The downward vertical movement experienced by structures or a soil surface as the underlying supporting earth compresses.

Shear strength.   The ability of a soil to resist shearing stresses developed within a soil mass as a result of loading imposed onto the soil.

Sieve. Pan or tray-like equipment having a screen or mesh bottom; used in laboratory or field work to separate particles of a soil sample into their various sizes.

Silt. The category of fine-grained soil particles (individual soil grains whose particle size is smaller than 0.07 mm or too small to be seen without visual aid) whose mineralogical composition remains similar to the rock they were derived from.

Soil sampler. The equipment used to extract soil samples from borings or test pits made in a subsurface investigation.

Soil stabilization. Treatment of soil to improve its properties; includes the mixing of additives and other means of alteration such as compaction or drainage.

Sump. Small excavation or pit provided in the floor of a structure, or in the earth, to serve as a collection basin for surface water and near-surface underground water.

Till. Description given to glacially transported soil formations which consist of a heterogeneous mixture of fine-grained and coarse-grained material.

Unit weight. The weight per unit volume of a material such as soil, water, concrete, etc. Typically expressed as pounds per cubic foot, grams per cubic centimeter, or kilonewtons per cubic meter.

Void ratio. The total volume occupied by a soil mass includes the soil particles plus void spaces (which in nature always exist between the particles because of their irregular shape). The void ratio is the ratio of the void space volume to the volume of soil solids.

Water content. The ratio of the quantity of water in a soil (by weight) to the weight of the soil solids (dry soil), typically expressed as a percentage.

Well point. The perforated end section of a wellpipe which permits the ground water to be drawn into the pipe for pumping.

References

(1)   Abdun-Nur, E. A. *A Standard Classification of Soils as Prepared by the Bureau of Reclamation.* ASTM Special Technical Publication #113. 1950.

(2)   Acker, W. L. III. *Basic Procedures for Soil Sampling and Core Drilling.* Scranton, Pa.: Acker Drill Co., Inc., 1974.

(3)   Agerschou, H. A. "Analysis of the Engineering Pile Formula," *Journal of the Soil Mechanics and Foundations Division, ASCE*[1] Vol. **88,** No. SM5, Oct. 1962.

(4)   Akroyd, T. N. W. *Laboratory Testing in Soil Engineering.* London: Soil Mechanics Limited, 1964.

(5)   Aldrich, H. P. "Precompression for Support of Shallow Foundations," *Journal of Soil Mechanics and Foundations Division, ASCE,* Vol. **91,** No. SM2, March 1965.

(6)   Aldrich, H. P. "Selection of Foundation Systems," *Soil Mechanics Lecture Series,* Foundation Engineering, Soil Mechanics and Foundations Division, Illinois Section, ASCE and Dept. of Civil Engineering, Northwestern University, Evanston, Illinois, 1969.

(7)   *AM-9 Chemical Grout, Technical Data.* American Cyanamid Co., Wayne, New Jersey.

(8)   Anderson, R. D. "Foster Vibrator," *Proceedings, Design and Installations of Pile Foundations and Cellular Structures,* Lehigh University, Envo Publishing Co., Lehigh, Pa., 1970.

---

[1]American Society of Civil Engineers.

(9)   *(Armco) Handbook of Drainage and Construction Products.* Armco Drainage and Metal Products, Middletown, Ohio.

(10) *Asphalt Handbook.* The Asphalt Institute, College Park, Maryland.

(11) Baker, C. N. and F. Kahn. "Caisson Construction Problems in Chicago," *Journal of Soil Mechanics and Foundations Division, ASCE,* Vol. **97,** No. SM2, Feb. 1971.

(12) Barron R. "Consolidation of Fine-Grained Soils by Drain Wells," *Transactions American Society of Civil Engineers,* Vol. **113,** 1948.

(13) Basore, C. E. and J. D. Boitano. "Sand Densification by Piles and Vibroflotation," *Journal of Soil Mechanics and Foundations Division, ASCE,* Vol. **95,** No. SM6, Nov. 1969.

(14) Baver, L. D., *Soil Physics.* New York: John Wiley and Sons, Inc., 1956.

(15) Berezantzev, V. G., V. S. Kristoforov, and V. N. Golubkov. "Load Bearing Capacity and Deformation of Piled Foundations," *Proceedings of the Fifth International Conference on Soil Mechanics and Foundation Engineering,* Paris, 1961, V.2.

(16) Bertram, G. E. "Design Requirements and Site Selection," *Design and Construction of 'Earth Structures, Soil Mechanics Lecture Series,* Soil Mechanics and Foundation Division, Illinois Section, ASCE and Civil Engineering Department, Illinois Institute of Technology, Chicago, Illinois, 1966.

(17) Brown, D. R. and J. Warner. "Compaction Grouting," *Journal of Soil Mechanics and Foundations Division, ASCE,* Vol. **99,** No. SM8, Aug. 1973.

(18) Bussey, W. H. "Foundation Evaluation and Treatment," *Design and Construction of Earth Structures, Soil Mechanics Lecture Series,* Soil Mechanics and Foundation Division, Illinois Section, ASCE and Civil Engineering Department, Illinois Institute of Technology, Chicago, Illinois, 1966.

(19) Carpenter, J. C. and E. S. Barker. "Vertical Sand, Drains for Stabilizing Muck-Peat Soils," *Transactions, American Society of Civil Engineers,* Vol. **124,** 1959.

(20) Carson, A. B. *General Excavating Methods.* New York: McGraw-Hill Book Co., 1961.

(21) Casagrande, A. "Research on the Atterberg Limits of Soils," *Public Road,* Vol. **13,** Oct. 1932.

(22) Casagrande, A. "Classification and Identification of Soils," *Transactions, ASCE,* Vol. **113,** 1948.

(23) Casagrande, L. "Electro-Osmotic Stabilization of Soils," *Journal Boston Society of Civil Engineers,* January 1952.

(24) Cedergren, H. R. *Seepage, Drainage and Flow Nets.* New York: John Wiley and Sons, Inc., 1967.

(25) Chellis, R. D. "The Relationship Between Pile Formulas and Load Tests," *Transactions, ASCE,* Vol. **114,** 1949.

(26) Chellis, R. D. *Pile Foundations,* 2nd ed. New York: McGraw-Hill Book Co., Inc. 1961.

(27) Christian, J. T. and W. F. Swiger. "Statistics of Liquefaction and SPT Results," *Journal of Geotechnical Engineering Division, ASCE,* Vol. **101,** No. GT11, Nov. 1975.

(28) Coyle, H. M. and L. C. Reese. "Load Transfer for Axially Loaded Piles in Clay," *Journal of the Soil Mechanics and Foundations Division,* ASCE, Vol. **92,** SM2, March 1966.

(29) Coyle, H. M. and I. H. Sulaiman. "Bearing Capacity of Foundation Piles: State of the Art," *Highway Research Record No. 333,* Highway Research Board, National Academy of Sciences—National Academy of Engineering, Washington, D.C., 1970.

(30) Cummings, A. E., G. O. Kerkhoff, and R. B. Peck. "Effect of Driving Piles into Soft Clay," *Transactions, ASCE,* Vol. **115,** 1950.

(31) D'Appolonia, D. J., E. D'Appolonia, and R. F. Brissette, "Settlement of Spread Footings on Sand," *Journal of Soil Mechanics and Foundations Division, ASCE,* Vol. **94,** No. SM3, May 1968.

(32) D'Appolonia, D. J., R. V. Whitman, E. D'Appolonia, "Sand Compaction with Vibratory Rollers," *Journal of Soil Mechanics and Foundations Division, ASCE,* Vol. **95,** No. SM1, Jan. 1969.

(33) D'Appolonia, E. "Load Transfer-Pile Clusters," *Soil Mechanics Lecture Series, Foundation Engineering,* Soil Mechanics and Foundations Division, Illinois Section, ASCE and Dept. of Civil Engineering, Northwestern University, Evanston, Illinois, 1969.

(34) Davisson, M. T. "Design Pile Capacity," *Proceedings Design and Installation of Pile Foundations and Cellular Structures,* Lehigh University, Envo Publishing Co., Lehigh Valley, Pa., 1970.

(35) DeBeer, E. E. "Bearing Capacity and Settlement of Shallow Foundations on Sand," *Proceedings, Bearing Capacity and Settlement of Foundations Symposium,* Duke University, Durham, N.C. 1967.

(36) deMello, V. F. B. "The Standard Penetration Test," *Proceedings, Fourth Pan American Conference on Soil Mechanics and Foundation Engineering,* Vol. **1,** Puerto Rico, June 1971 (Pub. ASCE).

(37) deRuiter, J. "Electric Penetrometer for Site Investigations," *Journal of Soil Mechanics and Foundation Division, ASCE,* Vol. **97,** No. SM2, Feb. 1971.

(38) *Design of Small Dams.* U.S. Bureau of Reclamation, Denver, 1960.

(39) Dunham, C. W. *Foundations of Structures.* New York: McGraw-Hill Book Co., 1962.

(40) *Earth Manual.* Bureau of Reclamation, U.S. Department of the Interior, Denver, Co.

(41) *Earth Resistivity Manual.* Soiltest, Inc., Chicago, Illinois, 1968.

(42) *Engineering Seismograph Instruction Manual.* Soiltest, Inc., Chicago, Illinois, 1975.

(43) Fellenius, B. H. "Test Loading of Piles and New Proof Testing Procedure," *Journal of Geotechnical Engineering Division, ASCE,* Vol. **101,** No. GT9, Sept. 1975.

(44) Fletcher, G. "Standard Penetration Test, Its Uses and Abuses," *Journal of Soil Mechanics and Foundations Division, ASCE,* Vol. **91,** SM4, July 1965.

(45) *Foundation Piling.* Report No. 4 for the Federal Construction Council Building Research Advisory Board, National Research Council—National Academy of Sciences Pub. No. 987, Washington, D.C., 1962.

(46) Fowler, J. W. "Pile Installation Case Histories," *Proceedings, Design and Installation of Pile Foundations and Cellular Structures,* Lehigh University. Lehigh Valley, Pa.: Envo Publishing Co., 1970.

(47) *Frost Action in Roads and Airfields.* Highway Research Board, Special Report No. 1, National Academy of Sciences—National Research Council, Pub. 211, Washington, D.C. 1952.

(48) Frye, S. C. "The Protection of Piling," *Proceedings, Design and Installation of Pile Foundations and Cellular Structures,* Lehigh University, Lehigh Valley, Pa.: Envo Publishing Co., 1970.

(49) Gendron, G. J. *Pile Driving: Hammers and Driving Methods,* Highway Research Record No. 333, Highway Research Board, National Academy of Sciences—National Academy of Engineering, Washington, D.C., 1970.

(50)   Gibbs, H. J. and W. G. Holtz. "Research on Determining the Density of Sands by Spoon Penetration Testing," *Proceedings of the Fourth International Conference on Soil Mechanics and Foundation Engineering*, London, 1957.

(51)   Gibbs, H. J. "Standard Penetration Test for Sand Denseness," *Proceedings of the Fourth Pan American Conference on Soil Mechanics and Foundation Engineering*, Vol. II, Puerto Rico, June 1971 (Pub. ASCE).

(52)   Golder, H. Q. "State-of-Art of Floating Foundations," *Journal of Soil Mechanics and Foundations Division, ASCE*, Vol. **91,** No. SM2, March 1965.

(53)   Golder, H. Q. "The Allowable Settlement of Structures," *Proceedings, Fourth Pan American Conference on Soil Mechanics and Foundation Engineering*, Vol. **1,** Puerto Rico, June 1971 (Pub. ASCE).

(54)   Goodman, L. S. and R. H. Karol. *Theory and Practice of Foundation Engineering.* New York: The MacMillan Company, 1968.

(55)   Grand, B. A. *Types of Piles: Their Characteristics and General Use.* Highway Research Record No. 333, Highway Research Board, National Academy of Sciences—National Academy of Engineering, Washington, D.C., 1970.

(56)   Gray, H. "Field Vane Shear Tests of Sensitive Cohesive Soil," *Transactions, ASCE*, Vol. **122,** 1957.

(57)   Grim, R. E. *Clay Mineralogy,* 2nd Ed. New York: McGraw-Hill Book Co., Inc., 1968.

(58)   Grim, R. E. "Physico-Chemical Properties of Soils: Clay Minerals," *Journal of Soil Mechanics and Foundation Division, ASCE,* Vol. **85,** No. SM2, April 1959.

(59)   Harr, M. E. *Groundwater and Seepage,* New York: McGraw-Hill Book Co., Inc., 1962.

(60)   Hedges, C. S. "Standard Test Boring with Drilling Mud," *Proceedings, Fourth Pan American Conference on Soil Mechanics and Foundation Engineering*, Vol. II, Puerto Rico, June 1971 (Pub. ASCE).

(61)   Hirsch, T. J., L. L. Lowery, H. M. Coyle, and C. H. Samson. *Pile Driving By One Dimensional Wave Theory: State of the Art.* Highway Research Record No. 333, Highway Research Board, National Academy of Sciences—National Academy of Engineering, Washington, D.C., 1970.

(62)   Holtz, W. G. and H. J. Gibbs. "Engineering Properties of Expansive Clays," *Transactions, ASCE*, Vol. **121,** 1956.

(63)   Holtz, W. G. and H. J. Gibbs. "Discussion, Settlement of Spread Footings on Sand," *Journal of the Soil Mechanics and Foundations Division, ASCE,* Vol. **95,** SM3, May 1969.

(64)   Hough, B. K. "Compressibility as the Basis for Soil Bearing Value," *Journal of the Soil Mechanics and Foundations Division, ASCE,* Vol. **85,** No. SM4, Aug. 1959.

(65)   Hough, B. K. *Basic Soils Engineering.* New York: Ronald Press Co., 1969.

(66)   Housel, W. S. *Checking Up on Vertical Sand Drains.* Highway Research Board Bulletin 90, National Academy of Sciences—National Research Council, Washington, D.C., 1954.

(67)   Housel, W. S. "Pile Load Capacity: Estimates and Test Results," *Journal of the Soil Mechanics and Foundations Division, ASCE,* Vol. **92,** No. SM4, July 1966.

(68)   Housel, W. S., "Michigan Study of Pile Driving Hammers," *Journal of Soil Mechanics and Foundations Division, ASCE,* Vol. **91,** SM5, September 1965.

(69)   Hunt, H. W. *Piletips, Design and Installation of Pile Foundations.* New Jersey: Associated Pile and Fitting Corp., 1974.

(70)   Hvorslev, M. J. *Subsurface Exploration and Sampling of Soils for Civil Engineering Purposes.* U.S. Waterways Experiment Station, Vicksburg, Miss. 1949.

(71)   Johnson, S. J. "Precompression for Improving Foundation Soils," *Journal of Soil Mechanics and Foundations Division, ASCE,* Vol. **96,** No. SM1, Jan. 1970.

(72)   Johnson, S. J. "Foundation Precompression with Vertical Sand Drains," *Journal of Soil Mechanics and Foundation Division, ASCE,* Vol. **96,** No. SM1, Jan. 1970.

(73)   Jumikis, A. R. *Thermal Soil Mechanics.* New Brunswick, New Jersey: Rutgers University Press, 1966.

(74)   Jurgenson, L. "The Application of Theories of Elasticity and Plasticity to Foundation Problems," *Journal, Boston Society of Civil Engineers,* July 1954.

(75)   Kansas, University of, *Proceedings 18th Annual Soil Mechanics and Foundations Engineering Conference,* Lawrence, Kansas, 1969.

(76)   Karol, R. H. "Chemical Grouting Technology," *Journal of Soil Mechanics and Foundations Division, ASCE,* Vol. **94,** SM1, January 1968.

(77)   Kerisel, J. L. "Vertical and Horizontal Bearing Capacity of

Deep Foundations in Clay," *Proceeding, Bearing Capacity and Settlement of Foundations Symposium*, Duke University, Durham, N.C., 1967.

(78)   Kreb, R. D. and R. D. Walker. *Highway Materials*. New York: McGraw-Hill Book Co., 1971.

(79)   Krynine, D. P. and W. R. Judd. *Principles of Engineering Geology and Geotechnics*. New York: McGraw-Hill Book Co., Inc., 1957.

(80)   Lambe, T. W. *Soil Testing For Engineers*. New York: John Wiley and Sons, Inc., 1951.

(81)   Lambe, T. W., "The Structure of Inorganic Clay," *Journal of Soil Mechanics and Foundation Division, ASCE*, Vol. **79**, Oct. 1953.

(82)   Lambe, T. W., "The Engineering Behavior of Compacted Clay," *Journal of Soil Mechanics and Foundations Division, ASCE*, Vol. **84**, SM2, May 1958.

(83)   Lambe, T. W., et al. "Compacted Clay—A Symposium," *Transactions, ASCE*, Vol. **125**, 1960.

(84)   Leggett, R. F. *Geology and Engineering*, 2nd ed. New York: McGraw-Hill Book Co., Inc., 1962.

(85)   Leonards, G. A., et al. *Foundation Engineering*. New York: McGraw-Hill Book Co., Inc., 1962.

(86)   McCarthy, D. F. *Basic Soil Testing Procedures*. Utica, N.Y.: Mohawk Valley Community College, SUNY, 1970.

(87)   McClelland, B., J. A. Focht, and W. J. Emrich. "Problems in Design and Installation of Off-Shore Piles," *Journal of Soil Mechanics and Foundations Division, ASCE*, Vol. **95**, No. SM6, Nov. 1969.

(88)   McClelland, B. "Design and Performance of Deep Foundations," *Proceedings of the Specialty Conference on Performance of Earth and Earth-Supported Structures*, Purdue University, ASCE, June 1972.

(89)   McClelland, B. "Design of Deep Penetration Piles for Ocean Structures," *Journal of Geotechnical Engineering Division, ASCE*, Vol. **100**, No. GT7, July 1974.

(90)   Meyerhof, G. G. "The Ultimate Bearing Capacity of Foundations," *Geotechnique*, Vol. **2**, No. 4, London, 1951.

(91)   Meyerhof, G. G. "The Bearing Capacity of Footings Under Eccentric and Inclined Loads," *Proceedings of the Third International Conference on Soil Mechanics and Foundation Engineering*, Zurich 1953.

(92)   Meyerhof, G. G. "The Influence of Roughness of Base and Ground Water on the Ultimate Bearing Capacity of Foundations," *Geotechnique*, Vol. **5**, No. 3, London, Sept. 1955.

(93)   Meyerhof, G. G. "Penetration Tests and Bearing Capacity of Cohesionless Soils," *Journal of Soil Mechanics and Foundations Division, ASCE*, Vol. **82**, SM1, January 1956.

(94)   Meyerhof, G. G. "The Ultimate Bearing Capacity of Foundations on Slopes," *Proceedings Fourth International Conference on Soil Mechanics and Foundation Engineering*, London, 1957.

(95)   Meyerhof, G. G. "Compaction of Sands and Bearing Capacity of Piles," *Journal of the Soil Mechanics and Foundations Division, ASCE*, Vol. **85**, SM6, December 1959.

(96)   Meyerhof, G. G. "Some Recent Research on the Bearing Capacity of Foundations," *Canadian Geotechnical Journal*, Vol. **1**, No. 1, Sept. 1963.

(97)   Meyerhof, G. G. "Shallow Foundations," *Journal of Soil Mechanics and Foundations Division, ASCE*, Vol. **91**, No. SM2, March 1965.

(98)   Meyerhof, G. G. "Bearing Capacity and Settlement of Pile Foundations," *Journal of the Geotechnical Engineering Division, ASCE*, Vol. **102**, No. GT3, March 1976.

(99)   Millar, C. E., L. M. Turk, and H. D. Foth. *Fundamentals of Soil Science*. New York: John Wiley and Sons, Inc., 1958.

(100) Mitchell, J. K. "Fundamental Aspects of Thixotropy in Soils," *Journal of Soil Mechanics and Foundations Division, ASCE*, Vol. **86**, No. SM3, June 1960.

(101) Mitchell, J. K. "In-Place Treatment of Foundation Soils," *Journal of the Soil Mechanics and Foundations Division, ASCE*, Vol. **96**, No. SM1, Jan. 1970.

(102) Moorhouse, D. C. and G. L. Baker. "Sand Densification by Heavy Vibratory Compactor," *Journal of Soil Mechanics and Foundations Division, ASCE*, Vol. **95**, No. SM4, July 1969.

(103) Moorhouse, D. C., "Shallow Foundations," *Proceedings of the Specialty Conference on Performance of Earth and Earth-Supported Structures*, Purdue University, ASCE, June 1972.

(104) Moore, W. S. "Experiences with Predetermining Pile Lengths," *Transactions, American Society of Civil Engineers*, Volume **114**, 1947.

(105) *Moretrench Wellpoint System*. Moretrench Corporation, Rockaway, N. J. 1967.

(106) Mosley, E. T. and T. Raamot. *Pile Driving Formulas.* Highway Research Board, National Academy of Sciences—National Academy of Engineering, Washington, D. C., 1970.

(107) Nordlund, R. L. "Bearing Capacity of Piles in Cohesionless Soils," *Journal of Soil Mechanics and Foundations Division, ASCE,* Vol. **89,** No. SM3, May 1963.

(108) Nordlund, R. L. "Pressure Injected Footings," *Proceedings, Design and Installation of Pile Foundations and Cellular Structures,* Lehigh University, Envo Publishing Co., Lehigh Valley, Pa.: 1970.

(109) Osterberg, J. O. "Drilled Caissons—Design, Installation, Application," *Soil Mechanics Lecture Series, Foundation Engineering,* Soil Mechanics and Foundations Division, Illinois Section, ASCE and Dept. of Civil Engineering, Northwestern University, Evanston, Illinois, 1969.

(110) Parola, J. F. *Mechanics of Impact Pile Driving.* Ph.D. thesis, University of Illinois, Urbana, Ill., 1970.

(111) Parsons, J. D. "Piling Difficulties in the New York Area," *Journal of Soil Mechanics and Foundations Division, ASCE,* Vol. **92,** No. SM1, Jan. 1966.

(112) Peck, R. B. "Pile and Pier Foundations," *Journal of Soil Mechanics and Foundations Division, ASCE,* Vol. **91,** No. SM2, March 1965.

(113) Peck, R. B. and A. R. S. Bazaraa. "Discussion, Settlement of Spread-footings on Sand," *Journal of the Soil Mechanics and Foundations Division, ASCE,* Vol. **95,** SM3, May 1969.

(114) Peck, R. B., W. E. Hanson, and T. H. Thorburn. *Foundation Engineering,* 2nd ed. New York: John Wiley and Sons, Inc., 1973.

(115) Peurifoy, R. L. *Construction Planning, Equipment and Methods,* 2nd ed. New York: McGraw-Hill Book Co., 1970.

(116) *Procedures for Testing Soils.* American Society for Testing and Materials, Philadelphia, Pa.

(117) Proctor, R. R., "Fundamental Principles of Soil Compaction," *Engineering News Record,* August 31, September 7, September 21, September 28, 1933.

(118) Reese, L. C. and M. W. O'Neill. "Criteria for the Design of Axially Loaded Drilled Shafts," Research Report 189-11F, Center for Highway Research, Univ. of Texas, Austin, Texas, 1971.

(119) Reese, L. C., F. T. Touma, and M. W. O'Neill. "Behavior of Drilled Piers Under Axial Loading," *Journal of Geotechnical Engineering Division, ASCE*, Vol. **102,** No. GT5, May 1976.

(120) Reginatto, A. R., "Standard Penetration Tests in Collapsible Soils," *Proceedings, Fourth Pan American Conference on Soil Mechanics and Foundation Engineering*, Vol. II, Puerto Rico, June 1971 (Pub. ASCE).

(121) Richart, F. E. "A Review of the Theories for Sand Drains," *Journal of the Soil Mechanics and Foundations Division, ASCE*, Vol. **83,** SM3, July 1957.

(122) Roberts, D. V. "Notes on Predetermination of Pile Capacities," Dames and Moore Engineering Bulletin 5, Dames and Moore, Los Angeles, Ca.

(123) Robinsky, E. I. and K. E. Bespflug. "Design of Insulated Foundations," *Journal of the Soil Mechanics and Foundations Division, ASCE*, Vol. **99,** No. SM9, Sept. 1973.

(124) Rodin, S. "Experiences with Penetrometer, with Particular Reference to the Standard Penetration Test," *Proceedings, Fifth International Conference on Soil Mechanics and Foundation Engineering*, Paris, 1961.

(125) Rosenquist, I. Th. "Physico-Chemical Properties of Soils: Soil-Water Systems," *Journal of Soil Mechanics and Foundations Division, ASCE*, Vol. **85,** No. SM2, April 1959.

(126) Rutledge, P. C. "Construction Methods and Inspection," *Soil Mechanics Lecture Series, Foundation Engineering*, Soil Mechanics and Foundations Division, Illinois Section, ASCE and Dept. of Civil Engineering, Northwestern University, Evanston, Ill. 1969.

(127) Rutledge, P. C. "Utilization of Marginal Lands for Urban Development," *Journal of Soil Mechanics and Foundation Division, ASCE*, Vol. **96,** No. SM1, Jan. 1970.

(128) Sanglerat, G. *The Penetrometer and Soil Exploration*, (English). London-New York: Elsevier Publ. Co., 1972.

(129) Schmertmann, J. H. "Static Cone Penetrometers for Soil Exploration," *Civil Engineering, ASCE*, June 1967.

(130) Schmertmann, J. H. "Static Cone to Compute Settlement Over Sand," *Journal of Soil Mechanics and Foundation Division, ASCE*, Vol. **96,** No. SM3, May 1970.

(131) Schmid, W. E. "Low Frequency Pile Vibrators," *Proceedings, Design and Installation of Pile Foundations and Cellular Structures.*

Lehigh Valley, Pa.: Lehigh University, Envo Publishing Co., 1970.

(132)   Sherard, J. L., R. J. Woodward, S. F. Gizienski, and W. A. Clevenger. *Earth and Earth-Rock Dams.* New York: John Wiley and Sons, Inc., 1963.

(133)   Skempton, A. W. "Soil Mechanics in Relation to Geology," *Proceedings of the Yorkshire Geological Society,* Vol. **29**, Part 1, No. 3, April 1953.

(134)   Skempton, A. W. "The Colloidal Activity of Clays," *Proceedings, Third International Conference on Soil Mechanics and Foundation Engineering,* Zurich, 1953.

(135)   Smith, E. A. L. "Pile-Driving Analysis By The Wave Equation," *Transactions ASCE,* Vol. **127**, 1962.

(136)   Smith, R. E. "Guide for Depth of Foundation Exploration," *Journal of Soil Mechanics and Foundations Division, ASCE,* Vol. **96**, No. SM2, March 1970.

(137)   *Soil Cement Construction Handbook.* Portland Cement Association, Chicago, Illinois.

(138)   *Soil Cement Laboratory Handbook.* Portland Cement Association, Chicago, Illinois.

(139)   *Soil Density Control Methods.* Highway Research Board, Bulletin 159, National Academy of Sciences-National Research Council, Pub. No. 498, Washington, D.C., 1957.

(140)   *Soil Mechanics, Foundations and Earth Structures.* NAVDOCKS DM-7, Dept. of the Navy, Bureau of Yards and Docks, Washington, D.C.

(141)   Sorensen, T. and B. Hansen. "Pile Driving Formulae—An Investigation Based on Dimensional Considerations and a Statistical Analysis," *Proceedings, Fourth International Conference on Soil Mechanics and Foundation Engineering,* London 1957.

(142)   Sowers, G. F. "Soil Stress-Strain, Strength and Earth Pressure," *Design of Structures to Resist Earth Pressures, Soil Mechanics Lecture Series,* Soil Mechanics and Foundations Division, Illinois Section, ASCE and Civil Engineering Department, Illinois Institute of Technology, Chicago, Illinois, 1964.

(143)   Sowers, G. F. "Fill Settlement Despite Vertical Sand Drains," *Journal of Soil Mechanics and Foundations Division, ASCE,* Vol. **90**, No. SM5, Sept. 1964.

(144)   Sowers, G. B. and G. F. Sowers. *Introductory Soil Mechanics and Foundations,* 3rd ed. New York: The McMillan Co., 1970.

(145)   *Stresses and Deflections in Foundations and Pavements.* Soil Mechanics
        and Bituminous Materials Research Laboratory, Dept. of Civil
        Engineering, Univ. of California, Berkeley, Ca. 1965.

(146)   Strom, J. A. "Development of the Earth Deformation Re-
        corder," Dames and Moore Engineering Bulletin 21, Dames
        and Moore, Los Angeles, Ca.

(147)   Taylor, A. W. "Physico-Chemical Properties of Soils: Ion Ex-
        change Phenomena," *Journal of Soil Mechanics and Foundations
        Division, ASCE,* Vol. **85,** No. SM2, April 1959.

(148)   Taylor, D. W. *Fundamentals of Soil Mechanics.* New York: John
        Wiley and Sons, Inc., 1949.

(149)   Teng, W. C., *Foundation Design.* Englewood Cliffs, N.J. Pren-
        tice-Hall, Inc., 1962.

(150)   Terzaghi, K. *Theoretical Soil Mechanics.* New York: J. Wiley and
        Son, Inc., 1943.

(151)   Terzaghi, K. *From Theory to Practice in Soil Mechanics.* New York:
        John Wiley and Sons, Inc., 1960.

(152)   Terzaghi, K. and R. B. Peck. *Soil Mechanics in Engineering
        Practice,* 2nd ed. New York: John Wiley and Sons, Inc., 1968.

(153)   Tomlinson, M. J. "The Adhesion of Piles Driven in Clay," *Pro-
        ceedings, Fourth International Conference on Soil Mechanics and Founda-
        tions Engineering,* London, 1957.

(154)   Tomlinson, M. J. *Foundation Design and Construction.* New York:
        Wiley-Interscience, 1969.

(155)   Touma, F. T. and L. C. Reese. "Behavior of Bored Piles in
        Sand," *Journal of Geotechnical Engineering Division, ASCE,* Vol. **100,**
        No. GT7, July 1974.

(156)   Tschebotarioff, G. P., *Foundations, Retaining and Earth Structures,*
        2nd ed. New York: McGraw-Hill Book Co., Inc., 1973.

(157)   Turnbull, W. J., J. R. Compton, and R. G. Ahlvin. "Quality
        Control of Compacted Earthwork," *Journal of Soil Mechanics and
        Foundations Division, ASCE,* Vol. **92,** No. SM1, Jan. 1966.

(158)   Turnbull, W. J. and C. I. Mansur. "Compaction of Hy-
        draulically Placed Fills," *Journal of Soil Mechanics and Foundations
        Division, ASCE,* Vol. **99,** No. SM11, Nov. 1973.

(159)   Tuttle, J., "Experiences with Piles in Coarse Granular Soils,"
        Dames and Moore Engineering Bulletin 6, Dames and Moore,
        Los Angeles, Ca.

(160)   Vesic, A. S. "Ultimate Loads and Settlement of Deep Founda-

tions in Sand," *Proceedings, Bearing Capacity and Settlement of Foundations Symposium,* Duke University, Durham, N.C., 1967.

(161) Vesic, A. S. "Test on Instrumental Piles, Ogeechee River Site," *Journal of Soil Mechanics and Foundations Division, ASCE,* Vol. **96,** No. SM2, March 1970.

(162) Vesic, A. S. "Load Transfer in Pile-Soil Systems," *Proceedings, Design and Installation of Pile Foundations and Cellular Structures,* Lehigh Valley, Pa.: Lehigh University, Envo Publishing Co., 1970.

(163) Vesic, A. S. "Analysis of Ultimate Loads of Shallow Foundations," *Journal of Soil Mechanics and Foundations Division, ASCE,* Vol. **95,** No. SM1, Jan. 1973.

(164) Vijayvergiya, V. N. and J. A. Focht. "A New Way to Predict Capacity of Piles in Clay," Fourth Annual Offshore Technology Conference, Houston, Texas, May 1972.

(165) Webb, D. L. and R. I. Hall. "Effects of Vibroflotation on Clayey Sands," *Journal of the Soil Mechanics and Foundations Division, ASCE,* Vol. **95,** No. SM6, Nov. 1969.

(166) Whitman, R. V. "Hydraulic Fills to Support Structural Loads," *Journal of Soil Mechanics and Foundation Division, ASCE,* Vol. **96,** No. SM1, Jan. 1970.

(167) Woodward, R. J., W. S. Gardner and D. M. Greer. *Drilled Pier Foundations.* New York: McGraw-Hill Book Co., 1972.

(168) Wu, T. H. "Relative Density and Shear Strength of Sands," *Journal of Soil Mechanics and Foundations Division, ASCE,* Vol. **83,** No. SM1, Jan. 1957.

Answers To Selected Problems

2-1. w = 13.2 percent
2-2.   (a) w = 20 pct; (b) γ_{wet} = 116.7 pcf, γ_{dry} = 97.5 pcf
2-3. G_s = 2.63
2-5. e = 0.58
2-6. G_s = 2.68, e = 0.62
2-10. w = 23.5 pct, e = 0.62, S = 100 pct
2-11. e = 2.31, S = 100 pct, w = 85 pct
2-13. G_s = 2.57
2-15. γ_{wet} = 135.4 pcf, γ_{sub} = 73.5 pcf

4-3. γ_{wet} = 131 pcf, γ_{dry} = 114 pcf
4-5. e_o = 0.55, D_r = 74 pct
4-6.   (a) water content; (b) PI = 32 pct; (c) MH
4-7.   (a) LI = 0.33; (b) CL–CH
4-8.   (a) LI = -0.22; (b) CH
4-11.   Activity = 0.88, probably kaolinite

5-1. $R_H = r/2$
5-2. $R_H = 1.2$
5-7. k = .022 cm/sec
5-10. k = 9.2 cm/min
5-11. k = .021 ft/min $\cong$ .01 cm/sec
5-14. k = 9.6×10^{-4} ft/min
5-17. p_w = -375 psf
5-19. $h_c \cong$ 1500 cm

6-1. $Q = 32 \ \text{ft}^3/\text{hr} = 5700 \ \text{gal}/\text{day}$

6-3. (a) 8.5 gal/day/ft. wide; (b) 5.6 gal/day/ft. wide

6-4. $i = 1.03$

6-6. drawdown slope approx. 5 pct, distance approx. 350 ft

7-1. (b) $\tau_{max} = 1500$ psf, $\sigma_n = 3500$ psf; (c) $\sigma_n = 2750$ psf, $\tau = 1300$ psf

7-3. (b) $\sigma_n = 30$ psi, $\tau_{max} = 50$ psi; (c) $\sigma = 55$ psi, $\tau = 43$ psi

7-4. (b) $\sigma_1 = 8600$ psf, $\sigma_3 = 1400$ psf; (c) $\theta = 16.9$ deg; (d) $\tau_{max} = 3600$ psf

7-7. $\sigma_1 = 130$ psi

7-9. (a) $\sigma_1 = 1200$ psf, $\sigma_3 = 600$ psf; (b) $\tau_{max} = 300$ psf

8-1. (a) $\bar{\sigma}_v = 1476$ psf; (b) $\bar{\sigma}_v = 1100$ psf, $u = 375$ psf

8-4. $\Delta \sigma_h = 372$ psf

8-6. (a) center, $\Delta \sigma_v = 1.08$ ksf; edge, $\Delta \sigma_v = 0.90$ ksf;
      (b) center, $\Delta \sigma_v = 0.72$ ksf; edge, $\Delta \sigma_v = 0.54$ ksf

8-9. Boussinesq, $\Delta \sigma_v = 2.20$ ksf; sixty-deg approx, $\Delta \sigma_v = 1.54$ ksf

8-12. net $\Delta \sigma_v = +760$ psf

9-6. (a) $a_v = 9.5 \times 10^{-5} \ \text{ft}^2/\text{lb}$; (b) $C_c = 0.64$

9-9. $C_c = 0.63$

9-10. (a) $\Delta H = 1.9$ in.; (b) $\Delta H = 3.3$ in ;(c) $\Delta H = 1$ in.

9-13. ΔH approx $\frac{3}{4}$ in

9-16. (a) $t = 5.8$ yrs; (b) $\Delta H = 0.85$ in; (c) $t = 1.37$ yrs

10-1. $\phi = 31$ deg

10-3. $\phi = 37$ deg

10-5. $\sigma_1 = 3400$ psf $(\pm)$

10-8. (a) $c = 1000$ psf; (b) axial load = 2000 psf

11-11. $V_1 = 2000$ ft/sec, $V_2 = 10,000$ ft/sec, $H = 82$ ft

12-4. (a) $\gamma_{max} = 115$ to 116 pcf, $w_{opt} = 15$ pct$(\pm)$; (b) 7 pct to 20 pct

12-5. $\gamma_{max} = 120$ pcf, $w_{opt} = 13$ to 14 pct

12-8. (a) $\gamma_{dry} = 112$ pcf, $w = 15$ pct $(\pm)$; (b) 97.5 pct compaction

12-9. (a) $\gamma_{dry} = 109$ pcf; (b) 93 pct compaction

14-1. $q_{des} = 6.3$ ksf

14-2. $q_{des} = 4.2$ ksf, $Q = 10.5$ kip/ft of length

14-3. 7.75 ft × 7.75 ft

14-4. 9 ft × 9 ft

14-7. $Q_{des} = 96$ kips

14-9. $L = 26$ ft, $B = 6.2$ ft

14-10. $Q_{des} = 8$ kip/ft of length

14-12. Q_{des} = 8.7 kip/ft of length

14-14.   (a) q_{des} = 3.3 tsf, B = 7 ft; (b) q_{des} = 2.15 tsf, B = 8.5 ft

14-17. Q_{down} = 85 kips (for K = 3)

14-19. Q_{down} = 43 kips

14-21. Q_{down} = 106 kips; Q_{up} = 94 kips

14-23. Q_{down} = 295 kips

14-25. Q_{des} = 50 tons

14-26.   set = 0.33 in

Index